Genetics and Criminal Behavior

In this volume, a group of distinguished philosophers address basic conceptual, methodological, and ethical issues raised by genetic research on criminal behavior. Their contributions fill a large gap between popular accounts of "crime genes" and technical discussions of heritability, linkage, and genetic variation. They explain the scientific debate about behavioral genetics in lucid but precise terms, and place it in the context of broader issues about causation, moral responsibility, and political justice. The book will be of great value to philosophers, legal scholars, scientists, and policy makers interested in the potential of genetic research to predict, understand, and modify human behavior, and to educated laymen curious or perplexed about the recent controversies surrounding behavioral genetics.

David Wasserman and Robert Wachbroit are both research scholars at the Institute for Philosophy and Public Policy at the University of Maryland.

Cambridge Studies in Philosophy and Public Policy

General editor: Douglas MacLean, *University of Maryland, Baltimore County*

Other books in series

Mark Sagoff: *The Economy of the Earth*
Henry Shue (ed.): *Nuclear Deterrence and Moral Restraint*
Judith Lichtenberg (ed.): *Democracy and Mass Media*
William Galston: *Liberal Purposes*
Elaine Draper: *Risky Business*
R. G. Frey and Christopher W. Morris: *Violence, Terrorism, and Justice*
Douglas Husak: *Drugs and Rights*
Ferdinand Schoeman: *Privacy and Social Freedom*
Dan Brock: *Life and Death*
Paul B. Thompson: *The Ethics of Aid and Trade*
Jeremy Waldron: *Liberal Rights*
Steven Lee: *Morality, Prudence, and Nuclear Weapons*
Robert Goodin: *Utilitarianism as a Public Policy*
Bernard Rollin: *The Frankenstein Syndrome*
Robert K. Fullinwider (ed.): *Public Education in a Multicultural Society*
John Kleinig: *The Ethics of Policing*
Norman Daniels: *Justice and Justification*
James P. Sterba: *Justice for Here and Now*
Erik Nord: *Cost-Value Analysis in Health Care*

Genetics and Criminal Behavior

Edited by

DAVID WASSERMAN
Institute for Philosophy and Public Policy, University of Maryland

ROBERT WACHBROIT
Institute for Philosophy and Public Policy, University of Maryland

CAMBRIDGE
UNIVERSITY PRESS

PUBLISHED BY THE PRESS SYNDICATE OF THE UNIVERSITY OF CAMBRIDGE
The Pitt Building, Trumpington Street, Cambridge, United Kingdom

CAMBRIDGE UNIVERSITY PRESS
The Edinburgh Building, Cambridge CB2 2RU, UK
40 West 20th Street, New York, NY 10011-4211, USA
10 Stamford Road, Oakleigh, VIC 3166, Australia
Ruiz de Alarcón 13, 28014 Madrid, Spain
Dock House, The Waterfront, Cape Town 8001, South Africa

http://www.cambridge.org

© Cambridge University Press 2001

First published 2001

Printed in the United States of America

Typeface Palatino 10/13 pt.   *System* QuarkXPress™ 4.04 [AG]

A catalog record for this book is available from the British Library.

Library of Congress Cataloging in Publication data

Wasserman, David, 1953–
    Genetics and criminal behavior / David Wasserman, Robert Wachbroit.
    p. cm. – (Cambridge studies in philosophy and public policy)
    Includes index.
    ISBN 0-521-62214-X – ISBN 0-521-62728-1 (pbk.)
    1. Mental illness – genetic aspects. 2. Behavior genetics. 3. Criminal behavior –
genetic aspects. 4. Criminal behavior, Prediction of. I. Wachbroit, Robert Samuel. II.
Title. III. Series.

RC455.4.G4.W38 2000
616.89′042 – dc21

                                                                    00-037821

                    ISBN 0 521 62214 x hardback
                    ISBN 0 521 62728 1 paperback

Contents

List of Contributors   *page* ix

Acknowledgments   xi

1 Introduction: Methods, Meanings, and Morals   1
   DAVID WASSERMAN AND ROBERT WACHBROIT

PART I

2 Understanding the Genetics-of-Violence Controversy   23
   ROBERT WACHBROIT

3 Separating Nature and Nurture   47
   ELLIOTT SOBER

4 Genetic Explanations of Behavior: Of Worms, Flies,
   and Men   79
   KENNETH F. SCHAFFNER

5 On the Explanatory Limits of Behavioral Genetics   117
   KENNETH A. TAYLOR

6 Degeneracy, Criminal Behavior, and Looping   141
   IAN HACKING

7 Genetic Plans, Genetic Differences, and Violence:
   Some Chief Possibilities   169
   ALLAN GIBBARD

PART II

8 Crime, Genes, and Responsibility   199
   MARCIA BARON

Contents

9   Genes, Statistics, and Desert   225
    PETER VAN INWAGEN

10 Genes, Electrotransmitters, and Free Will   243
    P. S. GREENSPAN

11 Moral Responsibility without Free Will   259
    MICHAEL SLOTE

12 Strong Genetic Influence and the New "Optimism"   273
    J. L. A. GARCIA

13 Genetic Predispositions to Violent and Antisocial
    Behavior: Responsibility, Character, and Identity   303
    DAVID WASSERMAN

    Index   329

Contributors

MARCIA BARON, Department of Philosophy, University of Illinois

J. L. A. GARCIA, Department of Philosophy, Rutgers University

ALLAN GIBBARD, Department of Philosophy, University of Michigan

P. S. GREENSPAN, Department of Philosophy, University of Maryland

IAN HACKING, Department of Philosophy, University of Toronto

KENNETH F. SCHAFFNER, Medical Humanities, George Washington University

MICHAEL SLOTE, Department of Philosophy, University of Maryland

ELLIOTT SOBER, Department of Philosophy, University of Wisconsin

KENNETH A. TAYLOR, Department of Philosophy, Stanford University

PETER VAN INWAGEN, Department of Philosophy, Notre Dame University

ROBERT WACHBROIT, Institute for Philosophy and Public Policy, University of Maryland

DAVID WASSERMAN, Institute for Philosophy and Public Policy, University of Maryland

Acknowledgments

This anthology arose out of a project on behavioral genetics and crime, funded by a grant from the Ethical, Legal, and Social Implications (ELSI) Branch of the National Institute for Human Genome Research (R13 HG00703). As it turned out, our proposed discussion of genetic research and criminal behavior proved to be as controversial as the research itself. We wish to thank the many friends, colleagues, and scholars who supported us in the face of this controversy. We are particularly grateful to Eric Juengst, ELSI branch chief at the time the project was funded, Jacob Goldhaber, acting provost of the University of Maryland, College Park, and Victor Medina, then the University's director of sponsored programs.

The contributions to the anthology were the product of two working groups that examined the meaning and significance of research claiming to find genetic influences on criminal behavior. One group focused on the interpretation and plausibility of claims of genetic influence, the other on the relevance of those claims to the assignment of moral and legal responsibility. We are indebted to several present and former colleagues at the Institute for Philosophy and Public Policy for their work on these groups. Alan Strudler helped recruit and direct the group on genetic influence and responsibility. Teresa Chandler provided skillful assistance both in conducting the working group meetings and in preparing summaries of the discussion for the authors. Carroll Linkins handled the arrangements for the meetings with great efficiency and finesse.

Chapter 1

Introduction: Methods, Meanings, and Morals

DAVID WASSERMAN AND ROBERT WACHBROIT

The ambition to explain and predict criminal behavior scientifically has defined the field of criminology and inspired a vast number of studies in psychology, sociology, and economics. As central as this ambition has been to a wide variety of research programs, it has been called into question by those who doubt that criminal behavior is susceptible to scientific explanation or prediction or who worry that the desire to explain and predict masks a desire for pervasive social control. The recent pursuit of this ambition by human geneticists has inspired particularly strong hopes and fears. The success of genetics in understanding human disease suggests that it could be a powerful tool in the scientific investigation of human behavior, including criminal behavior. At the same time, the checkered history of human genetics suggests that it can easily be abused, misrepresented, or misunderstood, regardless of the validity of the studies or the motivation of the researchers.

At present, a variety of research programs investigates genetic influences on human behaviors, dispositions, and mental traits. Heritability studies seek to tease out genetic from environmental effects on human behavioral differences, largely by examining twins and adoptees; molecular researchers look for markers, and ultimately genes, associated with crime and violence; neurobiologists explore causal pathways by which genetic variations may affect aggressive and impulsive behavior (see generally Raine 1993; Reiss and Roth 1994; Stoff and Cairns 1996; Sherman et al. 1997). Although specific research programs differ greatly in aim and method, they share the assumption that it makes scientific sense to look for genetic contributions to important mental and behavioral differences among people.

These programs have had to wrestle with the legacy of human genetic research from the first half of the twentieth century, which studied heritable differences among people with the goal of improving the genetic stock of humanity (Kevles 1985; Duster 1990; Paul 1995). The confidence of many early researchers in their moral authority to control human reproduction was matched only by their ignorance of the complex patterns and mechanisms of human inheritance. Not only did their research programs yield little scientific insight, but they also lent scientific prestige to the restrictive immigration policies and sterilization laws of many Western nations at that time as well as to the campaigns of mass sterilization and "euthanasia" in Nazi Germany.

After a period of understandable quiescence following World War II, human genetic research on mental and behavioral traits has made a dramatic comeback. Researchers attribute this comeback in large part to the advent of sophisticated techniques for isolating and manipulating genetic material and statistically assessing patterns of inheritance. These techniques, they believe, will eventually make it possible to identify the specific genes and causal mechanisms that underlie the heritability of many psychological traits and behavioral dispositions, including some of those associated with criminal behavior. Although no genetic variations have yet been identified that can explain any significant proportion of criminal behavior, researchers see ongoing studies of the genetics of psychiatric and behavioral disorders as encouraging preliminaries (see, e.g., Carey and Gottesman 1996; Goldman 1996a).

Researchers also claim that the renewed interest in genetic contributions to personality and behavior represents a sensible retreat from the dogmatic environmentalism of the postwar era, which made unrealistic demands on families and institutions and promoted "one-size-fits-all" social interventions, doomed to failure by their neglect of individual differences. An understanding of the genetic and biological contributions to such differences, researchers insist, is critical for humane and effective social policies (e.g., Carey and Gottesman 1996; Fishbein 1996).

Critics contend that the comeback of genetic research on human behavior is largely due to the public's obsession with, and credulity toward, genetic explanations, and to a blanket repudiation of the optimistic environmentalist policies that enjoyed a brief ascendancy in the postwar years. They argue that the search for genetic factors involves the "medicalization" of social behavior and thus diverts attention and resources from the social and economic conditions largely responsible for crime. Although most mainstream researchers do not look for, or expect to find, racial differences in genetic predispositions to antisocial behavior, opponents argue that their work reinforces public perceptions of criminal behavior as an essentially biological problem, affecting some races more than others (see Wasserman 1995; Miller 1996).

The public debate on current research has been characterized by the invocation of competing precedents. Researchers place behavioral genetics in the Enlightenment tradition of scientific progress, which has alleviated untold suffering by overcoming ignorance and superstition; critics place it in the more recent tradition of eugenics, which has justified terrible social inequalities by the invention or exaggeration of biological differences. Important as this debate is, its focus on historical precedents has tended to obscure critical issues about the scientific potential and moral relevance of claims of genetic influence on criminal behavior. While it would be naive to ignore history in appraising current human behavioral genetics, it would be equally mistaken to assume that an adequate appraisal can be made on that basis alone. Contemporary behavioral genetics appears to conform to prevailing norms of scientific inquiry and offers detailed and highly qualified findings, not the vague, sweeping assertions of causal influence that discredited eugenic research earlier in this century. We cannot judge whether contemporary research is likely to correct or repeat a history of scientific folly and social abuse without examining its assumptions, methods, results, and prospects. Similarly, we can hardly assess the social significance of current research, the motivation of those pursuing it, or the socially appropriate uses it may have without a clearer understanding of its potential for actually predicting, controlling, and explaining behavior, and its implications for the ascription of blame and punishment.

The chapters in this volume were originally commissioned for an interdisciplinary conference that sought to examine the scientific prospects and social implications of genetic research into crime and violence. The proposed conference was funded in 1992 by the Ethical, Legal, and Social Implications (ELSI) Program of the National Center for Human Genome Research, at the National Institutes of Health (NIH). The same year that the grant was awarded, a leading federal research psychiatrist announced plans for a comprehensive "Violence Initiative" that would investigate the genetic and biological sources of "individual vulnerabilities" to criminal and antisocial behavior. That announcement provoked a furor over public funding for behavioral genetic research, and the proposed conference soon became part of the controversy it had been organized to explore. Although it had no connection to the Violence Initiative, and its agenda featured some of the leading critics of human behavioral genetics, the conference was assailed as lending support to racist assumptions about crime and repressive programs of crime control. In response, the NIH suspended, then terminated, the grant. Funding was restored, however, and the conference rescheduled, after an administrative review by the Public Health Service, the parent agency of the NIH, held the termination "arbitrary and capricious." Although it was conducted in a harsh limelight (the entire proceedings were filmed by C-SPAN and attended by dozens of print and television journalists) and briefly interrupted by a demonstration, the conference succeeded in getting entrenched and often bitter adversaries to talk to each other with civility and mutual respect.

That success was due in part, we believe, to the circulation of early drafts of the commissioned papers. The relevance and quality of those papers were widely recognized by the participants, and several were repeatedly cited during the conference, providing a common framework for adversaries whose very terms of discourse had often been at odds. The papers did not, for the most part, directly address the issues on which the conference focused: the justifications for government funding of genetic research on violence and the impact of such research on law enforcement, criminal justice, and vulnerable minorities. Rather, they offered a scholarly assessment of the potential of such research for predicting, explaining, and controlling crime, and for assigning responsibility for its commission. That assessment was recognized by many conference participants as critical to the resolution of the issues they were debating. If genetic research had little prospect of predicting, explaining, or reducing crim inal and antisocial behavior, and little relevance for the imposition of punishment or blame for such behavior, its pursuit and its funding, would be misguided, based on unreasonable expectations or ulterior motives. Furthermore, even if the research had substantial value, its explanatory potential and social implications may have been greatly exaggerated or distorted in the public debate.

The chapters collected in this volume are largely a subset of the commissioned papers, significantly revised and supplemented by additional contributions. They focus on two sets of issues central to the appraisal of current research into the genetics of criminal behavior:[1]

1. What are the assumptions about criminal behavior, causation, and scientific explanation underlying research programs that seek genetic factors in criminal behavior, and are they justifiable? Given the widely acknowledged importance of environmental factors, and the

fact that crime is a social category, what can a genetic investigation possibly tell us about criminality?

2. How would credible evidence of genetic influence on criminal behavior affect our practices of blaming and punishing? Would such evidence compel us to revise our conceptions of moral and legal responsibility or modify our assessment of particular agents?

Unfortunately, much of the discussion regarding genetics and violence has taken place without serious reflection on these issues. This volume presents such reflection by a diverse group of philosophers, many of whom had not previously written in this area. Because a great deal of bad philosophy pervades the public debate – in arguments for and against behavioral genetic research – it was important to elicit papers from contributors who could subject these arguments to close critical scrutiny.

For the remainder of this Introduction, we briefly describe the basic methods and assumptions of recent behavioral genetics, highlighting the claimed findings and the controversies most relevant to the prediction, explanation, and appraisal of criminal behavior. We do not intend this to be a freestanding review, but rather an introduction to the essays that appear in this volume, many of which offer their own careful exposition of particular aspects of the research. We then describe how the essays in this volume approach the issues we have just outlined.[2]

CLASSICAL AND CONTEMPORARY
BEHAVIORAL GENETICS

Before summarizing the methods that behavioral geneticists employ, it is important to specify what it is that they investigate. Although the obvious answer is "behavior," we must distinguish the study of behavior from the study of behavioral *differences*. The latter is the primary focus of behavioral genetics. Unfortunately, this distinction and its significance have not been generally appreciated by nonscientists.

Conceptually, the contrast is clear enough; it is as clear as the difference between asking "Why is this person behaving criminally?" and "Why is this person behaving more criminally than this other person?" Presumably a complete answer to the first type of question can lead to answers to the second type; but the converse is not necessarily true: factors that account for behavioral differences may play a minor role in accounting for the behavior itself. By way of analogy, think of two people, one six feet tall and the other six feet, one inch tall. The factors that explain the one-inch difference may play hardly any role in explaining the six feet in stature that each has attained. For this reason, it would be a mistake to take the answer to one question as the answer to the other. Behavioral geneticists may hope that an explanation of differences will provide, or at least lead to, an explanation of the behaviors they study. But they cannot assume this connection at the outset, and there may be reasons to suspect that behavioral commonalities and differences have distinct sources.

The reason for the scientific emphasis on behavioral differences may be largely historical: earlier generations of human geneticists were preoccupied with the question of why people *vary* with respect to significant traits and behaviors. Perhaps as a result, the research methods adopted by behavioral genetics are generally better suited for investigating behavioral differences than the behavior itself. This is particularly true of heritability research.

Heritability Research

Until techniques were developed in the 1970s for isolating and manipulating genetic and other molecular material, human behavioral genetics was largely confined to heritability studies. Heritability research begins with the scarcely debatable observation that many behaviors and psychological traits travel in families. Parents usually confound the assessment of genetic influence, however, by providing their children with rearing environments as well as genes. Thus, it may be difficult to say whether their genes or their parenting is responsible for the transmission of behavior from one generation to the next. Behavioral genetics exploits two processes that can highlight the differential impact of genetic and environmental contributions: twinning and adoption. Twinning produces offspring that share either half their genes (dizygotic/DZ), the same proportion as in normal siblings, or all their genes (monozygotic/MZ). If the rearing environments of DZ twins can be assumed to be as much alike as those of MZ twins, and if other, more technical, assumptions are satisfied, then any greater similarity in the behavior or traits of the MZ twins can be attributed to their greater genetic commonality.

The second process that can highlight the contrast between genetic and environmental contributions is adoption. A true experiment would randomly assign children immediately at birth (or, better yet, at conception) to other parents, and compare their traits and behaviors with those of their biological and adoptive parents. Social practice very roughly approximates such an experiment by assigning children somewhat fortuitously, if nonrandomly, to adoptive parents, at some point after birth. The greater the similarity of the children to their biological parents in the trait or behavior studied, in comparison to their adoptive parents, the greater the estimated genetic contribution to that trait or behavior. A hybrid approach combines twinning with adoption, by studying the similarities of MZ twins who are reared apart. Although the logic of these comparisons has been understood and debated for over a century, the implementation of careful studies only became possible with the bureaucratic record-keeping and refined statistical techniques of the early and mid-twentieth century.

Studies conducted over the past thirty years have reported significant heritabilities for a variety of psychiatric and behavioral conditions, including schizophrenia, intelligence, neurosis, antisocial behavior, and property crime. Interestingly, they have failed to find significant heritabilities for other behavior, such as violent crime, which the public assumes to have a substantial genetic component (Raine 1993; Plomin, Owen, and McGuffin 1994; Carey 1994, 1996).

There has been much discussion about the validity of the assumptions on which these findings rest: to what extent does the departure from randomness in these studies introduce bias in the results? To what extent must, or can, we control for ways in which genetic similarities can affect rearing environments? For example, are the rearing environments of DZ twins as much alike as those of MZ twins, or are identical twins environmentally as well as genetically more alike, down to their identical wardrobes? Does the time adoptees spend with their biological parents, or do the adoption agencies' nonrandom placement practices, introduce systematic biases in the studies' results? The researchers themselves are keenly aware of these challenges to the validity of previous studies, although they tend to be more optimistic than their critics about the prospects for dispensing with controversial assumptions or controlling for their violation (Plomin et al. 1997, chs. 5, 11).

Beyond these specific methodological concerns, researchers must confront the general limitation of heritability studies we noted earlier: these studies are designed to investigate only behavioral differences, not behavior. Indeed, even with regard to behavior differences, they can reveal only variation, not causation. They proceed by calculating the variation in a population of some measure of a trait or behavior (e.g., IQ score, number of arrests) and comparing it with the genetic variation in that population. Thus, a typical finding of an MZ/DZ twin study might be that 60 percent of the variance in IQ score in a particular population is due to the variation in genetic constitution – IQ has a heritability of 60 percent. This finding emphatically does not mean that 60 percent of the IQ of an individual is caused by his genes. In this respect, the term "heritability" can be very misleading. Nevertheless, researchers would argue that heritability studies make a valuable contribution by identifying traits and behaviors that are good candidates for genetic influence in light of the strong evidence that *something* is being genetically transmitted.

Molecular Genetics Research

Researchers and critics agree that heritability research will play a diminishing role in the behavioral genetics of the twenty-first century. But they disagree about the scientific potential of the techniques that are superseding it (Wasserman 1996). These techniques, which attempt to link behavioral and other mental traits to specific genes, advance beyond heritability research in two respects: they seek to identify the genes being transmitted, and they attempt to trace causal pathways from genes to traits or behavior, thereby moving part way from the population to the individual. On the other hand, these techniques maintain the focus of heritability research on the genetic basis for *differences* among individuals.

Researchers expect that molecular and neurogenetic research will resolve the ambiguities about causation by tracing the complex pathways through which specific genes affect traits or behavior (Carey and Gottesman 1996; Goldman 1996a, 1996b). By replacing global estimates of heritability with narrow and testable hypotheses, this work, it is claimed, will establish behavioral genetics on a solid biological foundation. Critics, on the other hand, fear that neurogenetic research will fall prey to the same kind of oversimplification as heritability research (Balaban 1996; Balaban, Alper, and Kasamon 1996).

Thus far, researchers have only identified one genetic variation closely associated with criminal behavior, in one family. It is instructive to compare the scientific issues raised by this apparent link between a *genetic* abnormality and criminal behavior with those raised almost thirty years earlier by the claimed association of a *chromosomal* abnormality with criminal behavior. The comparison suggests that old issues of interpretation will continue to confront the new molecular genetics.

From Extra Chromosomes to Mutant Genes

The first microbiological marker thought to be linked with human mis-behavior was the XYY karyotype (possessed by males born with an extra Y chromosome). In 1965, researchers found an apparently high prevalence of that karyotype among prison inmates in Britain. That is to say, the percentage of prison inmates with XYY was higher than the percentage of XYY males in the general population (Jacobs et al. 1965). Unfortunately, many people quickly took this finding to be evidence of a direct link between an extra Y chromosome and a tendency to hyperaggressivity and violence. That assumption was eventually rejected by genetic researchers, but it held sway in the popular imagination long enough to stigmatize a generation of XYY males and (reportedly) lead to the abortion of a significant number of fetuses with that karyotype. It is now widely believed that if an extra Y chromosome leads to incarceration, it is by a indirect route. XYY individuals are no more aggressive than average, but they may be taller and less intelligent, hyperactive, and generally more impulsive (see Rutter, Griller, and Hagell 1998). Their increased risk of arrest or conviction may stem from an increased likelihood of getting caught, or of committing crimes more likely to be detected, rather than from heightened aggressiveness or greater disregard for social norms.

Both researchers and critics regard the XYY story as a cautionary tale, but in different ways. Where critics see an illustration of the risks inherent in any inquiry into biological markers for social behavior, researchers see a modest triumph of scientific self-correction. Critics observe that the early XYY investigators, in their rush to find a direct link between genes and behavior, assumed that an extra male chromosome would make a specific contribution to violence or aggression, instead of having the generally impairing effects typically associated with an extra chromosome. Researchers, on the other hand, note that it was behavioral geneticists who ruled out any association between the XYY karyotype and violence or aggression (while confirming the high incidence of XYY in prisons and other institutions; Witkin et al. 1976). Those findings, researchers note, came from the very studies of XYY individuals that critics vehemently opposed, on the grounds that they stigmatized their subjects and created a significant risk of a self-fulfilling prophecy.

In the twenty-five years since the XYY controversy, the techniques

9

for identifying biological markers may have changed more than the issues concerning their interpretation. With the development of recombinant DNA technology in the late 1970s, researchers were able to identify and manipulate individual genes and genetic material. That technology became relevant to behavioral genetics with the discovery of genetic markers for a variety of diseases and traits. These markers – highly variable (polymorphic) but functionally inert DNA segments – were found to be associated with various phenotypic traits, presumably because they were located in close proximity to genes that actually contributed to those traits. (In most cases, the markers were assumed to vary by family, such that a particular DNA variation [allele] at a given location on the genome would mark a given trait in one family, while a different allele at that location would mark that trait in another family. In other cases, it was hypothesized that the same allele would be linked with a trait in all individuals bearing it, greatly simplifying the task of identifying that allele.)

In the late 1980s and early 1990s, markers, and in some cases genes, were identified for a number of diseases known by inheritance patterns to have a significant genetic component. Behavioral geneticists were quick to adopt the same methods, hoping to replicate the dramatic success of medical genetics. They were soon reporting markers for a number of psychiatric and behavioral conditions, including bipolar disorder and alcoholism, though the first finding was retracted and the second has remained mired in controversy (Plomin et al. 1994; Holden 1994).

In 1993 researchers finally found a marker, then a gene, associated with violence and aggression, in the male members of a Dutch family (Brunner et al. 1993). Although the family was atypical in several relevant respects, the study had enormous impact. In part, this was because the affected gene was known to be responsible for a protein, MAO, involved in regulating the metabolism of serotonin, one of the neurotransmitters thought to play a critical role in mediating between genes and behavior and implicated in psychiatric and behavioral conditions ranging from depression to impulsive violence.

A comparison of the MAO and XYY studies suggests both significant advances in scientific technique and similar issues of interpretation. The connection between genotype and phenotype was closer in several respects for MAO than for XYY. First, a statistically significant association was found between MAO and aggressive and criminal behavior in one family; in contrast, no correlation was established between XYY and any form of criminal behavior until a decade after the karyotype was identified (Rutter et al. 1998). Second, the reason for suspecting a causal link was considerably stronger for the MAO gene, which helps to regulate the inhibitory mechanisms of the central nervous system, than for a chromosome associated with male gender. More than fifty studies have found an association between aggressive, criminal, antisocial, or suicidal behavior and the serotonin metabolite CSF 5-HIAA, which MAO helps produce (Balaban et al. 1996). Third, the measure of behavior – observation by the researchers or reports from close relatives, as opposed to inferences from official records – was more direct in the MAO study than in the original XYY study.

Yet critics have argued that the link between MAO and violence and aggression, even in this one family, is much more tenuous than the researchers claim, and that the study reveals some of the same inferential leaps that characterized early XYY research. Intent on finding a genetic cause for the high incidence of criminal and aggressive behavior among male family members – the presenting symptom – the Dutch researchers may have overlooked more global effects of the MAO mutation:

Since a primary characteristic of the affected subjects was lowered IQ, it is unclear why the subjects' aggression received more emphasis than their cognitive deficits, and why there was no mention of the possibility that these cognitive deficits may have contributed to behavioral pathologies. . . . Perhaps these acts of violence are secondary to some more widespread defect in affect or cognition. (Balaban et al. 1996, 18)

Although some of the men in the family studied engaged in clearly criminal and antisocial acts, that conduct, like the convictions on which the XYY researchers relied, may well have reflected a more general deficit. The mutant-MAO males, like the XYY males, may have been not more aggressive, but less intelligent, lacking constructive outlets for their aggression or clever ways of concealing it. The Dutch researchers acknowledged that their findings did not support "a simple causal relationship between the metabolic abnormality and the behavioral disturbance" they observed (Brunner 1996, 159) and that "borderline mental retardation" was also associated with the MAO mutation (156).

Critics, however, complain that the researchers did not fully explore alternative explanations for the observed effects of MAO on behavior, nor adequately examine the range of behavior that might have been affected. Moreover, some critics (e.g., Balaban et al. 1996) also question the claimed link between low serotonin and aggressive and antisocial behavior, which gives the MAO finding its plausibility. They argue that the misbehavior of the serotonin-deficient may reflect little more than the broadly debilitating impact of low serotonin levels on mental function, a breadth suggested by the sheer range of conditions in which serotonin deficits are implicated. They suggest that the fixation of serotonin researchers on specific psychiatric or behavioral pathologies has obscured these more general effects (Balaban et al. 1996).

Researchers might respond by citing other evidence, or other studies, which tend to rule out more global explanations of the association between serotonin and violence, or which support a more direct pathway. Or they might regard this challenge merely as posing a legitimate question for further research. But they might also insist that these concerns, though reasonable, are hardly unique to behavioral genetics.

Medical geneticists also confront the problem of genetic pleiotropy – the multiple effects of a single gene and the multiplicity of possible causal pathways from gene to trait. The same kinds of ambiguity arise in studies that search for the genes associated with various physiological effects, and in studies that probe the physiological effects of candidate genes (Culp 1997; Wachbroit 1998). Thus, the association between a gene mutation and a disease revealed in a linkage or targeting study may arise indirectly, from more global effects of the gene on the organism, or from the effort of other genes or bodily systems to compensate for the loss of the gene's standard function. Because researchers will rarely be able to track the full range of functions and interactions a gene can have, they will rarely be justified, on the basis of a linkage or targeting study alone, in claiming a direct causal relationship between that gene and a disease or other condition. Such claims will only be warranted when researchers have acquired sufficient knowledge about developmental and physiological processes to narrow the range of plausible causal pathways. But that knowledge cannot come entirely from molecular genetics.

If these inferential difficulties confront medical as well as behavioral genetics, why are critics so much more skeptical about the yield of the latter than of the former? For one thing, they believe that genes contribute less, and less directly, to behavior than to physiology; for another, they believe that the ways human behavior is defined and classified compound the difficulty of finding causal relationships. Because of the vast array of social and environmental forces that shape differences in human behavior and psychology, any genetic effect on these differences is both modest and difficult to track. And because many important behavioral and psychological categories are social in origin, they may not be amenable to genetic or biological explanation. Thus, researchers looking for genetic contributions to criminal behavior must begin with the fact that such behavior is defined by legislators and ascertained by police, prosecutors, judges, and juries. The difficulty is not that genes cannot affect voluntary behavior – most critics concede that they can – but that social types may not correspond with biological types. Hence there may be no one type of behavior to be explained. We should not expect much in common psychologically or neurobiologically, let alone genetically, between a child abuser, a pickpocket, a mob boss, and a political terrorist; between the bank robber prosecuted by one regime and the bank founder prosecuted by another. It is unlikely that any genetic feature distinguishes the members of such an eclectic rogues' gallery from the general population, and even if one does, it is unlikely to have much explanatory value.

Researchers acknowledge that the genetic contribution to behavior may well be more subtle and elusive than the genetic contribution to physiology, but they deny that there is less scientific value in finding smaller and less direct effects (Carey and Gottesman 1996; Goldman 1996a). Furthermore, they regard the heterogeneity and social construction of human behavior as part of the challenge of their work: either to find unity in heterogeneity, through such underlying traits as impulsivity, antisocial personality, or novelty seeking (some general tendency toward disobedience or risk) or to develop refined typologies of criminal behavior and look for different genetic influences on different types of crime (serial killing, leadership of an urban drug ring, or embezzlement or tax evasion to finance a second home or third car).

It should be apparent from this paraphrase of the debate that the issues dividing the researchers and critics are not narrowly scientific ones. The plausibility of particular methods or findings depends to a significant extent on assumptions regarding the appropriate classification and explanation of human behavior. The critical examination of such assumptions is an essentially philosophical undertaking.

CONCEPTUAL AND METHODOLOGICAL ISSUES

As this brief survey of the state of the research suggests, an appraisal of the conceptual assumptions and explanatory potential of the behavioral genetics of criminality is a complex undertaking. It requires more than the assessment of a particular research protocol, paradigm, or approach in a single discipline with well-established canons of inquiry. A variety of protocols and approaches falls under the rubric of behavioral genetics, and a variety of disciplines, from psychology to neurobiology, sponsors or appropriates that research. Assessing its assumptions and prospects demands a metascientific, philosophical perspective.

An important preliminary is an analysis of what precisely is in dispute. For some, the real question is the legitimacy or plausibility of any science of human behavior. For others, the central issue is the possibility of adequately characterizing the class of behaviors – criminal behavior – of interest; for yet others, the issue is how to integrate behavioral genetics with other legitimate scientific inquiries. The purpose of presenting a taxonomy of these issues is not merely to avoid misunderstandings or cross-purposes, but also to establish a logical structure for the debate, to determine whether the intelligibility of debating some issues requires a (particular) resolution of others. For example, whereas some criticisms in the debate are directed at the possibility of a scientific account of criminal behavior, others are directed specifically at a genetic or biological account, and may presuppose the legitimacy of some form of scientific explanation. It is also important to identify the vague or ambiguous terms that often fuel the debate while confusing the real issues. For example, although most people on both sides believe that behavior has both a genetic and an environmental component, they appear to disagree about the significance of each component. On several interpretations, a thesis becomes untenable; on others, uninteresting. These themes are examined in Robert Wachbroit's essay.

If a taxonomy of the dispute is critical to an assessment of the behavioral genetics of crime, so is a detailed examination of specific protocols and approaches. As we noted, there are two principal types of research protocols: one is based on the statistical methods of quantitative population genetics and heritability studies, the other on the techniques of molecular genetics. Regarding the first, Elliott Sober and Kenneth Taylor present clear and accessible accounts of what such research can and cannot show. In particular, they demonstrate that heritability – the chief measure of such studies – is not the same thing as inheritance: a trait with high heritability need not be a trait that is inherited. Although scientists are (usually) quite clear on this point – for example, recognizing that a high *statistical* correlation between genes and behavior in a given population need not reflect any particular *causal* connection between genes and behavior – they do not always communicate this to the lay consumers of their research or take account of it in discussing the social implications of their research.

The prospects for the second type of research protocol, studies based on molecular genetics, are the topic of Kenneth Schaffner's essay. Looking at the case of a much simpler organism – the small roundworm, *Caenorhabditis elegans* – he explains the immense challenge of providing genetic explanations of its (simpler) behavior. Even though we know the organism's complete neuroanatomy, including a circuit diagram of its nervous system, as well as much of the worm's neurophysiology, scientists have yet to provide a genetic explanation of any of the worm's behavior. Because we do not have any such comparable knowledge of the human nervous system, the task of providing a genetic explanation of any human behavior is more daunting still. Even if such explanations are possible, they are not likely to be achieved anytime soon. Schaffner's pessimism is reinforced by his equally careful consideration of the somewhat more complex and even more intensively studied organism, the common fruit fly, *Drosophilia melanogaster.*

Two other essays shift our attention from the general prospects of different research protocols in behavioral genetics to the specific issues raised by the behavioral genetics of criminal behavior. Kenneth Taylor and Ian Hacking examine the problems of treating the social category of criminal behavior as a biological category. Taylor discusses the difficulties in regarding criminal behavior as a biological abnormality, a pathology, without understanding the circumstances in which the behavior takes place. If these circumstances are themselves extraordinary (pathological), there may be no basis for regarding the behavior as pathological. In any event, explanations of individual criminal behavior rest on an understanding of the psychodynamics of such behavior. Suggesting a comparison with linguistic behavior, Taylor argues that a great deal of research in cognitive science needs to be done before research in genetics becomes relevant.

Hacking discusses the general significance of treating criminals or criminal behavior as a biological category – a "human kind" – and raises questions about the construction of such a category. (Of particular interest is his observation that the sheer knowledge of such categories can affect people's attitudes and behavior in a way that feeds back on the classification scheme itself.) According to Hacking, the basic issue is not so much the relationship between a social classification scheme and a biological one, but rather the more complex relationship between *four* classification schemes: common or ordinary, legal, psychiatric, and biological. In the end, he questions the relevance of the biological category to the public's interest in crime.

Finally, it is important to understand how behavioral genetics fits in with other sciences that attempt to link biology and behavior. Genes only produce proteins, so any link between certain genes and particular types of behavior will rest on the mechanisms or causal pathways that lead from the genes to the behaviors. The findings of other sciences – for example, developmental biology, neurophysiology, cognitive psychology – can identify the sorts of mechanisms or pathways that are possible or plausible, and so shape the significance of any empirical finding. Alan Gibbard discusses how behavioral genetics could fit in with evolutionary theory, particularly the claims of evolutionary psychology. Focusing on mechanisms that are shaped by evolutionary pressures and thus are specieswide, he finds some reason to be skeptical about whether genetic differences can account for much of the variation in criminal behavior *between* individuals. To return to the distinction we introduced at the beginning of this section, Gibbard suggests that selection pressures may have yielded genetic features that predispose the species as a whole (or at least the males of the species) to violence or aggression in some circumstances, but that they would be unlikely to result in genetic differences that predispose some individuals more than others to violence or aggression.

ASSIGNING BLAME AND IMPOSING PUNISHMENT

The second part of this volume considers some of the moral stakes in the empirical and conceptual debates presented in Part I. There are two general ways in which evidence of genetic influence on behavior might affect the ascription of responsibility: first, in helping to fill out a deterministic account of human behavior, by providing causal explanations for differences in harmful behavior that cannot be attributed to differences in environment; second, by revealing that some people are constitutionally more predisposed to, or have greater difficulty avoiding, harmful conduct. Although it might seem that the former would have more radical implications than the latter, a generation of "compatibilists," influenced by Peter Strawson (1962), would be inclined to conclude just the opposite. The truth of determinism, they would claim, poses little threat to the social practices of praising and blaming. We cannot give up the "reactive attitudes" underlying these practices without losing much of our humanity; nor do we need to do so, since those attitudes do not require the "panicky metaphysics" of indeterminism. On the other hand, claims of genetic predisposition might well affect our ascription of responsibility in specific cases – mitigating blame to the extent that such claims were seen as establishing volitional hardship or disability, aggravating blame to the extent that they were seen as revealing an antisocial character rather than a lapse in self-control.

Two of the philosophers writing in Part II reject what they see as the prevailing complacency about the threat posed by determinism, but differ sharply about the mitigating potential of genetic predisposition evidence. Peter Van Inwagen and Jorge Garcia agree that the truth of determinism would undermine the moral foundations for blame. Van Inwagen, however, insists that to the extent genetic predispositions leave the agent free to do otherwise, they lack, with narrow exceptions, the potential to excuse or even mitigate misconduct. Garcia, to the contrary, believes that even nondeterministic claims of "strong genetic influence" would be powerfully, and disturbingly, extenuating. Evidence that "our behavior was almost entirely a matter of genetics" would seriously threaten prevailing practices of assigning blame and punishment.

Michael Slote views this threat with far greater equanimity, because he favors a significant revision in the prevailing practices of assigning blame and punishment. Quite independently of the possible yield of behavioral genetic research, he rejects the assumption that punishment and blame should rest on a notion of preinstitutional desert. Rather, he finds the warrant for punishment and blame in the justice of the society that assigns them. Slote argues for a conception of justice as collective benevolence: the motivation of a society's institutional agents to promote the well-being of its members. He contends that the only reason benevolent legislators and judges would inflict harm on wrongdoers would be to prevent greater harm. Evidence that particular wrongdoers were genetically predisposed to act as they did would not undermine this motivation for imposing punishment, but merely refine it to take account of the greater threat of harm posed by such individuals as well as the heightened difficulties and risk they faced as a result of their predispositions. Garcia is sympathetic toward Slote's project of grounding justice in the virtue of benevolence, but finds his conceptions of benevolence and justice radically incomplete: justice as benevolence requires the notion of preinstitutional desert that Slote rejects, and demands not only concern but also respect, an attitude we can only adopt toward other people if we regard them as free, undetermined agents.

For those more liberal than Van Inwagen about the mitigation of harmful conduct, or more conservative than Slote about the revision of our moral practices, the relevance of genetic predisposition evidence for blame and punishment depends to some extent on the specific ways in which genes affect behavior, a central issue in Part I of this volume. Credible research may be less likely to find the kind of "strong genetic influences" discussed by Garcia and Van Inwagen (except in rare cases like the Dutch family described earlier) than weaker, subtler, more oblique ones. Although such influences may help to fill in a deterministic account of behavior, they will not by themselves pose a wholesale threat to the ascription of responsibility or the imposition of blame. They may, however, significantly alter our moral assessment of specific agents or specific conduct – including our own.

Patricia Greenspan argues that the main significance of genetic predispositions lies not in their contribution to a deterministic account of human agency but in the threat to self-control they are likely to pose. She suggests that genetic variations may typically contribute to antisocial or criminal behavior not by giving rise to exceptionally strong impulses, but by depriving the agent of the volitional resources for resisting normal impulses. This suggestion is informed by research indicating that genetic variations associated with criminal behavior are more likely to affect the inhibitory than the excitory system. Greenspan compares "inherited impulsivity" with more familiar volitional and cognitive deficiencies and explores some of the interesting complexities in our reactive attitudes toward these deficiencies.

One striking feature of genetic predisposition evidence is its moral ambiguity. It may make the agent appear more dangerous but less culpable, a persistent threat just because he is the victim of a chronic behavioral disorder. But it may also make him appear *more* culpable, by revealing an antisocial character. David Wasserman's chapter explores this ambiguity. He suggests that genetic predispositions will be seen as less extenuating than such predisposing social factors as childhood abuse and neglect, because we are more inclined to regard the agent as a victim of his upbringing than of his genes. This inclination, however, may reflect either the unwarranted assumption that genetic predispositions are immutable or essential to personal identity, or a conflation of genetic influence and early onset. Greenspan observes that the genetic predispositions she characterizes as inherited impulsivity may elicit pity as well as anger and underwrite strong claims for remedial assistance. Thus, Wasserman and Greenspan see genetic predisposition claims as capable of lending support to a variety of reactive attitudes and social policies.

Marcia Baron presents a valuable overview of ways in which the "strength" and character of genetic predisposition claims may affect their legal as well as moral implications. She suggests that genetic pre-dispositions to crime and violence are unlikely to be stronger or more subversive of free will than environmentally induced predispositions. If the former have a more significant impact on our ascription of re-sponsibility and imposition of punishment, it is likely to be because they look more scientifically credible or objective. Even then, the differences are likely to be matters of slight degree.

Baron's doubts about the normative significance of genetic predis-position claims are reinforced by her examination of the legal environ-ment in which genetic predisposition defenses will be introduced in this country. American courts are increasingly hostile to claims for any kind of volitional incapacity. The wholesale judicial rejection of XYY defenses a generation ago suggests that even more credible and causally specific claims of biological incapacity or deficiency will meet with very limited success. Finally, Baron considers the broader social impact of predispo-sition claims, which she fears may be on balance harmful.

Within the next few years, there will undoubtedly be a number of new studies implicating heredity in general, or specific genes, in criminal be-havior. We hope that this volume offers useful guidance in assessing the plausibility and the implications of that new research. The essays it contains analyze the complexities in tracing any genetic influence on human behavior, the variety of interpretations to which almost any claim of genetic influence on such behavior is subject, and the resilience of doubts about the causal importance and moral relevance of genetic influences, even in the face of the most carefully conducted, technically sophisticated research. Cumulatively, these essays do not depict human behavioral genetics as an inherently suspect enterprise. But they do sug-gest that its implications for understanding and evaluating human be-havior are often debatable, and that resolving the debate requires a broader understanding of human behavior and moral responsibility than behavioral genetics can itself supply.

NOTES

1. A note on terminology: in describing the subject matter of the research these essays examine, we often refer to "criminal behavior" or "violent and anti-social behavior," and sometimes use the terms interchangeably. There are, of course, important differences between these descriptions, some of which are discussed by our contributors, especially Ian Hacking. For one thing, crimi-nal and antisocial behavior are constructed in different ways, by different

people, for different purposes. For another, the evidence of genetic influence is different for different kinds of behavior classified as criminal or antisocial, a point also noted by several contributors. But despite these important caveats, we need to refer to several bodies of research without the obnoxious use of quote marks or the numbing repetition of qualifications, so we have left the burden of qualification to this note and the essays.

2 We commissioned several fine papers that we were unfortunately not able to include in an anthology of limited scope and size. They included essays by historians, social scientists, and biologists, some assessing possible historical precedents, others debating the current state of knowledge in the behavioral genetics of criminality.

REFERENCES

Balaban, E. (1996). Reflections on Wye Woods: Crime, biology, and self-interest. *Politics and the Life Sciences* 15 (1): 86–88, 87.

In press. Behavior genetics: Galen's prophecy or Malpighi's legacy? In *Thinking about evolution: Historical, philosophical, and political perspectives,* ed. R. Singh, C. Krimbas, D. Paul, and R. Beatty. Cambridge: Cambridge University Press.

Balaban, E., Alper, J., and Kasamon, Y. L. (1996). Mean genes and the biology of human aggression: A critical review of recent animal and human research. *Journal of Neurogenetics* 11: 1–43.

Brunner, H. G. (1996). MAOA deficiency and abnormal behavior: Perspectives on an association. In *Genetics of criminal and antisocial behavior,* ed. G. R. Bock and J. A. Goode, 155–167. New York: John Wiley and Sons.

Brunner, H. G., Nelen, M., Breakfield, X. O., Ropers, H. H., and van Oost, B. A. (1993). Abnormal behavior associated with a point mutation in the structural gene for monoamine oxidase. *Science* 262: 578–580.

Carey, G. (1994). Genetics and violence. In *Understanding and preventing violence: Biobehavioral influences on violence,* vol. 2, ed. A. J. Reiss Jr. and J. A. Roth. 21–58. Washington, D.C.: National Academy Press.

Carey, G., and Gottesman, I. (1996). Genetics and antisocial behavior: Substance vs. sound bytes. *Politics and the Life Sciences* 15 (1): 88–90.

Culp, S. (1997). Establishing genotype/phenotype relationships: Gene targeting as an experimental spproach. *Philosophy of Science* 64 (proceedings): S268–S278.

Duster, T. (1990). *Backdoor to eugenics.* New York: Routledge.

Eaves, L. J., Martin, N. G., and Heath, A. C. (1990). Religious affiliation in twins and their parents: Testing a model of cultural inheritance. *Behavior Genetics* 20: 1–22.

Fishbein, D. (1996). Prospects for the application of genetic findings to crime and violence prevention. *Politics and the Life Sciences* 15 (1): 91–94.

Goldman, D. (1996a). Interdisciplinary perceptions of genetics and behavior. *Politics and the Life Sciences* 15 (1): 97–98.

(1996b). The search for genetic alleles contributing to self-destructive and aggressive behaviors. In Stoff and Cairns 1996, 23–40.

Gottesman, I., and Goldsmith, H. H. (1994). Developmental psychopathology of antisocial behavior: Inserting genes into its ontogenesis and epigenesis. In *Threats to Optimal Development: Integrating Biological, Psychological, and Social Risk Factors,* ed. C.A. Nelson, 69–104. Hillsdale, N.J.: Erlbaum.

Holden, C. (1994). A cautionary genetic tale: The sobering story of D2. *Science* 264: 1696–1697.

Jacobs, P., Bruton, M., Melville, M. M., Brittain, R. P. and McClermont, W. F. (1965). Aggressive behavior, subnormality, and the XYY male. *Nature* 208: 1351–1352.

Kevles, D. J. (1985). *In the name of eugenics: Genetics and the uses of human heredity.* New York: Knopf.

Miller, J. (1996). *Search and destroy: African-American males in the criminal justice system.* Cambridge: Cambridge University Press.

Paul, D. (1995). *Controlling human heredity: 1865 to the present.* Atlantic Highlands, N.J.: Humanities Press.

Plomin, R., DeFries, J. C., McClearn, G., and Rutter, M. (1997). *Behavioral genetics,* 3rd ed. New York: Freeman.

Plomin, R., Owen, M. J., and McGuffin, P. (1994). The genetic basis of complex human behaviors. *Science* 264: 1733–1738.

Raine, A. (1993). *The psychopathology of crime: Criminal behavior as a criminal disorder.* New York: Academic Press.

Reiss, A. J., Jr., and Roth, J. A., eds, (1994). *Understanding and preventing violence: Biobehavioral influences on violence.* Vol. 2. Washington, D.C.: National Academy Press.

Rutter, M., Giller, H., and Hagell, A. (1998). *Antisocial behavior by young people.* Cambridge: Cambridge University Press.

Sherman, S. L, DeFries, J. C., Gottesman, I. I., Loehlin, J. C., Meyer, J. M., Pelias, M. Z., Rice, J., and Waldman, I. (1997). Recent developments in human behavioral genetics: Past accomplishments and future directions. *American Journal of Human Genetics* 60: 1265–1275.

Stoff, D., and Cairns, R. (Eds). (1996). *Aggression and violence: Genetic, neurological, and biosocial perspectives.* Mahwah, N.J.: Erlbaum.

Strawson, P. F. (1962). Freedom and resentment. *Proceedings and Address of the British Academy* 48: 1–25.

Wachbroit, R. (1998). The question not asked: The challenge of pleiotropic genetic tests. *Kennedy Institute of Ethics Journal* 8 (2): 131–144.

Wasserman D. (1995). Science and social harm: Genetic research into crime and violence. *Report from the Institute for Philosophy and Public Policy* 15 (1): 14–19.
  (1996). Research into genetics and crime: Consensus and controversy. *Politics and the Life Sciences* 15 (1): 107–109.

Witkin, H. A., Mednick, S. A., Schulsinger, F., et al. (1976). XYY and XXY men: Criminality and aggression. *Science* 196: 547–555.

PART I

Chapter 2

Understanding the Genetics-of-Violence Controversy

ROBERT WACHBROIT

The promise of behavioral genetics – especially the prospects of discerning a relationship between genes and violent behavior – is surely one of the most contentious issues in genetics, if not in all of biology. The development of ever more powerful and sophisticated research protocols and techniques of data analysis by some researchers has been greeted with a barrage of methodological complaints by others; whereas some scientists express optimism over the scientific fruitfulness of recent discoveries, others make pessimistic, if not dismissive, assessments of the likelihood of genetic explanations of behavior. And the controversy has not been confined to the lecture room, laboratory, or scientific conference. Many sectors of the public have followed this dispute keenly, some welcoming, others being alarmed by the news – which is often reported with little qualification – of what scientists claim to have demonstrated.[1] While the dispute has sometimes been intense, even confrontational, it has not been generally clear to everyone that several types of issues are in play in the controversy. Indeed, they form a structure of layers, a conceptual hierarchy, in which issues raised at one level presuppose that issues on a different level are not in dispute.

I want to set out this structure not only as a roadmap for understanding the controversy but also as an aid in sharpening the debate. The aim is not merely to clarify conceptual priorities and logical entailments, important as they are to get straight, but also to identify other problems that can arise from mixing levels of criticism. There is an old joke about the restaurant critic who claimed in his review that "not only was the food bad but the portions were small." It vividly illustrates how two criticisms, even though each might be true and make a forceful point, can undermine each other when combined. Though logically consistent, the criticisms are, so to speak, pragmatically inconsistent. We want to avoid having a counterpart of this joke arising in the more serious controversy over the possibility of genetic explanations of behavior.

FIRST-LEVEL CONTROVERSIES

At the most general level, the dispute is over whether human behavior can be scientifically studied and characterized. Although there are some disagreements over what constitutes a science and a scientific inquiry, especially among the social sciences, for the most part the people engaged in the controversy at this level understand a science as something that aims for and, with luck and perseverance, yields explanations, predictions, characterizations of how phenomena could be controlled or modified, and identifications of general causal mechanisms or processes.

So understood, a science of human behavior raises the issue of determinism and, consequently, purported threats to claims of free will and moral responsibility. For some people, a science of human behavior is impossible because, so it is argued, such a science would be incompatible with the truth that people typically have free will and are typically morally responsible for their actions. This a priori argument has a long and distinguished past that predates genetics; the argument has been assessed, refined, exposed, and re-posed by some of the best philosophers in history. Nevertheless, it is not an argument much invoked by critics of behavioral genetics, probably because it constitutes such a broad assault, taking in much more than genetics.

Many critics of the possibility of a science of human behavior frame their arguments more narrowly. They deny that there can be a science of certain types of behavior, particularly a science of violent or aggressive behavior. Their rejection can turn on three different objections. The first objection questions whether violent behavior is a sufficiently well defined classification. Any vagueness or ambiguity in the classification will translate into a corresponding vagueness or ambiguity in explanations, predictions, and the like. The second objection questions the stability of the classification: are researchers changing their definitions of violent or aggressive behavior to suit the results of their experiments? Changing definitions can misrepresent the significance of an experiment through equivocation. The third objection questions whether violent behavior constitutes a natural kind – that is, a classification amenable to distinct scientific generalizations. Not every classification yields groups that figure in scientific explanations or laws. Although the class of people who only watch television on every fifth day is well defined, no one believes that there are distinctive scientific laws in which this class figures.

Although the first two concerns – the need for clear and stable classifications – are important, we should keep in mind that, while science may begin with ordinary classifications, with all their possible fuzziness and variability, it will typically develop a different scheme. A common response to vague or ambiguous concepts in science is to propose a definition and then to revise it in the light of research. This is a well-established and proper indication of scientific progress. For example, the ancients plainly did not – indeed, could not – define pure water as H_2O. Consequently, some of the stuff they identified as pure water, correctly by their lights (e.g., clear spring water), is not pure water by the modern chemical definition because it contains various trace impurities. And so the scientific laws characterizing pure water would not apply to some of the stuff that used to be called "pure water." Such redefining is perhaps better viewed as refining, indicating that the categories we start with may need revision in the light of scientific discoveries or theoretical developments.

The point of these refinements and revisions in definitions is not simply to develop categories that are clear and well defined, but, more important, categories that are amenable to scientific laws. We may start with ordinary or commonly used classifications, but the demands of scientific theories may well transform them into something quite different. This transformation often happens in medicine. For example, until the nineteenth century, a fever was considered to be a (type of) disease. However, the search for the cause of this "disease" proved fruitless. It became apparent that fever can arise in quite diverse ways, from viruses, bacterial infections, or traumas. The concept of fever did not enter into systematic theorizing in the same way as other diseases. That is to say, even though the concept of fever is reasonably well defined and can figure in some scientific generalizations ("Sustained high fevers result in death"), there was no systematic characterization of its causes or treatments. In some cases, quarantining can prevent the spread of fevers, but in other cases it cannot. In some cases, drugs can treat fevers; in other cases, they cannot. Consequently, modern disease classifications do not regard fever as a kind of disease but rather as a symptom of many different kinds of disease. It may well turn out that many of our ordinary classifications of behavior – most notably, for our discussion, violent behavior – will not be scientifically tractable: some may turn out to be too vague to be of any use or, even when well defined, may turn out to refer not to different kinds of behavior but to symptoms or features of different kinds of behavior.

There are two responses to the concern that the concept of violent behavior is not amenable to scientific theorizing. One might replace the category of violent behavior with that of something else – for example, hyperactivity-inattention, impulsivity, sensation seeking, or lack of behavioral control – which, it is hoped, is scientifically more tractable. Such replacements are not without costs, even if redefinitions and reclassifications are the stuff of scientific progress: we need to ask whether the redefined category is still sufficiently similar to the old category so that the interest in the scientific findings is not "defined away." This question becomes especially urgent if there are divergences between the scientific and ordinary classification – for example, if a behavior that would fall under the ordinary classification of "violent" does not fall under the scientific counterpart of that classification. Should we revise our ordinary classification – should we, for example, be prepared to say that, as science has shown, that sort of behavior is really not violent behavior? Or should we regard the scientific classification as really about something different and so not possibly in conflict with ordinary classifications? When the chemist declares that (ordinary) glass is a liquid and not a solid, we don't revise our ordinary classification, but note the limited applicability of the chemist's definition for ordinary purposes.

The other response is not to replace the concept of violent behavior but to cast it as an occasional feature or symptom of a different, scientifically tractable concept. For example, one might hold that the tractable concept is cognitive ability, and that there can be scientific explanations – indeed, genetic explanations – of cognitive ability. Violent behavior is a sometime aspect of this condition in certain environments, such as when the environment is particularly frustrating, given the individual's cognitive level. In that case, there is no more a science of violence than there is a science of fever, at least as far as a systematic account of causes is concerned.

The non-a-priori objections to the possibility of a science of (violent) behavior can only be met in a non-a-priori manner: we have to see what researchers come up with. Will the various efforts at defining and refining behavioral classifications yield interesting scientific generalizations? If they do, how should divergences between classifications be regarded? As we assess developments, it seems reasonable to adhere to this principle: a science of violence, including a genetics of violence, cannot be held to a higher standard than other sciences. We cannot demand stronger proof or more certainty than we demand in, say, clinical genetics. If there is a case for departing from this principle when it comes to a genetics of violence, that departure must be explicitly acknowledged and argued for.

SECOND-LEVEL CONTROVERSIES

The second layer of the controversy proceeds by assuming that there can be a science of human behavior, including something sufficiently approximating a science of violent behavior. The second layer need not depend on actually settling the first-layer controversy. It can arise from an agreement, perhaps implicit, that certain issues are not presently in dispute; the possibility of a science of human behavior might be assumed for the sake of argument or be understood as a working hypothesis. In any case, the second layer consists of a dispute over what kind of science this science is, what sort of explanations it provides. In other words, to which science – genetics, psychology, economics – should we look to provide explanations of violent behavior? As we will see, the second layer is peculiar, because most participants in the debate deny violent behavior is the exclusive province of any one science while imputing such claims of exclusivity to their opponents.

The dispute over which science should provide explanations of violent behavior is inseparable from a dispute over the appropriate research questions to investigate. Typically, people do not simply ask, Why is there violence? They have a particular contrast class in mind. If they ask why this particular person is violent, the contrast class, perhaps only implicit, shapes the research agenda. They may be asking why is this person violent now, in contrast to how he was years ago – or in contrast to his siblings, in contrast to his peers, in contrast to other members of his socioeconomic class, in contrast to certain other mammals, and so forth. Each of these contrast classes suggests locating the study of human behavior in a different science. Conversely, once we locate the study of violence in one of the particular sciences, certain questions will be the wrong questions to ask; they are not fruitful or do not lend themselves to a clear answer or may require special assumptions for their pursuit. For example, the question, Why are young men of one particular social class more likely to be violent than those of another? is not likely to be considered the right question to ask by a behavioral geneticist. The contrast class is not readily characterized in genetic terms, and the assumption that there is a clear genetic difference may be too difficult to maintain.

The question should therefore, according to the geneticist, be tabled; with more research we may develop tools to answer that question or learn that there is a better question to raise about group behavior.

The question, Which science? may not seem particularly pressing or of general interest. The division of sciences reflects human convenience and the vagaries of history more than a fundamental division of the world. The distinction between nuclear physics and nuclear chemistry or between social anthropology, sociology, and social psychology can sometimes be more a question of tone and emphasis than of subject matter or method. Some sciences such as astrophysics are explicit hybrids of other sciences, and the list of sciences continues to change, as indicated by such recent entries as sociobiology and biopsychology. These new disciplines arise for a variety of reasons, such as to institutionalize multidisciplinary efforts or to attract new sources of funding. Consequently, given the shifting and always tentative division of the sciences, the controversy over which science can best provide explanations of violent behavior may seem of only limited interest and significance.

What drives a broader interest in the question is that it seems to be an articulation of the controversy over "nature versus nurture." The line separating nature from nurture is now drawn roughly and perhaps arbitrarily at conception: nature is what is present at conception, leaving everything else subsequent to be understood as nurture, including the random events that characterize the development of neural pathways in the fetus. Genetic explanations are therefore about the individual's nature; all other accounts are about his nurture. What is the point of this contrast?

Some have suggested that the significance of the distinction is that one's nurture can be changed but one's nature is immutable. This claim is simply false, as many writers (e.g., Jencks 1987) have pointed out: the effects of nature are no less mutable than the effects of nurture. It is well understood in medicine, for example, that the etiology of a disorder does not determine the mode of treatment, so that genetic disorders are sometimes treated with nongenetic therapies. The treatment of phenylketonuria (PKU) with dietary restrictions is a clear case of this.

A different suggestion for the significance of the nature-nurture distinction is that it is a proxy for one between the internal and the external: nature refers to who you (really) are, nurture refers to what (merely) happens to you. Plainly, the concept of identity (who you are) at work here is not a social one. Childhood experiences, for example, can profoundly affect a person's understanding of who she is, but they are all a matter of nurture. The concept of identity invoked is more metaphysical: she would be the same person even if her experiences were different, but she would not be the same person if her genetic makeup were different. But how different? To treat every gene as essential to a person's identity would make gene therapy a bad joke, for then a treatment that altered a particular (lethal) gene would not be curing that person but destroying her and replacing her with someone else. Similarly, because genes can sometimes mutate due to environmental effects (e.g., radiation), we would have to characterize such cases not as mutagenic effects but rather as the destruction and replacement of a person. In order to avoid such absurdities, we need to acknowledge that some genes are not essential to a person's identity. But if only some genes are held to be essential to a person's identity, there is no longer a rationale for equating a person's nature with her genome.

Even if strict genetic essentialism is rejected, the related concern of genetic determinism is sometimes raised at this level of the controversy. We can understand this concern in two very different ways. On the one hand, we can understand the concern as a worry about genetic *determinism* – that is to say, a concern about the implications of a particular kind of determinism. So understood, it is really a first-level issue: it is nothing more than the concern determinism sans phrase raises regarding the possibility of a causal science of (violent) human behavior. On the other hand, we can understand the concern as a worry about *genetic* determinism – that is to say, a concern not so much with the implications of determinism but with the implications *of reductionism.*

Determinism and reductionism are quite distinct, though they are often confused. Determinism is a thesis about universal causation – every event has a cause sufficient for its occurrence. Therefore, the phrase "genetic determinism" would, strictly speaking, mean that every event has a genetic cause sufficient for that event's occurring. Without a plausible restriction on the class of events being considered – for example, excluding meteorological events – genetic determinism is obviously false; genes do not determine the weather. Consequently, there are different kinds of genetic determinism, if you will, depending on how the class of events is restricted: we could consider a genetic determinism with regard to protein formation, a genetic determinism with regard to hormone secretion, a genetic determinism with regard to body mass, with regard to certain diseases, with regard to certain behaviors, and so forth. One kind of genetic determinism may be true even though others may be false. Unfortunately, many writers use the term "genetic determinism" without being either clear or precise as to the restriction they have in mind. This makes assessments difficult and confusion likely.

Whereas determinism is about causation, reductionism is about explanation. Causation and explanation are plainly related: in many contexts, identifying a cause amounts to providing an explanation. This may account for why causation and explanation, and so genetic determinism and reductionism, are conflated in some cases. Nevertheless, they are distinct: the causal relation is transitive, whereas the explanation relation is not always transitive – if A causes B and B causes C, then A causes C, but if X explains Y and Y explains Z, it does not necessarily follow that X explains Z.

Reductionism is a thesis about explanation – that a certain thing is really nothing but something else, so that explanations of the latter can yield explanations of the former. Reductionism need not explicitly invoke or entail a particular explanation; nevertheless, the point of reductionism is to help us construct better scientific explanations. This general position can get specified in different ways, leading to different kinds of reductionism. One kind is a reductionism between objects – sometimes called "ontological reductionism" – in which one type of object is held to consist of nothing but an ensemble of other objects of a different type, as far as explanations are concerned: for example, light is nothing but electromagnetic radiation and so can be explained by regarding it as electromagnetic radiation, or biological entities are nothing but physicochemical entities and so can be explained by regarding them as physiochemical entities.[2] Commentators who say that we are nothing but our genes should not be understood as making a claim of ontological reductionism. That would be obviously false, because the human body is composed of more than just DNA. Any such remarks should be understood as figures of speech referring to a different kind of reductionism.

Reductionism can also be understood as a relation between theories – for example, physical optics is nothing but electromagnetic theory, thermodynamics is nothing but (statistical) mechanics. In such cases, the reductionist's claim is not so much that certain objects are nothing but (ensembles of) other objects, but rather that the phenomena of one subject can be entirely explained by (nothing but) the lawlike generalities of the other subject. (There are several other kinds of reductionism as well, but we don't need to spell these out here.) For our purposes, the relevant reductionism is the view that certain phenomena – say, violent human behavior – can be entirely explained by theories concerning apparently different phenomena. For example, the view that violence can be entirely explained by genetic phenomena is reductionist; the view that violence can be entirely explained by social and economic forces or by the individual's psychology or by evolution are each in their turn reductionist. So understood, the dispute at this level is not over reductionism versus antireductionism but rather over what theory the phenomena in question are being reduced to.

It is striking that the views expressed at this level of the controversy about reductionism and nature versus nurture are not really in dispute. Most researchers and other scholars eschew talk of nature versus nurture, partly because they recognize its vagueness, but also because they acknowledge that it is a false opposition. Indeed, almost all reject genetic reductionism, social reductionism, and the like, and assert that human behavior is a product of nature and nurture (cf. Sherman et al. 1997). That is to say, (violent) behavior is the result of an interaction of a number of factors – genetic, developmental, social, economic, cultural – so that reductionist views, according to which such behavior is nothing but the operation of just one of these factors, are false. The consensus that behavior is a product of genes and environment – interactionism – rests on a number of considerations. At a minimum, there is the evident truth of counterfactuals such as: if many of that individual's genes were different, his behavior would have been different; and if his environment had been quite different, his behavior would have been different. Furthermore, various neurological disorders have been linked to genes – most notably, Huntington's Disease – which plainly have an impact on behavior. And genes can affect the production of various chemicals in the body (e.g., hormones), which can affect the brain, which in turn can affect behavior. Therefore, because genes can affect behavior and environments can affect behavior, it would seem that no one of the special sciences can provide a complete explanation of (violent) behavior. The second layer is important to the debate not because actual claims of exclusivity are widely embraced but rather because they are widely alleged.

One might wonder at this point that if there is actually a consensus on interactionism, what then is in dispute, aside from the details of particular studies or research designs? Why doesn't interactionism lead to a pluralism in the study of human violence? We could have many sciences – genetics, economics, psychology, evolutionary biology – each engaged in a different research agenda regarding violent behavior. That there still is an area of disagreement leads us to the third layer of controversy.

THIRD-LEVEL CONTROVERSIES

In the third layer there is agreement that no one science can provide a complete explanation of violent behavior. Disputes at this level arise over two related issues, reflecting an intellectual and a practical concern. First, an acknowledgment of interaction does not entail the view that the individual components are of equal importance. There may therefore be a dispute over whether one of the components (e.g., genetic factors) constitutes the primary cause or explanation of violence, relegating other factors to the status of refinements to the primary explanation (or, as a mathematician might put it, to second- and higher-order effects). Second, the study of violent human behavior is not undertaken merely out of curiosity. In many circumstances, violence is seen as a problem. Because violent behavior is the result of an interaction, there are several ways, at least theoretically, of solving the problem. The technical feasibility of different solutions is addressed by different research questions and agendas. Opposition to particular solutions, which may be based more on ethical or symbolic concerns than on scientific commitments, can in turn lead to opposition to pursuing associated research questions. Making an objectionable solution feasible is simply too dangerous, it might be argued.

Because the practical concerns are not the topic of this essay, our discussion here must be brief and sketchy. Three immediate problems could arise with a solution encouraged or suggested by a science of violence: it could have adverse side effects, it could easily be abused, or it could be deemed inappropriate though effective. The first two problems are clear enough, especially if the solution involves the administration of drugs. A proposed medication, for example, might impair a wide range of cognitive functions or might likely be administered, say, in a tense situation, prematurely and indiscriminately. The third problem arises if a mismatch is seen between the characterization of the problem and the solution. If a particular class of violent acts is seen as stemming from a moral failing, responding to it with drugs, as if it were merely a disease, could be regarded as an inappropriate response in that it does not give the moral character of the problem its due. The individuals should not receive treatment, but punishment or censure; the injustice should be rectified.

Genetic investigations can add a fourth problem to these three. In the clinical arena, genetic investigations have promised – and sometimes realized – the prospect of identifying conditions in asymptomatic individ-

uals. The possibility that scientists could (claim to) identify violent individuals before any trouble has occurred worries some people. In part
the worry is linked to the fear of diagnostic judgments superseding
moral judgments. Classifying certain people as violent before the act
can seem tantamount to judging people on the basis of their genetic endowment. The resulting stigma or discrimination would cause obvious
individual and social harm. Such classifications could also be used to justify preventive detention. The same reasoning that warrants quarantining individuals solely because they carry an infection that threatens public health might also be invoked to warrant similar measures to prevent
individuals who have "violence genes" from threatening the public.

While these worries are certainly important, they are to some extent
speculative. Arguing that certain research on violence should not be
pursued because it might lead to an objectionable solution rests on the
assumption that policy makers are unable, unwilling, or unprepared to
identify or respond to and correct problems. Debates over that assumption are all the more inconclusive because both pessimists and optimists
can point to history. There has certainly been a history of abuses in the
name of science, but there has also been a history of such abuses being
exposed, often by scientists, and corrected. The optimists are encouraged
by science's apparent ability to self-correct, but the pessimists, even if
they acknowledge this ability, worry about the (human) costs incurred
until the self-correction takes place. But short of a showing that research
on violence would yield "forbidden knowledge" – information that
would definitely lead to harm and mischief exceeding any possible
benefit – the sensible conclusion should be neither to stop the research
nor to ignore the risks, but to proceed cautiously.

Let us now turn to the more theoretical issue: even though there
could be a consensus that violent behavior reflects a complex interaction of various factors, there need be no agreement over the relative importance of each of these factors, or whether some particular factor
points to the primary or main explanation of the phenomenon. Thus,
some might say that even though behavior is a product of genetic and
social factors, the genetic explanation is the primary one; it explains
most of the phenomenon. What does that mean?

In many sciences, the idea of a primary or main explanation can be
understood in terms of a linear approximation. The explanation of the
complex phenomenon can be approximated by arranging a sequence of
simpler explanations, where each component explanation can be regarded as a refinement or modest correction of the previous component

explanation, and where the component explanations framed in terms of just one factor come before explanations framed in terms of two factors, and so on. For example, suppose we wanted to explain the trajectory of an object. The primary explanation of its path would be a simple one from classical mechanics in which only gravitational forces are invoked. But this is only an approximate explanation. If we wanted a more complete explanation (i.e., greater accuracy), we would refine this explanation by adding the effect of air friction on the velocity of the object. This would be the second term of the approximation – and so forth, as we added further and further refinements and modifications in the service of a more and more complete account of the object's trajectory. In a similar fashion, we might picture the environmental explanation of some particular piece of behavior as a refinement or filling-in of some details of the genetic explanation. The effect of the interaction between genetic and social factors – for example, the effect of environmental factors brought about by genetic factors, or of genetic factors that arise because of environmental factors – constitutes even finer detail. We might represent this as follows:

(Account of violent behavior) = (Account arising solely from genes) + (Account arising solely from environmental factors) + (Account arising from genes affected by environmental factors) + (Account arising from environmental factors affected by genes) + ...

The assumption of linearity is represented by the plus sign, meaning that the components can be simply summed. The claim that this equation is a reasonable approximation rests on each succeeding term representing a smaller contribution than the previous term.

As this representation should make clear, although the idea of explanations by linear approximation can be made precise in the case of quantitative explanations – in which an explanation can be seen as a convergent sequence of terms (or perturbations, as these would be characterized by the physicist) – it is admittedly vague in the context of explanations of behavior. What would it mean to say that the third term in the preceding representation is less important, makes less of a contribution, than the second? Consequently, it is not at all clear what remarks such as "Genetics plays a *significant* role in violent behavior" could mean. For example, Hamer and Copeland (1998), after duly acknowledging the truth of interactionism, assert that certain environmental factors, such as rearing or education, are not as important as certain biological factors, such as a childhood case of measles, for the explanation of behavior, and that genes "largely" determine behaviors such as alcoholism. But they don't really explain what "important" or "largely determine" means. Perhaps the intuitive idea is that some factors constitute causes, whereas others constitute merely influences or background conditions. But there are several ways one might try to understand such contrasts, regardless of whether the claim is that genetic factors or social factors constitute the primary or most important causal factor.

For example, someone who claims (contra Hamer and Copeland) that social factors are the most important might be claiming that although in any particular instance of violent behavior there are specific genetic factors that are causally implicated, there is no general, systematic story to tell. It might be like the weather. Although precipitation probably has some influence in particular instances of violence, we do not expect significant patterns such that level of precipitation could provide an explanation of violence. This response harkens back to the first level of controversy in that, although it acknowledges genetic factors in violence, it denies that these factors can be systematized into a scientific account – that a classification of violent behavior in terms of these factors forms a natural kind.

A different approach proceeds by invoking a contrast between direct and indirect causation. For example, it might be argued that genes affect behavior only indirectly – they merely influence it – and so are not as important as social factors in explaining violence, if one assumes that the latter affect behavior directly. The problem with this approach lies in understanding the contrast between direct and indirect causation. The distinction seems to rest on whether there are mediating steps in the causal sequence. That is to say, the relationship between A and B is an example of indirect causation if there is a C such that A causes C and C causes B. Thus, we might claim that genes don't bring about behavior; they manufacture proteins. Hence, genes are only indirectly related to behavior. Similarly, we might claim that any particular social factor (e.g., socioeconomic status) does not affect behavior directly but is mediated by other intervening social or psychological variables (e.g., peer pressure). But if that is what direct versus indirect causation means, the distinction is of little use here. Arguably, any factor could be characterized as representing indirect causation. It is just a matter of acknowledging intermediate steps. That is probably why the contrast between direct and indirect causation is not found in any scientific theory.

A more promising approach would be to say that the most important factors are those that are sufficient for the behavior to occur; next in importance are those that are necessary though not sufficient for the behavior; and, finally, the least important factors are those that are neither necessary nor sufficient for the behavior. Because we are assuming interactionism, we can rule out any factor being sufficient. (We could still talk about factors being conditionally sufficient or sufficient in the circumstances – that is, they might be considered sufficient when certain other factors are in place. But this is much weaker than straightforward sufficiency.) It would also seem that many researchers who hold to interactionism would deny that any particular factor is necessary for violent behavior. (I am, of course, putting aside trivial factors, such as those necessary for being alive, because they would be nondistinguishing – that is, necessary for any behavior.) There does not appear to be any distinguishing genetic, social, or other factor that people who behave violently have in common. This pattern could easily change: behaviors could be reclassified so that certain factors are seen to be common, and so necessary, for certain types of behavior. As we noted earlier, such reclassifications can be legitimate, in biology and elsewhere. Cancer is a case in point. Not only is cancer regarded as a category encompassing a number of different diseases, but even the common classifications in terms of affected organ or body part (lung cancer, breast cancer, etc.) are giving way to reclassifications in terms of genetic etiology (P53-gene cancers, BRCA1-gene cancers, etc.). As we noted earlier, the justification of such reclassifications rests on whether they yield fruitful scientific explanations and generalizations.

Another approach some have used for identifying the important factors has been the use of quantitative genetics – so-called heritability studies. By systematically observing people with varying degrees of genetic similarity (MZ twins, DZ twins, adoptees) in appropriately different settings, one can determine for a specific population the extent to which variations in a particular behavior are due to variations in genetic or nongenetic factors. If we put aside reservations about how well particular studies have been carried out, the difficulty with this approach, as Sober and Taylor argue in this volume, is that heritability tells us little about causation in individuals. Because heritability depends upon the particular population selected – that is to say, it reflects how much genetic factors, nongenetic factors, and types of behavior vary within a particular population – altering the population can easily change the results. Some factors may vary more (or less) in the new population than in the old, even though the different populations may contain many of the same individuals. Furthermore, the fact that a particular factor correlates highly with the variation of a particular trait does not entail that that factor plays an important role in the existence of that trait. Christopher Jencks gives a nicely vivid illustration of this: short hair is highly correlated with the Y chromosome, but plainly the length of people's hair is primarily due to social factors. (It is therefore a source of confusion to use the term "heritability," because a trait with high "heritability" needn't be heritable or even to a large extent genetic. As the American Society of Human Genetics [ASHG] notes in its statement on behavioral genetics, "heritability is a descriptive statistic of a trait in a particular population, not of a trait in an individual" [Sherman et al. 1997]. Nevertheless, the terms are too well entrenched for us to expect researchers to adopt clearer terminology soon.) Consequently, heritability measures are not useful for determining which factors are (causally) more important than others as far as a (causal) explanation of violent behavior in an individual is concerned.

It is unfortunate that many nonscientists do not understand or appreciate these points about heritability measures versus causation, but particularly troubling is that some prominent behavioral geneticists do not either. Consider a recent textbook on behavioral genetics (Plomin et al. 1997). In it there is a straightforward discussion of the contrast between investigating behavior and investigating behavioral differences as well as a discussion of heritability being a measure of variation in a particular population rather than a causal claim about individuals. Nevertheless, a discussion of antisocial behavior includes a figure and caption (p. 212), reproduced here as Figure 2.1. While the text talks about correlations, the figure caption talks about causes, suggesting causal influences on individuals. Given the heated controversy over the genetics of antisocial behavior, such inferential leaps are not only unjustified but especially irresponsible in a textbook.

It is perhaps worth noting that twin and adoptee studies also raise a first-level issue about scientifically tractable concepts. Insofar as such studies are investigating causes, it would seem that their subject is the causes of individual differences in violent behavior, rather than the causes of violent behavior itself. That is to say, the aim of such studies will often be to investigate not the question of why A is violent but rather

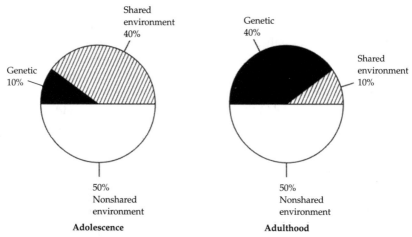

Figure 2.1. The causes of antisocial symptoms change from adolescence to adulthood, with genetics becoming more important and shared environment becoming less important (based on Lyons et al. 1995). From *Behavioral genetics* by Plomin et al. © 1997, 1990, 1980 by W. H. Freeman and Company. Used with permission.

the question of why A is more violent than B. There is no guarantee, however, that if the general category of violent behavior is scientifically tractable, the differential category – individual differences in violent behavior – will be so as well. To illustrate this, suppose there were a trait that was almost entirely biologically determined, though some small details arose from random events. Further, suppose that this trait were universal. In that case, individual differences would not be explicable (since they would be random) and so the category of individual differences in this trait would not be scientifically tractable. Nevertheless, for the most part, the trait itself, by hypothesis, could be explained biologically and so correspond to a scientifically tractable category. Consequently, research on differences, which is what family, twin, and adoptee studies investigate, needs to make special assumptions about the tractability of the categories.

A different approach to distinguishing the various factors would be to assign importance to causal factors in terms of the magnitude of their effect. For example, if a small change in the economic factor does not have as big an impact as a small change in genes, one might conclude that genetic factors are more important than economic ones. Despite its intuitive appeal, the problem with this approach is the lack of a common metric. What does it mean to say that some change in a factor is small, especially when we are dealing with factors as diverse as genetic and social factors? Consider an admittedly simple example. Suppose a particular genetic mutation has a bigger impact on the incidence of violence than a 10 percent increase in income but less than a 50 percent increase in income. Can we say whether genetic or economic factors are the more important? And how should the genetic change be represented: as a 100 percent change in a nucleotide or as a minuscule percent change of an individual's genome? Comparisons based on the magnitude of effect are not promising because we cannot give a nonarbitrary content to such questions as, Which is the bigger change – a point mutation or a 10 percent increase in income?

Another approach that one might consider is to appeal to a contrast between causes and conditions. Some factors should be characterized more as background conditions; the important factors are those that can properly be referred to as causes. There have been several proposals for how to draw the line between causes and conditions. Some have claimed that causes are factors that we can control or manipulate, whereas conditions are factors beyond our control. Thus, the blow from the bat swing caused the baseball to land in the bleachers; gravity was a (background) condition. Others have identified causes as the unusual or salient factors. Thus, the driver being drunk caused the accident last night; that the road was also dark was a (background) condition. Still others have identified causes as those factors that address the interests we may have in the inquiry. Thus, what the road engineer regards as the cause of a car accident (the banking of the turn) might be different from what the automotive engineer regards as the cause of the accident (the way the power is distributed on each of the car wheels). Despite the extent to which these suggestions are in line with ordinary talk about causes versus conditions, or even with the use of such concepts in the law, they are not the basis for determining which factors are primary or most important for a scientific account of violent behavior, because they rest on context-dependent considerations – which are subjective in the sense that they depend on what we find salient or what interests motivate the inquiry rather than solely on objective features of the world.

Finally, one might assign importance to various factors on the basis of a "scientific world view" which identifies the entities and processes that are more "fundamental" than others. For example, someone who holds that we are "fundamentally" social creatures (or biological entities, or products of evolution) – that is to say, we are primarily shaped by social and economic forces (biological drives or evolutionary pressures) – will likely hold that the proper (i.e., fundamental) explanation of violent behavior requires using the best methods and theories of the social sciences (or the biological sciences or evolutionary psychology). When the American Public Health Association calls violence a public health problem, such a charge amounts for some to seeing violent behavior as a disease whose explanation and treatment (control) is primarily to be found in the public health sciences. Such world views are not so much discoveries as working hypotheses: it is not that we have at present compelling scientific evidence that, for example, genetic factors merit the most (research) attention; rather, that belief fits so well with our other beliefs, scientific and otherwise, that we should proceed with it as a working hypothesis. The apparent success in using the theories and methods of genetics in the investigation of disease or other biological conditions can be highly suggestive.

The obvious worry with a working hypothesis is that its status can become forgotten or exaggerated, so that it is illegitimately construed as an established scientific claim. This can easily happen when scientists want to hype their work in order to attract funding or enhance their prestige, or when journalists want to dramatize their stories for similar reasons. It is plainly the responsibility of scientists to monitor each other and expose such errors and excesses.

CONCLUSION

There are three levels to the controversy over the prospects of a genetics of violent behavior. The first level, putting aside a priori arguments against the possibility of a causal science of (violent) behavior, consists of disputes over whether there is a way of classifying violent behavior suitable for scientific investigation. Determining the answer to that question is as much a part of the business of science as testing hypotheses and designing experiments. As things stand now, there is no consensus among scientists that such a classification has been identified. Whether a suitable classification can be found remains to be seen. In the meantime, given the all-too-human temptation of many scientists and journalists to hype or dramatize, the public should be wary of the latest "findings" reported in the media.

The second level of the controversy, in which it is assumed that there can be a science of violent behavior, is over what kind of science it should be. At this level, the choice is understood to be exclusive, so that the issue is over a choice of different reductionisms – genetic, psychological, social, evolutionary. Because most scientists reject reductionism, disputes at this level are often not against real opponents.

The third level of the controversy, in which it is agreed that (violent) behavior is the result of a complex interaction of factors, is over which science (or factor) is the most important or significant in explaining violent behavior. This level is beset by vagueness and unclarity over what "most important or significant" means. There are various ways of explicating the claim of significance, but such explication is unfortunately not made by the disputants. Although we have surveyed a range of possible ways of understanding what is at issue, these turn out to be either themselves unclear, hypothetical, context-dependent, and nonobjective, or to involve an illegitimate inference from the analysis of variance in a population to a claim of causation in individuals.

One obvious conclusion to draw from this taxonomy is that the various objections or worries about behavioral genetics are not all compatible with each other. Some objections can be raised only when others are not. A less obvious conclusion concerns the elusiveness of the controversy. Because most defenders of behavioral genetics subscribe to interactionism, only those critics who raise either first- or third-level objections are on target. But first-level disputes can only be settled by the progress of science, and third-level objections appear to rest on vague and unclear views about interactionism.

The taxonomy thus yields a classification of the various criticisms of behavioral genetics. Plainly it is not, however, a taxonomy of the various critics of behavioral genetics. Not only do many critics invoke a variety of arguments and considerations, but some appeal to arguments from several different levels simultaneously, resulting in an argumentative stand that may not be consistent. As an illustration of this problem, as well as of others raised in the course of this discussion of how the controversy should be structured, let us examine a particular set of objections. Steven Rose (1995), in an oft-cited article in *Nature,* objects to behavioral genetics in part because it engages in "reification" and "arbitrary agglomeration." The latter is a "lumping together [of] many different reified interactions as all exemplars of the [*sic*] one thing. Thus, aggression becomes the term used to describe processes as disparate as a man abusing his partner or child, fights between football fans, strikers resisting police, racist attacks on ethnic minorities, or wars" (380).

So understood, the charge of arbitrary agglomeration is a first-level criticism. Rose does not seem to be aware, however, that any reclassifi-

cation, proposed by any science, that diverges from an ordinary classification will appear to be an "arbitrary agglomeration," at least initially.
But that is the stuff of science – to discern patterns and similarities; its
dramatic successes occur when things ordinarily viewed as dissimilar
are shown to be similar on a deeper level. For example, up until at least
the sixteenth century, celestial motion was regarded as completely different from terrestrial motion. Newton's dramatic achievement was to
"agglomerate" them, successfully showing that they were in fact the
same type of phenomenon, governed by the same principles of mechanics. A similar story can be told about disease classifications: is the
viewing of certain diseases of the lung, of the breast, and of the brain
as cancers an *arbitrary* agglomeration? It is an empirical matter, a matter of discovery, whether any particular agglomeration works. To claim,
apparently as a matter of principle, that behavioral genetics cannot use
anything other than ordinary classifications is to claim that it cannot in
principle be a science, by holding it to a standard no other science is
held to.

The same considerations apply to Rose's earlier complaint about
"reification." The problem here, according to Rose, is that "reification
converts a dynamic process into a static phenomenon. Thus violence,
rather than describing an action or activity between persons, or even a
person and the natural world, becomes instead a 'character' – aggression – a thing that can be abstracted from the dynamically interactive
system in which it appears and studied in isolation. . . . [T]o reify the
process is to lose its meaning" (380). Again, Rose seems unaware that investigating interactions between objects by studying the *properties* of the
individual objects is a well-established procedure in the sciences. In
physics, we investigate the gravitational interaction between objects (a
dynamic process?) by studying the character of the gravitational field of
the individual objects (a static phenomenon?). In medicine, we might investigate a contagious disease by studying the character of the infectious
vector and the vulnerabilities of the host. Of course, there is no a priori
guarantee that such procedures will work, but in any case their working
does not depend on their entirely capturing the "meaning" of the interaction. In each case, their success depends on the particular scientific
theories of which they are a part. Again, behavioral genetics is being held
to a higher standard than other sciences.

Indeed, not just behavioral genetics. Because these alleged problems
are criticisms at the first level, they are quite general. It would seem
therefore that any science, including the various social sciences, would

also be subject to these criticisms insofar as they utilize the familiar science techniques of reclassification and the like. But this is not a road Rose wants to go down. He acknowledges that, to some extent, there is a science ("neuroscientific knowledge") of behavior. But this is an acknowledgment that forms the backdrop of third-level concerns; it is incompatible with first-level concerns. (Discerning inconsistency in Rose's attack may have a certain irony, since one of his first criticisms of behavioral geneticists is that they are inconsistent in that they claim to be interactionists but fail to act like ones.)

Behavioral genetics, especially a genetics of violence, needs to be closely scrutinized. As things stand now, in many cases legitimate concerns as well as hopes cannot be properly advanced without clearer arguments. By framing the arguments as the taxonomy of this chapter suggests, we will have a better understanding of what more needs to be said and – given the joke mentioned in the opening paragraph – what more should not be said.

NOTES

1. Several popular histories of the recent dispute have lately appeared (Wright 1998 and Wingerson 1998) as well as more scholarly efforts (Paul 1995 and Selden 1999).
2. Consequently, a claim about mere composition, with no eye toward explanation – for example, the human body consists primarily of H_2O and a few other chemicals – would not be reductionist. For a somewhat different view, see Sarkar 1992.

REFERENCES

Hamer, D., and Copeland, P. (1998). *Living with our genes: Why they matter more than you think.* Garden City, N.Y.: Doubleday.

Jencks, C. (1987). Genes and crime. *The New York Review of Books,* February 12, 33–41.

Lyons M. J., True, W. R., and Tsuang, M. T. (1995). Differential heritability of adult and juvenile antisocial traits. *Archives of General Psychiatry* 52: 906–915.

Paul, D. (1995). *Controlling human heredity.* Atlantic Highlands, N.J.: Humanities Press.

Plomin, R., DeFries, J. C., McClearn, G., and Rutter, M. (1997). *Behavioral genetics.* 3rd ed. New York: Freeman.

Rose, S. (1995). The rise of neurogenetic determinism. *Nature* 373: 380–382.

Sarkar, S. (1992). Models of reduction and categories of reductionism. *Synthese* 91: 167 -194.

45

Selden, S. (1999). *Inheriting shame: The story of eugenics and racism in America.* New York: Teachers College Press.

Sherman, S. Lu, DeFries, J. C., Gottesman, I. I., Loehlin, J. C., Meyer, J. M., Pelias, M. Z., Rice, J., and Waldman, I. (1997). Recent developments in human behavioral genetics: Past accomplishments and future directions. *American Journal of Human Genetics* 60: 1265–1275.

Wingerson, L. (1998). *Unnatural selection: The promise and the power of human gene research.* New York: Bantam Books.

Wright, W. (1998). *Born that way: Genes, behavior, personality.* New York: Knopf.

Chapter 3

Separating Nature and Nurture

ELLIOTT SOBER

Plant and animal breeders routinely attempt to disentangle the contributions of nature and nurture when they think about what makes corn grow tall or cows produce more milk. To apply the same concepts to human characteristics such as intelligence and violence, however, is politically explosive.

The discipline of quantitative genetics separates the relative contributions of genes and environment by deploying a set of technical concepts. The main one is called *"variance,"* which measures how much a trait varies in a population. Nature and nurture are analyzed by discussing variance in its different forms; there is phenotypic variance, genetic variance, environmental variance, and variance due to gene-environment interaction. Is it solely political considerations that make some people resist applying these humdrum scientific concepts to human beings? Or are there purely scientific considerations that block the easy transfer of these concepts from one domain to the other? I do not pose this question to disparage the significance of political questions; science is a human activity, and whether a scientific question should be pursued depends on what the consequences for human welfare would be of pursuing it. However, my goal in this chapter is to explain why the issue is not purely political. I am not going to argue that human beings and their traits are somehow outside the scope of biology, whatever that might mean. A human being develops a level of intelligence and attitudes toward violence because of the genes he or she possesses and the environments he or she inhabits. The very same thing is true of height in corn plants and milk yield in dairy cattle.

I am grateful to Andre Ariew, Ned Block, Richard Lewontin, and Steve Orzack for useful discussion.

No, the problem is not that we are outside the biological realm. The concepts of phenotypic, environmental, and genetic variance apply to human beings just as much as they do to cows and corn. For any trait we care to name, there is a fact of the matter concerning what the heritability of that trait is within this or that human population, whether we want to know about that fact or not. The problem is that our current level of knowledge frequently prevents us from ascertaining what the relevant facts are.

In what follows, I explain the relevant mathematical concepts in a way that is both simple and general. The math involved never goes beyond the arithmetic you learned in elementary school. The explanation, however, is quite general, applying as it does to any trait in any population of organisms. I am hoping that this chapter will make the meaning of heritability and related concepts completely transparent even to those with serious cases of math phobia. After defining these quantities, I explain two procedures that are frequently used for estimating the heritability of traits in populations of organisms. Once again, the discussion of estimation procedures involves nothing more than simple arithmetic. At this point in the exposition I explain why it is hard to find out how heritable many traits are.

The upshot of this discussion is not that there is a vitally important property of a trait – its degree of heritability – that we unfortunately are cut off from apprehending. Quite the contrary. In addition to explaining what heritability is, I discuss the question of why it matters and why the everyday concept of a trait's being "inherited" differs in several respects from the technical concept of its having a high degree of "heritability." Once we see heritability for what it is, the question arises of why the heritability of a trait is interesting; another question I ask is whether the heritability of a trait will become less interesting as science learns more about its underlying genetics.

THE ANALYSIS OF VARIANCE

Nuts and Bolts

Consider a farmer who grows two fields of corn. In the first, the corn plants are genetically identical – all have genotype G_1 – and they receive one unit of fertilizer (E_1). In the second field, the corn plants also are genetically identical, but they have genotype G_2; in this second field, the plants receive two units of fertilizer (E_2). At the end of the growing season, the farmer sees that the plants in the first field are one unit tall (on average), whereas those in the second field average four units of height. These observations can be recorded as two entries in a two-by-two table:

		Genes	
		G1	G2
	E1	1	-
Environment			
	E2	-	4

The farmer wants to answer a question about nature and nurture: Do the corn plants in the two fields differ in height because they are genetically different, because the plants grew in different environments, or for both these reasons? And if both genes and environment are responsible for the difference between the two fields, which mattered more? With the data described so far, the farmer has no way to answer these questions. The reason is that the genetic and the environmental factors are perfectly *correlated*; G1 individuals always inhabit E1 environments and G2 individuals always live in E2 environments.

The way for the farmer to make headway on this problem is to break the correlation. Let him plant a third field in which corn plants have G1 genotypes and receive two units of fertilizer; let him plant a fourth field in which G2 plants receive one unit of fertilizer. This will allow him to enter data in the other two cells in the two-by-two table just displayed. From these data, the farmer can make an inference concerning how genetic differences and differences in fertilizer treatment contributed to variation in plant height.

The experiment just described might generate different observational outcomes. Here are four possibilities to consider:

	G1	G2		G1	G2		G1	G2		G1	G2
E1	1	1	E1	1	2	E1	1	3	E1	1	4
E2	4	4	E2	3	4	E2	2	4	E2	1	4
	(i)			(ii)			(iii)			(iv)	

In outcome (i), the genetic factor makes no difference; whether the plants have genotype G1 or G2 does not affect their height; it is the environmental factor – the amount of fertilizer the plants receive – that explains all the observed variation.[1] Outcome (iv) is the mirror image of (i). In (iv), the fertilizer treatment makes no difference; the genetic variation explains all the variation in height. Outcomes (i) and (iv) support *monistic* explanations of the variation in plant height; each suggests that only one of the factors considered made a difference in the observed outcome.

Outcomes (ii) and (iii), on the other hand, support *pluralistic* conclusions. Both suggest that genetic and environmental factors made a difference. They disagree, however, as to which factor mattered more. In outcome (ii), changing the fertilizer treatment yields two units of change in height, whereas changing from one genotype to the other produces only a single unit of change. In this case, the environmental factor makes more of a difference than the genetic factor. By the same reasoning, we can see that outcome (iii) suggests that genetic variation was more important than the environmental factor considered.

Although the four possible outcomes described so far differ in various respects, they have something in common. In each of these data sets, the change effected by moving from G_1 to G_2 does not depend on which environmental condition one considers. Similarly, changing from E_1 to E_2 has the same impact on plant height, regardless of which genotype the plant possesses. Results (i)–(iv) are thus said to be "additive" (or to show no "gene-environment interaction"). This is not the case for the following possible results:

	G_1	G_2		G_1	G_2
E_1	1	7	E_1	1	1
E_2	7	4	E_2	1	4
	(v)			(vi)	

In outcome (v), going from one unit of fertilizer to two increases plant height for plants with genotype G_1, but reduces height for plants that have genotype G_2. In (vi), changing genotype has an effect on plant height within one fertilizer treatment but not within the other.

This simple two-by-two experiment illustrates the method that statisticians call the analysis of variance (ANOVA). By generalizing the scheme just described, we can introduce some terminology to label the relevant concepts. In the example just discussed, genotype and fertilizer treatment combine to make a plant grow to a certain height. Height is a phenotypic trait; amount of fertilizer is an environmental condition. There are two possible causes and one effect:

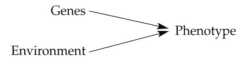

What, in general, do the words "gene," "environment," and "phenotype" mean in ANOVA investigations?

It is hard to answer this question with precision and without circularity. Phenotypes are often described as any feature of an organism's morphology, physiology, or behavior. However, psychological and cultural characteristics also are parts of the phenotypes of some organisms; knowing Korean and liking rock and roll are phenotypic traits, just as much as an individual's height and blood type. What is excluded from the phenotype is the sequence of nucleotides found in the strands of DNA in each cell of an organism's body. In a sense, we can view the organism's phenotype as all the traits it has that are caused by its genes and/or its environment. This way of defining "phenotype" entails that having a certain set of genes is not part of the organism's phenotype, and living in a certain environment isn't either. It will be convenient in what follows to think of phenotypes as quantitative features; they may come in integer values (like number of fingers), or they may be continuous (like height and weight).

Genes are possible causes of phenotypes. But there is also the contribution of the environment. What does "environment" mean? Once again, we must realize that the concept of environment is used in ANOVA as a garbage can category; an environmental factor is anything that is not genetic. The most important point to recognize here is that "genetic" is not synonymous with "biological." This is a common confusion in discussions of nature and nurture. Fertilizer is a "biological" cause of corn plant height, but it counts as an environmental, not a genetic, factor.

If phenotype is defined by contrast with what is genetic, and if the environment is defined by contrast with what is genetic, how are phenotype and environment distinguished? Well, an organism's environment causally contributes to the phenotype it has, but that is not a sufficient answer. After all, it also is true that the organism lives in a particular environment in part because of the phenotype it has. For example, lizards, being cold-blooded creatures, seek out warm environments and shun cold ones. Where the organism lives is a consequence of its physiological makeup.

It suffices for our purposes to distinguish phenotype and environment in the following rough-and-ready way: the organism's phenotype includes only the traits it has in virtue of what is going on inside its own skin (Sober 1984, sec. 1.5).[2] The organism's environment includes only the traits the organism has in virtue of what is going on outside its own skin. Living in a warm place is an environmental trait; being cold-blooded is a feature of the organism's phenotype (Dawkins 1982).[3]

In the corn plant example, there were two genetic conditions and two environmental conditions. Let's generalize. Each organism studied is in one of m possible genetic states and experiences one of n possible environmental conditions. An experiment that examines all combinations of these genetic and environmental conditions will have n-times-m treatment combinations. Organisms within each treatment cell are measured for some phenotypic trait. As in the simpler two-by-two example, cell entries record the average phenotype of individuals in each treatment. To simplify exposition, I'll assume that each cell contains the same number of individuals:

	G1	G2	G3	...	Gm	
E1	x_{11}	x_{12}	x_{13}	...	x_{1m}	$x_{1.}$
E2	x_{21}	x_{22}	x_{23}	...	x_{2m}	$x_{2.}$
	.		.		.	
	.	.	.	.	.	
	.	.	.	.	.	
En	x_{n1}	x_{n2}	x_{n3}	...	x_{nm}	$x_{n.}$
	$x_{.1}$	$x_{.2}$	$x_{.3}$	...	$x_{.m}$	M

In addition to the x_{ij} entries in the m-by-n table itself, the table also provides some numbers that are written down along the margins. These are called, appropriately enough, the *marginal averages*. They describe the average phenotype for individuals who experienced the same environment (but different genes), or the same genes (but different environments). The table also states, as a final entry in the lower right-hand corner, the grand mean M – the average phenotype in the entire population of individuals.

The phenotypes represented in the m-by-n table – the different x_{ij} entries – display a certain amount of variation; the numerical values may be tightly clustered or they may be spread out. The standard mathematical measure of the amount of "spread" a phenotype has in a population is given by the phenotypic *variance* (V_p). To compute this quantity, one finds the difference between each x_{ij} and the grand mean, squares this difference, and then computes the average of these squared differences:

$$V_p = \Sigma_{ij} (x_{ij} - M)^2 / nm.$$

Intuitively, the amount of variation present in this population can have two sources. First, there is the fact that individuals live in different environments; second, there is the fact that individuals have different genes. This idea is captured by the fact that we can decompose the total (phenotypic) variance in the population into two parts – the genetic variance and the environmental variance:[4]

$$V_g = \Sigma_j \, (x_{.j} - M)^2 / m.$$

$$V_e = \Sigma_i \, (x_{i.} - M)^2 / n.$$

Each of these variances is computed by seeing how much the marginal averages (the $x_{.j}$'s and the $x_{i.}$'s) vary from the grand mean.

Just to demystify this way of representing and decomposing variation, let's analyze a very simple data set, which the farmer might obtain in his two-by-two experiment:

	G1	G2	
E1	1	3	2
E2	5	7	6
	3	5	4

The marginal averages and the grand mean are duly recorded, from which we can compute the three variances:

$$V_p = [(1-4)^2 + (3-4)^2 + (5-4)^2 + (7-4)^2]/4 = 5$$

$$V_g = [(3-4)^2 + (5-4)^2]/2 = 1$$

$$V_e = [(2-4)^2 + (6-4)^2]/2 = 4.$$

Note that in this example $V_p = V_g + V_e$. This defines what it means for a data set to be additive.

However, the sums do not come out this way in the following data set, which involves an interaction:

	G1	G2	
E1	1	3	2
E2	5	11	8
	3	7	5

Here are the values for the three variances:

$$V_p = [(1-5)^2 + (3-5)^2 + (5-5)^2 + (11-5)^2]/4 = 14$$

$$V_g = [(3-5)^2 + (7-5)^2]/2 = 4$$

$$V_e = [(2-5)^2 + (8-5)^2]/2 = 9.$$

In the present case, we introduce a quantity I to represent the difference between V_p and $V_g + V_e$. In this example, the gene-environment interaction term has a value of unity.

As the various hypothetical data sets I have described make clear, we should not assume in advance that the data produced in an experiment will turn out to be additive. Rather, we should describe the total phenotypic variance as decomposing into three parts:

$$V_p = V_g + V_e + I. \tag{1}$$

The data we obtain may show us that $I = 0$. Indeed, it may turn out that $V_g = 0$ or that $V_e = 0$, as was true in data sets (i) and (iv). These are empirical matters, which will vary with the population studied and the trait considered.

Because V_g and V_e are quantities that describe how much variance is associated with the different genes and the different environments considered in the experiment, we may compare these two numbers to say which of them induced the larger amount of variation. It is customary to do this by taking ratios of each of these quantities, relative to the total phenotypic variance. If we divide both sides of (1) by V_p, we obtain:

$$1 = V_g/V_p + V_e/V_p + I/V_p. \tag{2}$$

The three right-hand terms describe, respectively, the proportion of the phenotypic variance that is due to genes, to environment, and to gene-environment interaction; they total 100 percent.

The first of the ratios in proposition (2) defines the concept of "heritability"[5] $h^2 = V_g/V_p$. A phenotype's heritability is the proportion of its variance that is caused by genetic variance. Notice that heritability is a property of phenotypes, not genes. In ordinary parlance, we talk of genes as well as phenotypes (like eye color) being "inherited"; however, "heritable" and "inherited" are not synonymous. Genes are not heritable. We examine other differences between the concepts of "heritable" and "inherited" in the next section.

Notice that proposition (1) cites *three* possible causes of phenotypic variation. This means that the genetic contribution to variation cannot be defined as the contribution that is not environmental. Thus, the analysis of variance provides a finer-grained representation of causal contribution than the intuitive one with which we started. We began with the idea that phenotypes are caused by genes and by environment, where "gene" and "environment" are defined so that there is no other type of factor that can cause an organism's phenotype. Although this intuitive picture is good enough when it comes to talking about the traits of an individual *organism*, it is not adequate as a description of what can produce phenotypic variance in a *population*. For the latter task, the mathematics of ANOVA requires that we recognize genetic variance, environmental variance, and variance due to gene-environment interaction. We have moved from *two* causes to *three*.[6]

Some Philosophical Comments

The type of experiment just described can provide information about the relative causal contributions of genes and environment to the phenotypic variation found in a population. However, that information must not be misinterpreted.

The analysis of variance can provide information about causality only if the population exhibits variation in the effect term studied. Consider a population of human beings in which everyone has exactly two hands. The grand mean is 2, and each genotype/environment treatment has 2 as its average number of hands. There is no variation to explain; and there is no phenotypic variation due to genes nor any due to environment; $V_p = V_g = V_e = 0$. This example shows another respect in which "heritability" and "inherited" differ in meaning. Hand number may seem like an obvious example of an "inherited" characteristic; however, its heritability in the population just described is not defined, since $V_g/V_p = 0/0$.

What would happen if we considered a larger population in which the number of hands does vary? We can construct an example of this sort by augmenting a population of two-handed individuals with individuals who are born with a smaller number of hands. These individuals may have one hand or none solely because they have some genetic defect; alternatively, they may have been born without hands solely because of a feature of their fetal environment (e.g., perhaps their mothers took a certain drug while pregnant). If the enlarged population includes handless individuals solely of the first type, then hand number will turn out to be highly heritable; if the enlarged population includes handless individuals solely of the second type, then hand number will have zero heritability. And if both types of individuals are included, the resulting phenotypic variation will have both a genetic and an environmental explanation.

This example provides an interesting lesson. It is possible to assess the heritability of a trait without having any understanding of the developmental processes that lead individuals to exhibit the trait. Even if we know nothing about how an individual's genes and environment conspire to insure that he or she develops hands, we nonetheless can tell whether *variation* in hand number is genetic or environmental or both. This point was already visible in the farmer's experiments discussed in the previous section. He need not understand *why* a particular combination of genes and environment yields plants that average four units of height. It suffices for him to observe that this has happened. The analysis of variance permits one to infer *how much* a cause contributes to an outcome without understanding *how* the cause manages to have its effects. In part, this is because ANOVA aims to explain the variation of traits in a population, not to explain why individual members of the population have the traits they do.[7] It is a fact about development that the genes in an organism's body help explain why that individual ends up with two hands; it is a quite separate matter whether genetic differences in a population help explain differences in hand number.

Perhaps the most important point about interpreting ANOVA data as evidence about causal contribution is that the inferences are specific to the phenotypic trait considered, the range of environments and genotypes studied, and the population studied. In the two-by-two experiments contemplated before, we considered the results of a specific pair of genetic traits (G_1 and G_2) and a specific pair of environmental variables (E_1 and E_2). The results obtained in that restricted domain say nothing about what would happen if some new genotype G_3 were taken into account, or if some new environment E_3 were brought into the problem. For example, it could turn out that the difference between G_1 and G_2 makes no difference in plant height, but that G_3 makes all the difference in the world. The same could happen for the environmental effect; the difference between one unit of fertilizer and two might not matter, even though three units cross a threshold that matters a lot. In short, an ANOVA experiment does not ascertain how much genes *in general* matter, or how much the environment *in general* matters. What the experiment investigates is a *specific* set of genetic factors and a *specific* environmental treatment.

Another respect in which ANOVA yields specific results, not general ones, is that the relative importance of genes and environment to a phenotype can change as a population ages. Suppose the farmer conducts his two-by-two experiment, computes the environmental and genetic effects after the corn plants are three weeks old, and finds that most of the variation is due to genetic differences. If he then follows the plants for an additional three months, it may turn out that variation in height at that later date is mainly due to environmental variation. Here we have another difference between the technical concept of heritability and the commonsense idea of a trait's being inherited. According to the commonsense concept, if you inherit a trait, it remains an inherited trait as long as you have it; however, the heritability of a trait, because it is a property of a population, can change as the population changes.

Two more details are worth mentioning in connection with the fact that ANOVA is specific to the range of environments and genotypes considered and the specific population under study. Different subgroups in the same overall population may show very different patterns of genetic and environmental variation. Imagine a four-by-four experiment; environmental treatments E_1, E_2, E_3, E_4 are paired with genetic conditions G_1, G_2, G_3, G_4. Let us consider two subsets of this entire experiment. In the upper left-hand corner of the four-by-four data table we are imagining, we find a description of what happens when E_1 and E_2 are paired with G_1 and G_2. In the lower right-hand corner, we find E_3 and E_4 paired with G_3 and G_4. It is entirely possible that genes make a great deal of difference to the resulting phenotype in the first case, but little or no difference in the second. Nor is this possibility merely hypothetical. Suppose we were studying variation in skin color among various populations in North America. Variation in skin color will be mainly genetic, if we focus on the people who live in New York City. But if we compare people who live in North Dakota with people who live in Utah, the variation will be mainly environmental (Block and Dworkin 1976).

The last wrinkle of this sort that I want to describe is this: even if genes matter a lot and environment matters only a little within each of two populations, this does not mean that the difference between the two populations is mainly due to their being genetically different. Lewontin (1970) illustrated this point by describing an experiment in which a genetically heterogeneous collection of corn seeds is used in two experiments. In the first, seeds are drawn from this collection and are planted in standard potting soil; in the second, seeds are drawn from the same collection and are planted in potting soil from which various trace elements have been removed. Within each experiment, all the variation in height will be due to genetic differences; however, the difference between the two experimental populations will be entirely environmental. This point was central to the controversy in the 1960s and 1970s concerning how the observed IQ difference between American whites and American blacks should be explained. Jensen (1970) argued that the between-group difference is partly due to genetic differences. Lewontin (1970) replied that Jensen's reasoning was fallacious – that one can't conclude that the between-group difference has a genetic component just from the fact that there is a genetic explanation of within-group variation. Jensen (1972) replied that his reasoning committed no such fallacy.

I now want to make a point that is specifically about the concept of heritability. High heritability does not imply that a trait is difficult or impossible to manipulate by changing the environment – an especially important point, because many individuals who have argued that IQ is highly heritable have inferred from this that it is futile to look for environmental interventions that boost IQ. One reason this is a fallacy connects with a point I made before. Because ANOVA is specific to a given range of environments, nothing follows about how the trait would respond to a *new* environmental variable, one not covered in the initial analysis. For example, before the invention of eyeglasses, poor vision was highly heritable. However, this did not mean that environmental interventions were bound to fail. As it turned out, eyeglasses have done wonders. In effect, this invention created a new environment, one that had a profound effect on the ability to see (Goldberger 1979).

Another example of this sort is provided by PKU disease – phenylketonuria. Individuals with two copies of the relevant recessive gene are unable to digest phenylalanine. The result of accumulating this substance is a severe retardation. However, if individuals with the genetic condition are at birth placed on special diets that don't include phenylalanine, they develop normally. Before the disease was diagnosed and all individuals had diets that contained phenylalanine, the retardation was completely heritable; all the variation in phenotype was explained by variation in genes. Once the disease was understood, a new environment was created that had a profound effect on the phenotype. Notice that in this example understanding the *genetic* causes of a condition opened the door for constructing a new *environmental* manipulation.

There is an additional reason why high heritability does not imply that environmental change will make little difference in the resulting phenotype. Even when we consider just the environments analyzed in an ANOVA study, high heritability does not mean that shifting a particular person from one environment studied to another will have little effect. Consider the following hypothetical data set:

	G1	G2	G3	G4	
E1	1	2	3	4	1.75
E2	1	2	3	2	2
E3	1	2	3	3	2.25
	1	2	3	2	2

The phenotype described here is highly heritable; $V_p = 2/3$ and $V_g = 1/2$, so $h^2 = V_g/V_p = 0.75$. Notice that individuals with genotypes $G1$, $G2$, and $G3$ do not exhibit different phenotypes when their environments are changed. However, matters are quite different for genotype $G4$. It would be a mistake to infer that environmental manipulation has little effect on individuals with genotype $G4$ on the ground that the trait is highly heritable.

The reason that heritability provides little guidance about the effects that environmental change will have on particular individuals is that heritability is a summary statistic about the whole population; it distills the n-times-m pieces of data in an ANOVA table into a single number. A much better guide to the issue of malleability is provided by the data in the table itself. If you know an individual's genotype, you can look down the relevant column in the ANOVA table and obtain an estimate of how changing the environment will produce changes in phenotype, within the range of environments considered. This information is sometimes presented graphically, as in Figure 3.1, by plotting, genotype by genotype, how an environmental circumstance induces a particular phenotypic condition. These graphs represent what is called the genotype's *norm of reaction.*

Graphing norms of reaction provides a handy way of summarizing some of the concepts we have already described in ANOVA. Imagine an ANOVA experiment in which genotypes $G1$ and $G2$ are tested in a range of environmental settings. We can predict the outcome of this experiment from knowledge of the genotypes' underlying norms of reaction. Graph (a) in Figure 3.1 depicts a situation in which there is no genetic variance; all the phenotypic variation will be due to environmental variation. In (b) we have the opposite situation. Environmental variation will make no difference; all the variation will come from the difference

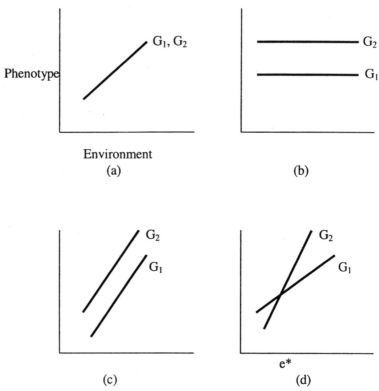

Figure 3.1. Hypothetical norms of reaction for two genotypes, G1 and G2.

in genotype. In (c), both genetic and environmental variations make a difference; because the norms of reaction are straight lines running in parallel, the situation is additive. In (d), the phenotypic variation will be due to genetic variation, environmental variation, and gene-environmental interaction. Whether G1 usually has a higher phenotypic score than G2, or the reverse is true, or they come out with the same average performance, will depend on how the environments investigated in the experiment are selected. If they come entirely from points to the left of e^*, G1 will do better; if they come entirely from points to the right, G2 will do better; and if they are equally spaced around e^*, then there will be a tie.

In an ANOVA experiment, it need not be true that all the individuals in the same environmental treatment experience *identical* environments. There is no way to ensure such uniformity. The plants in the corn experiment described earlier may be unequal distances from the barn; air currents in the experimental plots may differ slightly in direction and speed. No ANOVA study could be performed if perfect uniformity were required within environmental treatments. It might be thought that experimenters are able to attain the more limited goal of having plants within the same experimental treatment differ only in ways that have no effect on the phenotype under study. But here again, the demand is too strong. If we are studying the impact of fertilizer on height, we might set up the experiment so that each experimental plot receives the same amount of water. If there are 100 plants in a plot, however, it is practically certain that some plants will get a bit more water than others. If we are studying the effect of fertilizer, then what is required is that variation in *other* environmental factors that affect the phenotype be the same across plots.

It is at this point that theoretical understanding impinges on an analytic technique that in many ways seeks to get by without the backing of theory. I mentioned before that an ANOVA investigation can proceed without an understanding of the developmental processes that lead individual organisms to end up with their phenotypes. However, there is another type of causal information that is not so dispensable. You must know whether differences in environmental treatments besides the one under study are causally inconsequential for the phenotype under study.

There is no precise level of theoretical understanding that the experiment must satisfy – no cutoff point that marks the boundary between adequate and inadequate studies. Rather, the relevant consideration needs to be stated as a matter of degree. The less you know about which environmental factors influence the phenotype besides the environmental manipulation under study, the less certain you can be that your study has the proper controls in place. Far more is known about the environmental factors that influence the height attained by corn plants than is known about the environmental factors that influence the IQs attained by human beings. Because of this difference, we must be more circumspect in the human case when we say that two individuals grew up in environments that were "the same."

Symmetrical points apply to the concept of a genetic treatment in ANOVA. In an ANOVA experiment individuals said to have "the same" genetic characteristic need not be genetically *identical.* Although desirable, and characteristic of studies that focus exclusively on identical twins, the logic of ANOVA experiments does not demand this.

Individuals have many genes. Suppose $G_1, G_2, \ldots, G_m$ are alternative states of a single gene, and we are doing an ANOVA experiment in which these G conditions constitute the genetic variable under study. Individuals alike in the G trait they have may differ with respect to other genetic characteristics. Individuals who are all G_2 may differ in whether they have H_1 or H_2, whether they have J_1 or J_2, and so on. Some of these other genes may affect the phenotype under study. What is required is that the individuals in one genetic treatment have the same *distribution* of other genetic traits as the individuals in the other genetic treatments, for all the *other* genetic traits that may affect the phenotype.

In this section, I have made several points about how one goes about interpreting the components of variance inferred from an ANOVA experiment. These points do not show that ANOVA is useless as a device for apportioning causal responsibility; rather, the message is that ANOVA must be understood for what it is. ANOVA describes how much of the observed phenotypic variation is due to environmental variation and how much to genetic variation, for the phenotype considered, the range of environmental and genetic variation considered, and the population at hand.

ESTIMATING COMPONENTS OF VARIANCE

In the ANOVA procedure explained so far, investigators specify the various genetic and environmental traits they want to study. Once these are identified, the procedure is to find one or more individuals in each of the *m-by-n* treatment cells. To perform this type of study, we must know, not just the phenotype of each individual, but its genetic and environmental characteristics.

If we are interested in a phenotype as prosaic as height or weight, which genetic traits should we examine? If we already know that variation in $G_1, G_2, \ldots G_m$ influences height, or that variation in $H_1, H_2, \ldots, H_m$ does not, there is no point in doing a study that will tell us what we already know. On the other hand, if we don't know which genes matter, how should we proceed? Do we simply move arbitrarily from one array of genes to another, testing whether variation in a randomly selected gene helps explain variation in height? The same question arises with respect to environmental influences on height; if we know that an environmental factor matters, or that it does not, we will learn little[8] from an ANOVA study. On the other hand, if we focus only on environmental factors whose role is unknown, how do we decide which traits in that infinity of possibilities are worth studying? Naive empiricism is not a recipe for efficient inquiry. What alternative strategy will do better?

There are two standard alternatives to this strategy of hit-or-miss. The first is to examine identical twins reared apart. The second is to compare identical twins and fraternal twins, both reared by their biological parents. In these studies, one of course needs to decide which phenotype one wishes to study. But having decided this, one does not need to further specify which environmental variables and which genetic traits one wishes to consider. The reason is that these studies are designed to estimate the heritability of the trait with respect to the full range of genetic and environmental variation found in the population as a whole. This convenience, though enormous, also has its price. One need not know which genes affect the phenotype in question to do a heritability study, but the upshot is that the study does not tell you which genes make a difference. Although heritability is defined in terms of concepts drawn from ANOVA, there is a big difference between the way causal variables are treated in ANOVA studies and the way they are treated in heritability studies.

Let's now consider how heritability in the whole population is estimated by examining pairs of identical (i.e., monozygotic) twins who were "reared apart," meaning that the twins were separated at birth or shortly thereafter and raised in different environments. This procedure attempts to ascertain the genetic variance and the environmental variance of a trait in a population by looking at a very special subpopulation. Here is an approximate statement of the idea behind such studies: monozygotic twins have exactly the same genes. If they are reared apart, and if they differ from each other phenotypically, then this difference can be attributed entirely to environmental causes. If it turns out that these twins usually are more similar to each other than are two randomly selected individuals from the population, we can conclude that there is a genetic cause of the phenotypic variation in the population as a whole.

Let us now examine this line of reasoning with more care. No new mathematical concept will be introduced, and the only mathematics that will occur in what follows is some subtracting and replacing of equals with equals. I'll discuss the different types of variance that were explained before – the quantities V_p, V_g, V_e, and I, which describe the population as a whole. In addition, I'll talk about the quantities V_p(mono-twin), V_g(mono-twin), V_e(mono-twin), and I(mono-twin), which characterize the subpopulation of monozygotic twins reared apart.

Our goal is to infer the genetic and environmental variances represented in proposition (1). Let's begin by reminding ourselves of what we actually *observe* in such twin studies and what we must *infer* from our observations. Suppose the phenotype of interest is height. What we observe is how tall various people are in the whole population and how tall pairs of monozygotic twins are. From these numbers we deduce values for V_p and of V_p(mono-twin). V_p(mono-twin) describes how much difference there is in height in the average pair of twins. These two phenotypic variances are quantities that we know by observation. In addition, we know that monozygotic twins are genetically identical. We want to use this information to infer what the relative contributions are of genes, environment, and gene-environment interaction. As stated, the structure of this problem should be puzzling. Proposition (1) tells us that an observational quantity, V_p, is the sum of three quantities whose values we do not know by direct observation. How are we to infer three theoretical quantities from the observed phenotypic variance? It looks as if there are too few equations and too many unknowns.

We examine twins reared apart as a device for solving this problem about the full population. Although proposition (1) describes what is true in the population as a whole, the same set of relationships obtains within the subpopulation composed of the monozygotic twins in our study:

$$V_p(\text{mono-twin}) = V_g(\text{mono-twin}) + V_e(\text{mono-twin})$$
$$+ I(\text{mono-twin}). \qquad (3)$$

Notice that (3) by itself does not allow you to assign a value to V_e(mono-twin), based on the observed value of V_p(mono-twin). However, we should take note of the fact that

$$V_g(\text{mono-twin}) = 0. \qquad (4)$$

There is no genetic variation within pairs of monozygotic twins. Propositions (3) and (4) allow us to deduce that

$$V_p(\text{mono-twin}) = V_e(\text{mono-twin}) + I(\text{mono-twin}). \qquad (5)$$

Let us now assume that the relationship between gene and environment in this subpopulation of twins is additive – that is, that the interaction term is 0:

$$I(\text{mono-twin}) = 0. \qquad (6)*$$

Propositions (5) and (6) entail that

$$V_p(\text{mono-twin}) = V_e(\text{mono-twin}) \qquad (7)$$

We now have solved for one of the theoretical "unknowns" in proposition (3). We began by knowing only what the twins' phenotypic variance is; we now are able to assign a value to the environmental variance that exists within the twin population.

How do we use this result to estimate genetic and environmental variances in the whole population? We begin by making another assumption – that the variances are purely additive in the whole population:

$$I = 0. \qquad (8)^*$$

Proposition (8) combines with proposition (1) to yield:

$$V_p = V_g + V_e. \qquad (9)$$

We now further assume that twins reared apart live in environments that tend to be just as varied as the environments occupied by two randomly selected individuals in the whole population:

$$V_e(\text{mono-twin}) = V_e. \qquad (10)^*$$

Proposition (9) combines with (10) to yield:

$$V_p = V_g + V_e(\text{mono-twin}). \qquad (11)$$

Propositions (11) and (7) entail that

$$V_p = V_g + V_p(\text{mono-twin}). \qquad (12)$$

Proposition (12) rearranges to yield

$$V_g = V_p - V_p(\text{mono-twin}) \qquad (13)$$

and (13) and (9) imply that

$$V_e = V_p(\text{mono-twin}). \qquad (14)$$

We are done. The last two equations tell us that the values for the genetic and the environmental variances in the whole population are identical with phenotypic quantities that we can measure by observation. All we have to do is find out how height varies in the population as a whole and how it varies in the subpopulation of monozygotic twins reared apart, and we can calculate what the underlying theoretical quantities are.

As the reader may have guessed, I have placed asterisks besides two crucial assumptions that are used in this derivation. Propositions (6) and

(8) assert that the system is additive; proposition (10) says that twins reared apart tend to live in environments that are just as similar as the environments of two randomly selected people in the population. Both these assumptions are open to question.

Recall what the assumption of additivity means. It means that shifting from one environment to another "adds" the same change in phenotypic value for all individuals, regardless of their genotype. It is not hard to see why this condition can fail. Again, let us consider the example of height. Nutrition in early childhood affects height. But is it plausible to think that shifting from 1,500 calories to 1,600 calories per day will have the same effect on everyone's height, regardless of what their genotype is? This could easily fail to be true. Individuals of different genotype differ in their metabolism; genotypes therefore may differ in how efficiently they convert additional calories into additional height. Notice that the example of height is by no means unusual; just as additivity cannot be assumed a priori for a trait such as height, neither can it be assumed for psychological traits such as intelligence or propensity to violence. The claim of additivity is an empirical claim, and must be supported by evidence.[9]

I am not trying to propose an a priori argument for the claim that additivity always fails to obtain. There can be no a priori argument on this question, one way or the other. Maybe some traits in some populations are additive. Figure 3.2 illustrates the fact that height fails to be additive in a population of the plant *Achillea millefolium*. My main point is to warn against the idea that it is somehow a "safe" general assumption that the trait one is studying is additive.

The assumption that $V_e = V_e$(mono-twin) is also problematic, but for a reason that derives from the specifics of how human adoption agencies work, not from general biological considerations. When monozygotic twins are separated from each other, how are their new environments selected? Sometimes they are adopted into the homes of relatives. At other times, an adoption agency places one or both into a new home. It is well known that adoption agencies give strong preference to adults seeking to adopt who have high socioeconomic status. Both these considerations suggest that V_e(mono-twin) will be smaller than V_e, but by an amount that is difficult to estimate.

The assumptions I have singled out for criticism are very common ones in twin studies; however, the derivation I have described does not depend absolutely on their being true. By this, I mean not that the assumptions could simply be *removed* and the derivation would still go

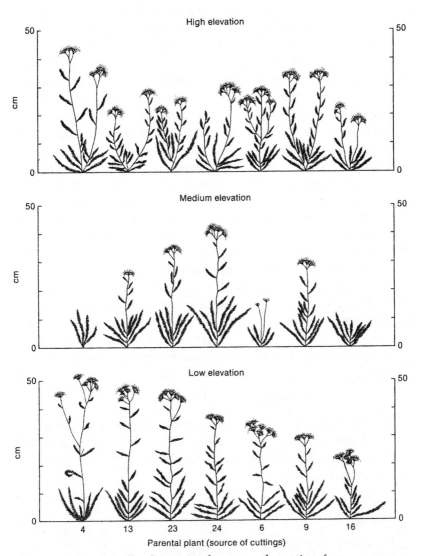

Figure 3.2. Empirically determined norms of reaction for seven geno-types of the plant *Achillea millefolium* grown at three different elevations (reprinted from Griffiths et al. 1993, 14). From *An introduction to genetic analysis* by Griffiths et al. © 1996, 1993, 1989, 1986, 1981, 1976 by W. H. Freeman and Company. Used with permission.

through but that the assumptions could be *replaced* by other assumptions that are equally substantive. The assumption of additivity could be replaced by assigning a value other than zero to the interaction term. The same holds for the assumption that $V_e = V_e$(mono-twin); this could be excised from the argument and replaced by some other characterization of how the two environmental variances are related. However, the problem is not solved by *stipulating* a couple of new assumptions; rather, it must be demonstrated empirically that the assumptions are plausible for the case under analysis (Layzer 1972).

I now want to consider a second popular methodology for inferring genetic and environmental variances. Instead of looking at monozygotic twins reared apart, one compares fraternal (dizygotic) twins reared together with monozygotic twins reared together. The rough idea in this inference procedure is as follows: monozygotic twins have all their genes in common, whereas fraternal twins on average share half their genes. It is taken to follow that if monozygotic twins are more similar to each other than dizygotic twins, then this difference furnishes a valid estimate of the genetic contribution to the phenotypic variance in the population as a whole.

Let us now consider the details. As before, we assume that the interaction term for monozygotic twins reared together is zero; this, and the fact that V_g(mono-twin) = 0, allow us to write

$$V_p\text{(mono-twin)} = V_e\text{(mono-twin)}. \tag{15}$$

Because V_p(mono-twin) is a quantity we can observe, proposition (15) tells us what value we should assign to the underlying parameter V_e(mono-twin). If we assume that the interaction term is zero for dizygotic twins as well, then

$$V_p\text{(di-twin)} = V_g\text{(di-twin)} + V_e\text{(di-twin)} + I\text{(di-twin)}.$$

reduces to

$$V_p\text{(di-twin)} = V_g\text{(di-twin)} + V_e\text{(di-twin)}. \tag{16}$$

The question is how we can use (15) and (16), which concern subpopulations of twins, to draw conclusions about the components of variance in the population as a whole.

We now introduce a further assumption – that the environments experienced by monozygotic twins reared together are just as similar as the environments experienced by dizygotic twins reared together:

$$V_e\text{(mono-twin)} = V_e\text{(di-twin)}. \tag{17}*$$

Propositions (16) and (17) entail that

$$V_p(\text{di-twin}) = V_g(\text{di-twin}) + V_e(\text{mono-twin}). \tag{18}$$

If we combine (15) and (18), we obtain

$$V_p(\text{di-twin}) = V_g(\text{di-twin}) + V_p(\text{mono-twin}), \tag{19}$$

which rearranges to yield

$$V_g(\text{di-twin}) = V_p(\text{di-twin}) - V_p(\text{mono-twin}). \tag{20}$$

Because the two phenotypic variances in (20) are observable, (20) tells us what value to assign to the genetic variance for dizygotic twins. We now need to appeal to a genetic claim – that dizygotic twins differ genetically from each other on average half as much as do randomly chosen individuals from the population at large:

$$V_g(\text{di-twin}) = (1/2)V_g. \tag{21}*$$

Propositions (20) and (21) combine to yield a formula that shows how the genetic variance in the whole population can be computed from the difference between the observable phenotypic variances for the two classes of twins:

$$V_g = 2[V_p(\text{di-twin}) - V_p(\text{mono-twin})]. \tag{22}$$

Finally, (22) combines with (9) to yield a formula for estimating the environmental variance from the observed phenotypic variances:

$$V_e = V_p - 2[V_p(\text{di-twin}) - V_p(\text{mono-twin})]. \tag{23}$$

In addition to the familiar assumption of additivity, propositions (17) and (21) are noteworthy. Proposition (21) depends on the assumption of random mating in the parental generation; this will often be false. For example, for many phenotypic traits, similar individuals tend to pair up to have children; this is true for height, socioeconomic status, IQ, and so on. The effect of assortative mating is to make dizygotic twins more than twice as genetically similar as pairs of individuals drawn at random; how much more depends on the intensity of the assortative process.

Another complication that affects the assessment of proposition (21) is that dizygotic twinning occurs more frequently in some genotypes than it does in others (Falconer 1981, 160). This is a problem, because the basic idea of twin studies is that the population of twins provides a representative sample of the genetic composition of the population as a whole.

Another problem arises in connection with proposition (17). For many psychological traits, it is questionable whether the environments of identical twins reared together are just as similar, on average, as the environments of fraternal twins reared together. Perhaps parents treat identical twins more similarly than they treat fraternal twins. For example, identical twins have the same sex, but fraternal twins often do not, and this may affect the way the twins are reared.[10]

There is an observation that throws light on this question. We know by observation that fraternal twins reared together are more similar in IQ than are nontwin siblings reared together (Plomin and Fries 1980). Because fraternal twins and nontwin sibs have the same degree of genetic similarity, it follows (if the system is additive) that fraternal twins experience more similar environments than do nontwin sibs. Of course, we cannot deduce from this that proposition (17) is incorrect; comparing fraternal and nontwin sibs is not the same as comparing fraternal and identical twins. Still, it would be naive simply to assume that (17) is correct.

Just as was true in the discussion of identical twins reared apart, the point about the present procedure is not that (17) and (21) are essential; other specific assumptions could be substituted for them and would allow the derivation to go through. Rather, the point is that assumptions of these types are needed, and must be defended empirically, before estimates of heritability can be obtained by comparing identical and fraternal twins.

In this section, I have emphasized the types of assumptions that need to be made in using data from twin studies to estimate heritability. However, it is well to remember, in addition, how little one would know, even if these assumptions were entirely correct. In a controlled ANOVA study in which each individual is measured for its environment, its genotype, and its phenotype, one can say which genes make a difference and how much difference they make; one also can say which environmental treatments raise phenotypic scores and which lower them. To be sure, the developmental processes that link causes to effects remain opaque. But at least one knows, from such a study, something about the identity of the causes.

None of this information is provided by a twin study that estimates heritability, even when that study is methodologically sound. If we infer that $h^2 = 0.6$, we have no idea which genes make a difference to the phenotype in question. Nor do we know which environmental variables are responsible for the fact that V_e/V_p has the value it does. What one knows is that certain *existence claims* are correct. There exist genetic differences that help explain phenotypic differences and there exist environmental differences that do the same thing. The fact that one can assign numbers to these causal contributions should not obscure the fact that these estimates say very little.

GENE-ENVIRONMENT CORRELATION

In my explanation of ANOVA, I assumed that each of the m-by-n treatment cells contains the same number of individuals. This setup is ideal in a controlled experiment, but the real world of natural populations rarely conforms to this tidy arrangement. In reality, there often are correlations between the genes that individuals have and the environments they tend to occupy. In this section, I want to explain how this fact further complicates the task of estimating heritability.

Let's begin with a simple example that illustrates how gene-environment correlation can affect the total phenotypic variance. Consider the following data that are drawn from a study of four hundred individuals.

	G1	G2
E1	1	3
E2	3	5

If there are 100 individuals in each of the four treatment cells, then the phenotypic variance is 2. But now suppose that 199 individuals are in the upper left cell and 199 are in the lower right cell; the phenotypic variance now has a value of approximately 4. On the other hand, if 199 individuals are in the upper right cell and 199 are in the lower left, the phenotypic variance will be close to 0.

In this example, individuals in $E2$ have a higher phenotypic value than individuals in $E1$; and individuals with $G2$ have a higher phenotypic value than individuals with $G1$. If higher-valued genotypes tend to occur in higher-valued environments, we have a *positive* association of genes and environment; this association tends to boost the total phenotypic variance. On the other hand if higher-valued genotypes tend to occur in lower-valued environments, there is a *negative* association of genes and environment, which tends to reduce the total phenotypic variance.

What this means is that proposition (1) provides an incomplete list of the possible sources of phenotypic variance. Proposition (1) said that

$$V_p = V_g + V_e + I.$$

It should be replaced with the following:

$$V_p = V_g + V_e + I + 2Cov(g,e). \qquad (1)^*$$

$Cov(g,e)$ is the gene-environment *covariance*. The covariance, which measures the strength of association of genotypes and environment, ranges from -1 to +1.

How does this complication affect the procedures for estimating heritability reviewed in the previous section? Just as twin studies often assume that the interaction term $I = 0$, they also often assume that there is no correlation between genes and environment. The point made earlier about interaction applies here as well. One cannot simply assume that the covariance is zero; one must estimate it empirically.

With respect to traits such as IQ, the gene-environment covariance is known to be positive (Falconer 1981, 121). Individuals with favorable genes tend to grow up in favorable environments, due to the fact that parents not only pass their genes along to their children but do a great deal to structure the environments in which the children are reared. However, if our goal is to estimate heritability, knowing that the covariance term is positive is not enough. We must be able to estimate its value. This requires a type of theoretical understanding that the simple data drawn from twin studies do not provide.

The so-called question of "nature" versus "nurture" or of "genes" versus "environment" suggests that the inferential problem involves saying how important *two* possible causes are. The idea that there are just two causes allows one to think that the contribution of genes can be viewed as a *remainder;* it is what is left unexplained by environmental factors. Proposition (1)* shows that this suggestion is doubly misleading and has the effect of inflating one's estimate of the importance of genes. Only when the interaction and the covariance terms are both zero will genetic variation explain everything that environmental variation fails to explain.

THE CONCEPT OF ENVIRONMENT

Earlier in this chapter, I adopted the rough-and-ready idea that an environmental factor is any property of the world that depends just on what is going on outside the organism's skin. The fertilizer treatment in the farmer's corn plant experiment was judged an environmental factor for just this reason. I now want to explain how quantitative geneticists assign a narrower meaning to the idea of an "environmental factor."

I begin with a simple example, due to Jencks et al. (1972). Suppose it is true in a given population of human beings that red-haired individuals have lower IQ scores than people with other hair colors and that this is true solely because they are treated badly when they are young. It may seem to follow from this description that the lower IQ is due to environment, not genes. For quantitative geneticists, this conclusion is not so straightforward.

I said that redheads have lower IQs because of how they are treated, but I did not say *who* treats them badly. Let's consider two scenarios. In the first, redheads tend to be born to parents who treat their children in ways that reduce their children's IQs. These parents, let us suppose, abuse *all* their children, regardless of what hair color the children happen to have. The point is that red-headed children are born disproportionately into such families. In the second scenario, society as a whole treats redheads in ways that reduce their IQ scores. The parents of redheads are not especially abusive toward their offspring; rather, society treats redheads badly. These two scenarios may be schematized as follows:

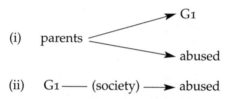

(i)   parents ⟨ G1 / abused

(ii)   G1 —— (society) ⟶ abused

The difference between these scenarios is that genotype causes an individual to be abused in (ii), but not in (i). In (ii), society abuses redheads *because* they are red-headed, and redheads have red hair *because* they have genotype G1; in (i), it is not true that parents of redheads abuse their children *because* the children have red hair.

In quantitative genetics, abuse counts as an "environmental" factor if it is provided by parents but not if it is provided by society generally. Notice that an adoption study will have quite different results in these two cases. If redheads are removed from their parents and placed in new homes, then they will suffer no deficit in IQ, if (i) is true. However, if scenario (ii) is in place, adoption will make no difference; redheads will continue to show lower IQs.

Quantitative geneticists do not regard abuse as an environmental factor, if the situation is of type (ii). The lower IQ of redheads in scenario (ii) is said to be genetic, rather than environmental, on the grounds that individuals experience abuse because of their genes (Falconer 1981). An environmental factor is not just something that occurs outside the organism's own skin; in addition, an environmental factor is defined as a factor that is not caused by genes.

Let's see how this mode of description applies to other examples. Suppose we observe that the women in a given population like to knit more than the men do. The question may then be posed of whether this pattern is due to the genetic difference between the sexes or to differences in how boys and girls are reared (or to both). One might have thought that the only way to separate nature and nurture in this case is to see what happens to *XX* individuals who are raised as boys and to *XY* individuals who are raised as girls. Only by breaking the correlation between genotype and rearing environment can nature and nurture be disentangled. This assessment is not correct, according to the practice of quantitative geneticists, if society tends to treat *XX* individuals one way and *XY* individuals another. If the causal relationships conform to pattern (ii), the correlation need not be broken. The conclusion will then be drawn that there is a genetic explanation for why women like to knit more than men.[11]

I hope the reader can infer from this discussion how societal racism will be classified in the quantitative geneticist's separation of genes from environment. Suppose that the difference in IQ scores between blacks and whites in the United States is entirely due to the fact that the United States is a racist society – that people treat blacks worse than whites. It might be thought that this hypothesis constitutes a purely environmentalist explanation of the IQ difference. Actually, this is not correct, according to the standard framework of quantitative genetics. If blacks are badly treated because of their skin color, and their skin color is genetic, then the lower IQ will be assigned to genes, not to environment.

It may be a bit surprising that quantitative geneticists use the terms "gene" and "environment" in this way. However, as odd as this usage may seem, it is not hard to understand why quantitative geneticists feel driven to adopt it. Every gene has the effect it does because of the ambient environment in which it acts. The so-called gene for eye color produces blue or brown eyes only as a result of the somatic environment in which it acts. To say that eye color is genetic means that genetic variation at a particular locus, *given the environment in which those genes exist*, causes variation in eye color.

The same point can be illustrated by returning to the farmer's two-by-two experiment in which fertilizer is the environmental variable. Suppose that G_1 has a positive effect on plant height because G_1 produces more of a particular gene product, one that repels an insect pest that happens to be present in the fields in which the experiment is performed. G_1 plants grow taller than G_2 plants because G_1 plants are treated differently by the insect pests. It is not an objection to the claim that genes make a difference in the determination of plant height that G_1 and G_2 make the difference they do only because the insect pest is present. Admittedly, it is quite possible that if a different pest had been present, the very opposite result would have transpired; maybe a different pest would have been *attracted* by larger quantities of the gene product.

In the plant example, we view the insect pest as a background condition and tend to focus on the difference in genotype as the genuine cause. In the hair color example, we view an individual's hair color as given, and tend to focus on societal abuse of redheads as the genuine cause. It is an interesting psychological question why we find some causal factors more salient than others. But from the point of view of quantitative genetics, an environmental factor is one that is not caused by genes. Abuse of redheads can arise by either pathway (i) or (ii). When (i) obtains, abuse counts as an environmental condition; when (ii) is in place, abuse of redheads is not part of the environment but is an effect of genes. Without this convention, it is hard to see how quantitative geneticists would be able to say that genes *ever* have any effect.

Quantitative geneticists differ from the rest of us in the way they tend to use the term "environment," and this difference in usage will probably persist. This means that when quantitative geneticists say that the variation in some phenotype has a genetic component, the rest of us must be very careful. The reason is that a genetic cause, in the quantitative genetics sense, may be what the rest of us would regard as an environmental cause. As I explained earlier, a so-called genetic cause may be changed just by changing the environment. If societal abuse of redheads, women, or blacks is changed, the "genetic" causes of the resulting phenotypes may entirely disappear.

CONCLUDING REMARKS

The criticisms I made of the methodology used in twin studies does not mean that such studies will never underwrite reasonable inferences about components of variance and heritability. As we learn more about the issue of gene-environment interaction, and as we learn more about the ways in which environmental factors influence various phenotypic traits, the sophistication of twin studies will improve. The criticisms reg-

75

istered here apply only to a certain "naive," though pervasive, approach in twin studies; these studies rest on assumptions that are nothing more than assumptions. However, there is no reason in principle why a groundless assumption should not someday be replaced by a statement that is empirically well attested. The new knowledge that will facilitate this improvement in twin studies will not come from surveys, which is what twin studies are, but from theoretical work on how genes and environment work together to produce phenotypes.

The points I made about interpreting ANOVA and heritability have a quite different status. For example, I have emphasized that high heritability of a trait does not mean that it cannot be altered much by the environment. I also explained why we must be careful to separate the task of explaining phenotypic variation *within* groups from the task of explaining variation *between* groups. These points will not disappear once we learn more about genetics. They are permanent features of the conceptual landscape of heritability and ANOVA.

As we learn more about genetics, what will happen to questions about heritability? I have explained how ANOVA studies do not depend on understanding the developmental processes whereby genes and environment conspire to produce phenotypes. The farmer can examine the average height of corn plants in his *m*-by-*n* experiment and obtain a heritability estimate without knowing why one unit of fertilizer and genotype G_1 combine to produce plants that average one unit in height. Twin studies, which are stopgap measures that scientists use when they cannot manipulate organisms or identify the relevant genotypes that they want to study, also attempt to draw conclusions with little or no information about developmental processes. My suspicion is that as we learn more about these developmental issues, questions about heritability will become increasingly marginalized. For example, why would a population statistic about the heritability of IQ matter to us, if we understood why some interventions in the lives of individual children boost their IQs while others do not? It is already perfectly clear, as a conceptual point, that high heritability does not mean that a trait cannot be modified much by environmental change. I suggest that as we learn more about the norms of reaction of different genotypes, and *why* genotypes differ from each other in their norms of reaction, we will come to care less and less about assigning a number to a trait's heritability. Ironically, the reason that heritability studies will suffer this displacement is not that science will become convinced that radical environmentalism is true. It isn't that we will lose interest in h^2 because we think its value is lower than we thought before. Rather, heritability will move to the periphery of scientific interest as we learn more about the details of genetic processes. The real challenge to quantitative genetics is not the advocate of nurture over nature, but the developmental geneticist who provides insights into underlying processes.

NOTES

1. Here and in what follows, I am ignoring the way in which statistical inference enters into the interpretation of ANOVA data. One must ask whether the pattern of variation is due to the factors considered or should be attributed to chance. This depends on the number of individuals and on the amount of variation that there is *within* treatment cells. For purposes of getting clear on the bare bones of ANOVA, however, assume that there is very little variation within cells and that the cells each contain a large number of individuals.
2. The philosophical concept of supervenience plays a useful role here. Phenotypic traits supervene just on what is going on inside the skin; environmental traits supervene just on what is going on outside.
3. This proposal goes against the quite reasonable idea that the web a spider weaves is part of its phenotype. However, the distinction of phenotype and environment we are proposing is a convenient one for understanding the basics of ANOVA, and that is all that matters here.
4. There are other components of variance, which will be discussed in due course.
5. This is the so-called broad heritability. In some contexts, it is important to decompose broad heritability into a sum of terms, one of which is the so-called narrow heritability. The narrow concept will not be relevant to our discussion.
6. In the subsequent discussion, we will move from three causes to four, by introducing the concept of gene-environment correlation.
7. There are other contexts in biology in which a population configuration is explained without explaining why individual organisms have the traits they do. I discuss this pattern in connection with the concept of natural selection in Sober (1995).
8. Even if you know that a genetic or an environmental factor positively affects some phenotype, an ANOVA study will add at least some information: it will tell you how much each of them matters, relative to the other. However, the larger question of how important these two factors are, as compared with other traits that were not investigated, remains open.
9. More precisely, the additivity thesis must be *tested* statistically. The question is whether relevant data deviate sufficiently from the predictions of additivity for this to justify rejecting the additivity hypothesis. "Sufficient deviation" is defined to reflect both the size of the difference between predicted and observed values and the sample size.
10. Even when the dizygotic twins considered are of the same sex, there may be

77

other reasons why monozygotic twins tend to share environments that differ in their degree of similarity from those occupied by dizygotic twins – for example, parents may treat identical twins more similarly (or may encourage differences).

11. This example is due to Dawkins 1972; I discuss it in Sober 1984.

REFERENCES

Block, N., and Dworkin, G. (1974). IQ, heritability, and inequality. *Philosophy and Public Affairs* 3: 331–409, 4: 40–99. Reprinted in Block and Dworkin 1976, 410–540.

(1976). *The IQ controversy.* New York: Pantheon.

Dawkins, R. (1982). *The extended phenotype.* San Francisco: W. H. Freeman.

Falconer, D. (1981). *Introduction to quantitative genetics.* London: Longman.

Goldberger, A. (1979). Heritability. *Econometrica* 46: 327–347.

Griffiths, A., Miller, J. Suzuki, D., Lewontin, R., and Gelbart, W. (1993). *An introduction to genetic analysis.* New York: Freeman.

Jencks, C., Smith, M., Acland, H., Bane, M., Cohen, D., Gintis, H., Heyns, B., and Michelson, S. (1972). *Inequality: A reassessment of the effect of family and schooling in America.* New York: Basic Books.

Jensen, A. (1970). How much can we boost IQ and scholastic achievement? *Harvard Education Review* 39: 1–123.

(1972). Race and genetics of intelligence – a reply to Lewontin. *Bulletin of the Atomic Scientists* (May). Reprinted in Block and Dworkin 1976, 93–106.

Layzer, D. (1972). Science or superstition: A physicist looks at the IQ controversy. *Cognition* 1: 265–300. Reprinted in Block and Dworkin 1976, 194–241.

Lewontin, R. (1970). Race and intelligence. *Bulletin of the Atomic Scientists* (March): 2–8. Reprinted in Block and Dworkin 1976, 78–92.

Plomin, R., and Fries, J. (1980). Genetics and intelligence: Recent data. *Intelligence* 4: 15–24.

Sober, E. (1984). *The nature of selection.* Cambridge, Mass.: MIT Press.

(1994). Apportioning causal responsibility. In *From a biological point of view,* 184–200. Cambridge: Cambridge University Press.

(1995). Natural selection and distinctive explanation. *British Journal for the Philosophy of Science* 46: 384–398.

Chapter 4

Genetic Explanations of Behavior: Of Worms, Flies, and Men

KENNETH F. SCHAFFNER

INTRODUCTION

For several hundred years scientists and philosophers have speculated about the character and scope of explanations of a particular type, where that type might be, for example, "mechanical," "chemical," "electrical," and, more recently, "biochemical," "molecular," "selectional," "developmental," "adaptational," and "genetic."[1] Extensive discussions about "mechanical" explanations are an interesting and useful analogous problem to the one that is the subject of the present chapter, especially in the context of nineteenth-century debates about the reach of mechanics, because it was thought for much of that century that *all* of physics and, ultimately, all natural science were susceptible to a reductive "mechanical explanation."[2] In a sense, in biology, genetics now plays the part that mechanics did for nineteenth-century physics. Philosophy of science in the twentieth century has been less interested in general unificatory or in special-science explanations, though there are important exceptions, including some in recent philosophy of biology. E. Nagel's analysis of "What is a mechanical explanation?" (1979, 153–174) is one exception that discusses a special-science explanation,

This research was partially supported by National Institutes of Health Grant R13 HG00703 to the University of Maryland, and by the National Science Foundation's Studies in Science, Technology, and Society Program. I would also like to express my gratitude to the NIH for inviting me to a special August 1993 workshop convened under the auspices of the NICHD at which work in progress by Drs. Bargmann and Chalfie on *C. elegans*, Drs. Hall and Tully on *Drosophila*, Dr. Hamer's studies on humans, and Dr. Plomin's general methodological approaches were presented and discussed. I am also grateful to Robert Wachbroit and David Wasserman for comments on an earlier draft of this manuscript. This chapter draws partially on my 1998 and 1999 essays, which develop some of the themes in the present chapter in different directions. Those essays will be referenced as appropriate.

and that in-depth examination offers some useful methodological suggestions that can be generalized to the topic at hand.

Mechanics and Mechanical Explanations

Nagel's analysis of "mechanical explanation" points out that the terms "mechanical" and "mechanics" admit of a number of different senses, so many in fact that he concludes, preanalytically at least, "there is no core of precise meaning common to these usages" (1961, 156). The most direct and satisfactory method for ascertaining the scope of a science and the distinctive character of its explanations, and for constructing an explication of a "mechanical explanation," he then argues, is to turn to the comprehensive laws and theories – whenever such theories are available – that constitute at a given stage of the development the ultimate premises of its explanations." I do not review Nagel's discussion of mechanics since that is not the point of the present chapter. Suffice it to say that he claims his analysis reveals a core of common meaning of "mechanical explanation." Moreover, he contended, this exercise "illustrates a mode of approach for characterizing what is distinctive of various explanatory systems in different branches of science, and thereby for examining important methodological problems concerning relations of dependence between different explanatory systems" (1961, 174). I want to use these suggestions as a methodological point of view for approaching the question, What is a genetic explanation? by looking closely not at mechanics but rather at genetics, or, more accurately, at behavioral genetics as it is currently practiced.[3] First, however, I consider a few general points about the form of an explanation and a related issue of the nature of a science that may not in a prima facie way conform to the received view of scientific explanation – explanation by the referring of events to be explained by appealing to "laws" of the science.

Genetics and Genetic Explanations

WHAT IS THE GENERAL FORM OF GENETIC EXPLANATIONS? If I were to be writing this chapter circa 1900, and following the method of analysis recommended by Nagel, I would begin with Mendel's theory of inheritance, and perhaps start with Mendel's laws. To some extent this would be successful, but, as extensive debate over the philosophical interpretation of genetics has concluded (see Wimsatt 1976; Kitcher 1984; Culp and Kitcher 1989; Schaffner 1993, esp. ch. 9), there is much more even in

Mendel's "theory" than is captured in (only) Mendel's laws. Mendel's theory of heredity, properly understood, involves not only several general laws or principles, but also the fact that these laws are *intertwined* with their cytological and subcytological bases. Genetics (from its inception) is closely associated with cytological mechanisms, and thus what we mean by "genetics" is not only a set of general (though, in retrospect, restricted) "universal law" type of statements, but also a set of cellular and subcellular mechanisms that are associated with the hypothesized entities of genetics. Recall that even in his first paper Mendel proposes that "in the ovaries of the hybrids there are formed as many sorts of egg cells, and in the anthers as many sorts of pollen cells, as there are possible *constant* combination forms" ([1865], 1966, 136). This point, one that has been emphasized – though to somewhat different ends – by Wimsatt (1976) and Kitcher (1984), has implications for any general account of a "genetic" explanation, because contemporary genetics has further (and in extraordinary detail) amplified the "subcytological" basis of inheritance, going to the level of individual nucleotides in the genetic material, and to enormously interactive complex mechanisms of translation and regulation.

The minimalized role of "laws" in genetics raises a prima facie problem for an account of genetic explanation, because most accounts of explanation in recent philosophy of science are based on having a law or lawlike (set of) sentence(s) as part of the explanans – variations in a sense of what some refer to as the Popper-Hempel-Oppenheim or deductive nomological model. Without such "laws" it might be thought that the notion of a general form of explanation would be difficult to formulate. This problem can, I think, be addressed by three moves that will ultimately take us back to being able to appeal to models and theories of a fairly standard type. Later in the chapter, however, I argue that such standard models need to be embedded in a broader explanatory context.

First, we could try to interpret genetics as illustrating what Salmon (1984) and others (e.g., Brandon 1990) have called a "causal/mechanical" explanation; Salmon also calls this type of explanation "ontic" in contrast with an "epistemic" or "unification" approach to explanation. These two approaches were not necessarily deep contrasts however, and in his 1984 book Salmon wrote:

The ontic conception looks upon the world, to a large extent at least, as a black box whose workings we want to understand. Explanation involves laying bare the underlying mechanisms that connect the observable inputs to the observable outputs. We explain events by showing how they fit into the causal nexus. Since there seems to be a small number of fundamental causal mechanisms, and some extremely comprehensive laws that govern them, the ontic conception has as much right as the epistemic conception to take the unification of natural phenomena as a basic aspect of our comprehension of the world. *The unity lies in the pervasiveness of the underlying mechanisms* upon which we depend for explanation. (276)

Thus even in the absence of a strong role for "laws," we could take the Salmon view of causal-mechanical explanation as the basis of a model within which "genetic explanation" could be located. Genetic explanations can, in general, be conceived as appealing to mechanisms, often quite complex ones.

The next move alleviates the prima facie lack of laws in genetics, including their absence in contemporary molecular genetics, by arguing that, although appeals to biological mechanisms *seem* to be different from appeals to laws, the difference is somewhat illusory, as shown by several considerations. First, an appeal to mechanisms also requires appeals to the "*laws* of working" (see Mackie 1974, 221–223) that are implemented in the mechanisms, and which provide part of the analysis and grounding of the "causal" aspect of the causal-mechanical approach; also note Salmon's reference to the "the laws that govern" the causal mechanisms in the preceding quotation. These may be the laws of broad scope similar to those found in physics, for example, the hydrogen bonding found between DNA chains; or they may be narrower in scope, for example, the Nernst and Goldmann equations used to calculate membrane potentials that figure in explaining action potentials. But they will be generalizations capable of sustaining counterfactuals (Schaffner 1980; 1993, ch. 3). "Mechanisms," from this point of view, implicitly contain generalizations.

Second, if we appeal to *any* general notion of a theory (pick your favorite, whether it be a semantic or a syntactic analysis), it would seem that we would have to deal with "generalizations." For example, if we consider one of the standard semantic conceptions of a scientific theory (Suppes 1967), such "theories" involve set-theoretic (or other language) predicates, which are essentially constituted by component generalizations. Thus, if we were to think about attempting to characterize a causal-mechanical approach to explanation with the aid of a semantic construal of theories, we would *still* require recourse to traditional generalizations.

Third, in those cases in which we wish to verify the *reasoning* supporting some *causal consequence* of an explanatory theory (whether it be a general result or a particular one), we can do no better than utilize the principles of *deductive* reasoning, other forms of logic and reasoning being significantly more suspect. This consideration suggests that causal-mechanical explanation will be a variant of classic deductive nomological explanation of the Popper-Hempel-Oppenheim type, and not one in strong contrast with it. Inductive-statistical explanations, which will appear in cases where we have probabilistic premises and specific data to explain, are more difficult to find a consensus on regarding the type of logical inferences involved, but even here I think we can appeal to standard hypothesis-testing analyses or, better (so I have argued in 1993, ch. 5), Bayesian inference.

What we are left with then, if these arguments are correct, is an account of explanation that will typically be phrased in terms of mechanisms, and often very complex mechanisms, but one that is also congruent with more traditional models of explanation discussed in the philosophy of science. How the explanation is phrased, that is, in terms of mechanisms and models (including statistical models) or in terms of generalizations of narrow and broad scopes, will be a matter of convenience rather than one of substance. The main point of interest about genetic explanation then will *not* be in the *form* of the explanation but rather in the content – the types of mechanisms and/or generalizations found in genetics, and the extent to which they account for other findings, in our case, for behaviors.

THE INCOMPLETENESS OF GENETIC EXPLANATIONS. Genetic explanations are in all interesting cases *incomplete.* Genes, even at the DNA level, do not, *all by themselves,* explain much. The genes have to be translated into phenotypes, and typically have to function in a specified environment. Also, genes do not act in a solitary manner – they act in concert with other genes, often with many genes. Thus a *genetic* explanation will need to demarcate its limitations to explain a phenotype, including a behavioral phenotype, and be situatable in a broader explanatory context that identifies that there are missing elements, such as environmental factors, even if the explanation cannot cite specifics. It is probably these types of limitations, and the ignoring of them, that fuel much of the debate about genetic reductionism and determinism (see Schaffner 1998).

To indicate how genetic explanations work, particularly in the do-

main of behavior, I turn to some background history and terminologi-
cal material, and then to a review of three specific examples.

NATURE AND NURTURE

A 1991 review by Kupferman in Kandel, Schwartz, and Jessel's *Principles of Neural Science* – a book that is a "bible" of neuroscience – begins
by noting that "Behavior in all organisms is shaped by the interaction of
genes and environment. The relative importance of the two factors
varies, but even the most stereotyped behavior can be modified by the
environment, and most plastic behavior, such as language, is influenced
by innate factors" (1991, 987).[4] Kupferman then focuses on "*aspects* of
behavior" (my emphasis) that might be inherited, and on the processes
of interaction between genes and environment affecting behaviors.
Thus the point about the incompleteness of any exclusively "genetic"
basis for the explanation of behavior is taken as a general premise in the
scientific community.

Scientists examining inherited aspects of behavior now frequently refer to what the ethologists called "instinctive behaviors" as *species-specific behaviors* because these are inherited as characteristic of a species
(Kupferman 1991, 989). Ethologists such as Lorenz and Tinbergen introduced two theoretical concepts to describe such behaviors: the cause
(or releaser) of the behavior was termed the "sign stimulus" and the
stereotypical response of the organism was called the "fixed-action pattern," often abbreviated as FAP. A FAP can be quite complex, and in simple organisms the firing of a single *command neuron* can trigger activity
in over a thousand different neurons in different neuronal subsystems.
Such command neurons have been found in the crayfish and in *Aplysia*
(Kupferman 1991, 990–991). The input to command neurons is from sensory neurons that serve as quite specific feature detectors. Although
these types of behaviors are highly stereotypical, environmental factors
and learning history can modify them to some extent.

A simple model of the implementation of a FAP is suggested
by Kupferman, who shows an essentially linear-flow chart leading
from sensory input(s) to a sensory analyzer, to a command system,
to a motor pattern generator, resulting in a motor output (Kupferman
1991, 92). There are some examples of a sign stimulus and a FAP in
humans. Ahren's work on sign stimuli that elicit smiling in young
infants illustrates this. His work in the 1950s on sign stimuli indicates
that two large dots in an otherwise featureless face are more effective

84

in eliciting smiling behavior in young infants than either one large dot or a fully featured face. As the infant matures, the double-dot patterns become less effective, and a face image with features more effective (Kupferman 1991, 994).

Genes do not control even species-specific behaviors directly, and the synthesis of even a single neuron (say a command neuron) requires the coordinated action of *many genes*. I call this principle the rule of "many genes, one neuron." This principle is one that has several other close analogues to be discussed in the next section. Studies done on some simple animals, however, show that certain specific genes can be critical in producing a behavior, and that this can be experimentally demonstrated by laser ablation of neurons or by finding mutants with deletions of genes and correlating them with ablated neurons.

SIMPLE MODEL ORGANISMS

Much of what we know about the biological basis of behavior and learning is based on studies involving simpler organisms. In this section I summarize information about two such organisms, *Caenorhabditis elegans*, a small ground worm, and *Drosophila melanogaster*, the fruit fly. Because the study of the behaviors of both of these organisms has resulted in a large amount of literature, my summary will itself have to be selective, though I do provide bibliographic pointers to the larger literature.

Caenorhabditis elegans *as a Model Organism*

The nematode *Caenorhabditis elegans* is one of the model organisms targeted by the Human Genome Project as a source of potential insight into the working of human genes.[5] Although the organism has been closely studied by biologists since the 1870s (see von Ehrenstein and Schierenberg 1980, for references), it was the vision of Sydney Brenner that has made *C. elegans* the model organism that it is today. In 1963 Brenner had come to believe, as had some other molecular biologists including Gunther Stent, that "nearly all of the 'classical' problems of molecular biology" had been solved or soon would be solved, and that it was time to move on to study the more interesting topics of development and the nervous system.[6] Brenner argued that the nematode had a number of valuable properties, such as a short life cycle, small size, relatively few cells, and suitability for genetic analysis, that could make the nematode the *Escherichia coli* of multicellular organisms. By 1967 Brenner had isolated the first behavioral mutants of *C. elegans,* and in 1970 John White began detailed reconstruction of its nervous system (Thomas 1994, 1698). In 1974 Brenner published the first major study of the genetics of this organism. *C. elegans* has in the past twenty years been intensively studied, and a landmark collection of essays summarizing the field appeared in 1988 (Wood 1988).

In his pioneering article of 1974, Brenner laid out the rationale and general methodology for studying *C. elegans.* Of related interest is his comment within this general methodological framework citing the utility of a similar methodology for the study of *Drosophila.* Brenner wrote:

In principle, it should be possible to dissect the genetic specification of a nervous system in much the same way as was done for biosynthetic pathways in bacteria or for bacteriophage assembly. However, one surmises that genetical analysis alone would have provided only a very general picture of the organization of those processes. Only when genetics was coupled with methods of analyzing other properties of the mutants, by assays of enzymes or *in vitro* assembly, did the full power of this approach develop. In the same way, the isolation and genetical characterization of mutants with behavioral alterations must be supported by analysis at a level intermediate between the gene and behavior. Behavior is the result of a complex and ill-understood set of computations performed by nervous systems and it seems essential to decompose the problem into two: one concerned with the question of the genetic specification of nervous systems and the other with the way nervous systems work to produce behavior. Both require that we must have some way of analyzing a nervous system.

Much the same philosophy underlies the work initiated by Benzer on behavioral mutants of *Drosophila* (for review, see Benzer 1971). There can be no doubt that *Drosophila* is a very good model for this work, particularly because of the great wealth of genetical information that already exists for this organism. There is also the elegant method of mosaic analysis which can be powerfully applied to find the anatomical sites of genetic abnormalities of the nervous system. . . .

Some eight years ago, when I embarked on this problem, I decided that what was needed was an experimental organism which was suitable for genetical study and in which one could determine the complete structure of the nervous system. *Drosophila,* with about 10^5 neurons, is much too large, and, looking for a simpler organism, my choice eventually settled on the small nematode, *Caenorhabditis elegans.* (Brenner 1974, 72)

C. elegans is a worm about one millimeter long that can be found in soil in many parts of the world. It feeds on bacteria and has two sexes: hermaphroditic (self-fertilizing) and male. Figure 3 from Wood (1988) shows a photograph and a labeled diagram of each sex. The organism has been studied to the point where there is an enormous amount of detail known about its genes, cells, organs, and behavior. The adult hermaphrodite has 959 somatic nuclei and the male 1,031 nuclei. The haploid genome contains 8×10^7 nucleotide pairs, organized into five autosomal and one sex chromosome (hermaphrodites are XX, males XO), comprising about 19,000 genes (see *C. elegans* Sequencing Consortium 1998). The organism can move itself forward and backward by undulatory movements and responds to touch and a number of chemical stimuli, of both attractive and repulsive forms. More complex behaviors include egg laying and mating between hermaphrodites and males (Wood 1988, 1, 14). The nervous system is the largest organ, being comprised, in the hermaphrodite, of 302 neurons, subdivisible into 118 subclasses, along with 56 glial and associated support cells. The neurons are essentially identical from one individual in a strain to another (Chalfie and White 1988, 338; Bargmann 1993, 48), and form approximately 5,000 synapses, 600 gap junctions, and 2,000 neuromuscular junctions (Bargmann 1993, 48).The synapses are typically "highly reproducible" from one animal to another, but are not identical.[7]

In 1988, Wood, echoing Brenner's earlier vision, wrote that "The simplicity of the *C. elegans* nervous system and the detail with which it has been described offer the opportunity to address fundamental questions of both function and development. With regard to function, it may be possible to correlate the entire behavioral repertoire with the known neuroanatomy" (1988, 14).

Unfortunately there are some limitations that make this optimistic vision difficult to bring to closure easily. Chalfie and White noted in 1988 that "because of the small size of the animal, it is at present impossible to study the electrophysiological or biochemical properties of individual neurons" (1988, 338), but they add that the much larger neurons in another closely related nematode, *Ascaris suum*, permits some analogical inferences about *C. elegans*'s neurons. Only very recently have patch clamping and intracellular recordings from *C. elegans* neurons begun to be feasible (Thomas 1994, 1698; also see some results and recent references in Goodman et al. 1998). In her 1993 review article, Bargmann writes that "heroic efforts" have resulted in the construction of a wiring diagram for *C. elegans* that has "aided in the interpretation of almost all *C. elegans* neurobiological experiments." Bargmann goes on to say that:

However, neuronal functions cannot yet be predicted purely from the neuroanatomy. The electron micrographs do not indicate whether a synapse is excitatory, inhibitory, or modulatory. Nor do the morphologically defined synapses necessarily represent the complete set of physiologically relevant neuronal connections in this highly compact nervous system. (1993, 49–50)

She adds that the neuroanatomy needs to be integrated with other information to determine "how neurons act together to generate coherent behaviors," studies that utilize laser ablations (of individual neurons), genetic analysis, pharmacology, and behavioral analysis (50).

Various painstakingly careful studies that have been done comparing behavioral mutants' behaviors with neuronal ablation effects, in attempting to identify genetic and learning components of *C. elegans's* behaviors. In addition to Brenner's pioneering work already cited, Chalfie's classic and ongoing studies of movement and touch sensitivity have revealed a large number of mutants affecting these behaviors (see Chalfie and White 1988; and Huang and Chalfie 1994, for references). Some of the most interesting recent work that takes the analysis to a molecular sequence level is by Bargmann and her associates, who have examined the nematode's complex response to volatile odorants (Bargmann, Hartwieg, and Horvitz 1993; Sengupta, Colbert, and Bargmann 1994; Thomas 1994). *C. elegans* is able to distinguish among seven classes of compounds and react by movement toward (chemotaxis) the odorant-emitting compounds. These seven classes of odorants are distinguished using only two pairs of sensory neurons, named AWA and AWC. Laser ablation studies of these neurons and the identification of mutations in about twenty genes affecting very similar behaviors indicate that these genes are required for AWA and AWC sensory neuronal function.

Bargmann and her associates describe about half a dozen *odr* mutations that affect the AWA and AWC neurons (Bargmann et al. 1993; Sengupta et al. 1994), and in the latter study have focused on the *odr-7* mutation, which has an exceptionally specific effect on the two AWA neurons. The general "rules" regarding the relation of genes to behavior is stated by Avery, Bargmann, and Horvitz (1993):

One way to identify genes that act in the nervous system is by isolating mutants with defective behavior. However, the intrinsic complexity of the nervous system can make the analysis of behavioral mutants difficult. For example, since behaviors are generated by groups of neurons that act in concert, a single genetic defect can affect multiple neurons, a single neuron can affect multiple behaviors and multiple neurons can affect the same behavior. In practice these complexities mean that understanding the effects of a behavioral mutation depends on understanding the neurons that generate and regulate the behavior. (455)

A single neuron is the product of many genes, a point that I suggested be termed the rule of "many genes, one neuron." In the quotation from Avery et al. we encounter several other similar rules. These "rules," which have some additional generalizations about the relations of genes and behavior, can be further elaborated.[8] A summary is presented in Table 4.1 (from Schaffner 1998).

In my view, these rules, based on empirical investigations in the simplest model organism possessing a nervous system that has been studied in the most detail, should serve as the *default assumptions* for further studies of the relations of genes and behavior in more complex organisms. These eight rules are generalizations involving principles of genetic pleiotropy, genetic interaction, neuronal multifunctionality, plasticity, and environmental effects, and, like virtually any generalization in biology, they are likely to have exceptions, or near exceptions, but I think these will be rare.[9]

Taken together, what these rules tell us is that the relation between genes and behavior types will be "many-many" (compare Lewontin 1991, 27, on this point). There is also a prima facie stochastic component present that shows up in some developmental variation for the synapse-wiring diagram, even in simple systems in which the complete lineage of cells has been identified and is traceable.

Although the analysis of the *unc-31* gene reported by Avery et al. (1993) illustrates this complexity of effects, the 1994 *odr-7* study appears to be considerably more specific and almost supports a "one-gene, one type of behavior" analysis. As Sengupta et al. (1994) show, a null mutation in the *odr-7* gene causes *C. elegans* to fail to respond to all odorants detected by the AWA neuron pair. A missense mutation in this gene results in a specific defect in one odorant response (971). (Of related interest are the quite specific effects of a mutation in the *npr-1* gene on "social" feeding and movement behavior in the worm recently reported by de Bono and Bargmann 1998.)

Sengupta et al. (1994) were also able to map *odr-7* to the X chromosome, and further localize the gene by restriction fragment length polymorphism (RFLP) mapping as well as by "germline rescue of the *odr-7* diacetyl chemotaxic defect with cosmids from a defined interval" (973).

Table 4.1. *Some Rules Relating Genes (through Neurons)*
to Behavior in C. elegans

1. Many genes → one neuron
2. Many neurons (acting as a circuit) → one behavior
3. One gene → many neurons (pleiotropy)
4. One neuron → many behaviors (multifunctional neurons)
5. Stochastic (embryogenetic) development → different neural connections[a]
6. Different environments/histories → different behaviors[a]
   (learning/plasticity)
7. One gene → another gene → behavior (gene interactions, including epistasis and combinatorial effects)
8. Environment → gene expression → behavior (long-term environmental influence)

Note: The → can be read as "affect(s)," "cause(s)," or "lead(s) to."
[a]In prima facie genetically identical (mature) organisms.

These investigators were then able to clone the gene, sequence it, and to determine its transcript, which encoded a "predicted protein product" 457 amino acids in length. The *odr-7* gene sequence was also compared with other sequences in available data bases, a comparison that indicated that the gene is a member of "the [super]family of nuclear hormone receptors" (974). Sengupta et al. have speculated on the manner in which *odr-7* functions, writing:

> Three general classes of models could account for the phenotype of *odr-7* mutants. First, *odr-7* could be involved in the cell type-specific expression of receptor or signal transduction molecules in the AWA olfactory neurons. Second, *odr-7* could determine the cell fate or development of the AWA neurons. Third, *odr-7* could interact directly with odorants in an unusual signal transduction cascade. Our results favor the first possibility. (1994, 977)

Bargman and Mori (1997) provide further information on the relations of *odr-7* to *odr-10*.

The *odr-7* gene is clearly more specific in its effects than typical behavior-influencing genes. In further work by Bargmann and her associates not yet published, this specificity has been further confirmed by results using antisera against the endogenous gene product that shows that *odr-7* is only expressed in a single cell type (Bargmann, personal communication). Although in the account just described *odr-7* appears to be monofunctional, more recent unpublished work suggests that *odr-7* mutations have at least two effects. In addition to its chemotactic function, AWA also helps integrate olfactory information over time, and *odr-7* mutants are defective in this AWA function (Bargmann, personal communication). The results thus far suggest that *odr-7* is exceptional in its specificity, though not monofunctional, thus preserving in attenuated form the general principles of genetic pleiotropy and multifunctionality described here. The interesting specificity of a single mutation and its effect on feeding and locomotion reported recently should also be considered in connection with the *odr-7* (and, later, *ord-10*) mutations (see de Bono and Bargman 1998). The bottom line is, in my view, that the many-many genes, neuron-behavior paradigm articulated in the eight rules in Table 4.1 will be pervasive, but that rare single-gene, behavior links can trump those rules – but *rarely*.

In addition to Bargmann and her colleagues' ongoing research program, some of the most useful prospective work on the explanation of *C. elegans* behavior may come from Shawn Lockery's laboratory at the University of Oregon. In the early 1990s; Lockery worked with Terry Sejnowski to develop a sophisticated connectionist model of the bending reflex in the leech (see Churchland and Sejnowski 1992: 339–353 for a general "philosophically oriented" introduction to this work, and also Lockery and Sejnowski 1993 for an update). Recently Lockery has developed some clever techniques for putatively recording from single neurons in *C. elegans* and has embarked on a research program to develop and test connectionist models for *C. elegans* (Lockery 1994; 1995; 1999).

Interestingly, the most complex behavior of *C. elegans*, male mating behavior, has yet to be studied in depth, and until recently the main citations to this work are to analyses done more than ten years ago (Hodgkin 1983). In 1995 Liu and Sternberg, 1995 provided some newer insights in this area. But Chalfie and White (1988) provide perhaps the best summary of this behavior pattern – a highly stereotypical form of behavior, which relates to the additional two examples of mating behavior I discuss later.

Male mating is one of the most complex behaviors in *C. elegans*. . . . males exhibit a chemotaxis toward hermaphrodites that is thought to be mediated by the four CEM [cephalic companion sensory neuron] cells [found only in males]. When a male's tail touches a hermaphrodite, the male places its copulatory bursa in contact with the hermaphrodite and begins moving backward along it. (The males do not seem to recognize the hermaphrodites as such; contact with other males or even a male's own head [*sic*] will produce this same behavior.)

This backward movement, which may result in the male's encircling the length of the hermaphrodite many times, allows the male to find the vulva. Once the bursa has contacted the vulva, the male extends a pair of cuticle-covered projections called spicules. The attachment of the spicules secures the male to the hermaphrodite and opens the vulva. Sperm are then ejaculated into the hermaphrodite, the spicules are removed, and the male swims away from the hermaphrodite. (383)

Chalfie and White add that "mating behavior thus involves both chemosensory and mechanosensory input, as well as special motor control, and the male nervous system differs in a number of ways from that of the hermaphrodite" (383).

C. elegans is presently the organism about which we know the most concerning its genetics and the relation of the genetics to its nervous system and to behavior. As Chalfie and White (1988) note, "The structure and development of the *C. elegans* nervous system have been analyzed to a resolution and completeness that has not been possible for any other animal" (390). Although it is a comparatively simple organism, if my overview of the results and the limitations of an analysis of the genetics of the behavior of this model organism is correct, we are quite some distance from being able to relate genetics to behavior in any scientifically credible *complete* sense even for this organism. My own view is that progress will be made by following two routes, hopefully in a synergistic manner. The first route or research strategy is the more classical search for simple gene–neuron behavior interactions. These may, however, turn out to be quite atypical even in *C. elegans*, though the *odr-7, odr-10,* and *npr-1* results do seem to confirm the usefulness of this type of strategy. The second route is to develop biologically informed connectionist models of the neuronal circuits. Lockery is following this route, and I suspect it will turn out to be the method of choice as he (and others) transcend the technical problems of single neuron recording in this organism. The eight rules involving pleiotropic and multifunctional effects also appear in our next example, *Drosophila,* and also argue strongly against any simple "one gene, one behavior" explanation in still more complex organisms, such as humans.

Drosophila melanogaster *Courtship and Mating Behavior*

The common fruit fly, *Drosophila melanogaster* (or *Drosophila* for short, when species differences are not important to distinguish) is the same organism that T. H. Morgan and countless other geneticists have investigated to explore the relations between genes and phenotypes since the early twentieth century. As with *C. elegans, Drosophila* is another proto-typical organism whose 120 Mb nucleotide genome is being sequenced as part of the human genome project. Here I concentrate on the complex courtship and mating behavior of *Drosophila*, and on what has become recently known about the cellular and biochemical etiology of aspects of its behavior patterns in this domain.

Although *Drosophila* has a long history, since about 1910, as a model organism, our story begins in the 1960s when Seymour Benzer, a mo-lecular biologist, began to identify behavioral mutants of *Drosophila* (Greenspan 1995). In the 1970s and 1980s, a number of investigators, es-pecially Jeffery C. Hall at Brandeis, continued this work and used a tech-nique involving genetic mosaics, which permits the creation of male cells in localized regions of female flies.[10] Hall (1994) recently summa-rized his decades-long investigation in a detailed review article in *Sci-ence*. Feveur, Stortkuhl, and Greenspan's research was also reported re-cently in *Science* (1995) and summarized by Greenspan in *Scientific American* (1995).

We can begin by noting the extraordinary range of behavioral mu-tants in *Drosophila*, many with colorful and evocative names such as *couch potato, nerd, fruitless* (or *fru)*, and *ether-à-go-go;* the last name, coined appropriately enough in the 1960s, is based on a shaking-limb muta-tional behavior produced by the anesthetic. Table 4.2, reproduced from Hall's (1994) review article, summarizes these and other classical muta-tions, such as *white eye.*

In his *Science* article, Hall provides a series of photographs along with a detailed textual account of the courtship-mating, species-specific be-havior. In this book, it is not possible to reprint the color photos, and black-and-white renditions cannot provide sufficient clarity. The reader is encouraged to consult Hall's original article for the graphics, but his text is also quite specific, and references to Hall's original figures are in-cluded in the following:

Courtship in *Drosophila melanogaster* involves a series of behaviors, most of which were caught on film during the production of a certain blue movie (Fig. 1). Once the male and the female have come into some reasonable proximity (perhaps on a food source or when experimentally put together in a mating cell), they quickly sense each other. Primarily, this seems to be the male detecting the female by using more than one sensory modality. Soon after the male reveals that he has noticed the female (which one infers by observing the orientation of his body toward hers [Fig. IA]), he taps the female's abdomen (Fig. IB). If she is

Table 4.2. *Mutants, Manipulated Genes, and* Drosophilia *Courtship*

Mutant
General decrements in courtship and vigor and male mating ability
Yellow and ebony body-color mutants
Inactive
Couch potato
Cuckold
Minibrain and no-bridge brain-damaged mutants
Nerd
Visual mutants
White eye mutants, depleted of screening pigment
Optomotor-blind
No-receptor-potential-A blind mutants
Olfactory mutants
Smellblind (sbl) and olfactory-D (olfD) alleles of paralytic gene
Abnormalities of female receptivity
Spinster
sbl and olfD olfactorily defective mutants
Sex-peptide gene ectopically expressed in female transgenics
Rhythm variants
Period (per) mutants
Per gene from *D. simulans*
Learning and memory variants
Dunce
Rutabaga
Amnesiac
Shaker
Ether-à-go-go
CAM-kinase-depleted transgenic
Courtship song mutants
Cacophony
Dissonance allele of no-on-transient-A gene
Croaker
Fruitless
Behavioral male sterility and bisexual orientation
Fruitless
Sex-determination variants
Sex-lethal

Table 4.2. (*cont.*)

Transformer
tra⁻ in XX flies
Ectopic expression of TRAf protein in brains of XY transgenics
Transformer-2-temperature sensitive
Doublesex
Loss-of-function mutations in XX and XY flies
Constitutively expressed mutations in XX flies
Fruitless?

Source: J. C. Hall, "The mating of a fly," *Science* 164 (1994): 1704. Used with permission.

walking about, he follows her (Fig. IC) during most of the time that she is moving in this manner (no courtship occurs in flight, unlike the capabilities of some other dipterans [4]). As the male orients to a stationary female – including circling her (5) – or follows a mobile one, he frequently sticks out one wing or the other (Fig. ID). This extension of the wing is accompanied by its vibration (Fig. IE), which produces a "love song" that can be recorded with specialized microphones (6, 7); the measurable components of these sounds are among the more salient species-specific elements of fruit-fly courtship. Several seconds to a few minutes after the two flies have begun to interact, the male extends his proboscis and licks the female's genitalia (Fig. 1F). Licking is almost immediately followed by the male's first copulation attempt (not shown in Fig. 1), which involves an abdominal bending by the male; this can be viewed in more contorted form by looking at the posture accompanying copulation per se (Fig. IG).

Hall continues:

If an attempted copulation fails, the male may cease courting for some moments. Thus, overt courtship interactions occur only about 60 to 80% of the time when the male and female are together (called the Courtship Index). . . . When the male resumes courting, he almost always drops back to the orientation and following or singing stages (that is, not to tapping or licking) and continues through the rest of the sequence. This series of actions and inter-fly interactions is successful in more than 90% of short-term laboratory observations of wild-type pairs. "Success" means copulation, which has a species-specific duration (about 20 min in *Drosophila melanogaster*). (Hall 1994, 1702)

The "genetic explanation" of this complex choreography is incomplete, but some important mutants that exhibit alternative courtship behavior have been identified and cellular explanations sketched for them. On the basis of these, aspects of the genetic explanation for such behavior have been the focus of recent speculation. In his account of the genetics of courtship behavior, Hall focuses on the *fruitless* (or *fru*) mutation, writing that the *fruitless* mutation may define a sex determination factor as well as a courtship gene. This mutation is now distinguished into several variants, fru^1, fru^2, fru^3, and fru^4. The behavioral phenotype of the fru^1 mutation in males involves vigorous courtship of females but no attempts at copulation; a fru^1 male never executes the abdominal bending described in the quotation from Hall. Hall adds that "the most dramatic reproductive anomaly associated with fruitless is that the fru^1 mutant courts another male just as vigorously as he does a female. Moreover, groups of fru^1 males will snake around a chamber, forming "courtship chains" (Hall also calls these "conga lines") in which most individuals are simultaneously courters and courtees.

The fru^1 males are also mild song mutants and display other anomalies as well. One important finding was that severe forms of this mutation lacked an abdominal bending muscle, known as the muscle of Lawrence or MOL after its discoverer. This may account in part for the lack of abdominal bending, but fru^2 mutants lacking a MOL do exhibit both abdominal bending and copulatory behavior with females. The uncertainty about the behavioral causation pathway(s) from *fru* is underscored by several other facts reviewed by Hall, who cautions us against accepting the nihilistic view (this is my term, not his) that "everything is expressed everywhere and affects everything," although he admits that "some of the effects and inferred noneffects are not very clean" (1994, 1712). He also adds that "if any mutant type in *Drosophila* is a true behavioral one, then this [*fru*] is it."

In spite of the enormous amount of work already done, however, how *fru* might produce its effects is still speculative, as Hall makes clear:

The *fru* gene could therefore act in part within the PNS to influence the development or function of sensory structures that are known or suspected to be involved in initiating and sustaining court ship. . . . Inasmuch as one *fru* variant is a double mutant that is likely to include a pheromone mutation, this autosomal region points to the question of what actual dimorphisms underlie the sex-specific production of these courtship-modulating substances: Is the origin of something like the aphrodisiac pheromone a matter of female-specific chemistry or could it also involve an as yet unknown element of sexually dimorphic anatomy? Drosophila is underanalyzed with respect to the tissue sources of its pheromones and in terms of the male and female sensory structures responsible for inputting each of the relevant chemical stimuli. . . .

That a *fru* male so avidly courts other males and seems so utterly unwilling to mate with a female may imply a central defect in the mutant (whether or not there is also a peripheral one). For this to occur, an involvement of the gene in the differentiation of ganglia within the head can be tentatively predicted. As for the courtship song defect exhibited by *fru* males, this strongly suggests that the gene also influences the structure or function of the thoracic nervous system. *fru*'s male-muscle defect . . . and its neural etiology . . . imply that the gene is expressed in neurons within the abdominal ganglion as well as in the more anterior CNS regions just noted. These suppositions are based in part on the behaviors and MOL phenotypes exhibited by gynandromorphs, whereby the "foci" responsible for a fly's thinking it is a male, singing like one, and developing the male-specific muscle are located in the brain, thoracic ganglia, and abdominal ganglia, respectively. (Hall 1994, 1712)

This general research program is, as suggested earlier, not only genetic: it also involves an examination at the cellular and suborgan level of determining elements. I believe (in agreement with the methodological comments quoted from Bargmann earlier) that just as neuroanatomy needs to be integrated with other information to determine "how neurons act together to generate coherent behaviors," so the genetics of *fru* seems in need of a robust neuroanatomy in order to make a complete story that makes scientific sense. This approach is one that Churchland and Sejnowski (1992) have urged in the neurosciences, and it seems especially required in this domain as well. In point of fact, Greenspan (Hall's former student) provides some of these neuroanatomical dimensions, using an artificially induced feminization of male cells (gynandromorphs) in key parts of *Drosophila*'s nervous system. Greenspan and his associates have been able to develop a different kind of gyandromorph, which is primarily male but has small localized areas of female cells. These gynandromorps, though primarily male with small elements of a female neuronal system, had their courtship behavior changed: the males of some of these transformed strains courted male flies as well as female flies. The technique was an application of Brand and Perrimon's work, which was adapted by Jean-François Ferveur while working as a postdoctoral student in Greenspan's laboratory. By some clever breeding techniques, they were able to switch localized neuronal cells into a ferminized path of development (Greenspan 1995).

It needs to be added that even with the addition of this neuroanatomical dimension, the explanations offered by Greenspan (and

Hall) remain speculative. The pathways from genes through neuronal development to behavior are still quite "gappy." Just as we found with the simpler organism C. *elegans*, much more work is required before we can understand the complex mechanisms underlying behavior. Greenspan (1995) underscores this, and adds that learning phenomena, discussed also by Hall but largely ignored by me in this account, add to the complexity. Greenspan writes that "Just as the ability to carry out courtship is directed by genes, so, too, is the ability to learn during the experience. Studies of this phenomenon lend further support to the likelihood that behavior is regulated by a myriad of interacting genes, each of which handles diverse responsibilities in the body" (75–76).

A question that arises from these caveats about complexity is whether there may be behavioral "simplifications" of a sort that can be discovered – a kind of "final common pathway" in a complex network of interactions. Certainly that seems to be what researchers approaching the genetic basis of human behaviors in the area of psychiatric disorders hope to find, so that they may intervene at specific pharmacological points and treat those affected by these illnesses. In our final specific example I move beyond so-called simple organisms and examine a putative genetic cause of sexual orientation in men, reported by Hamer in 1993.

SEXUAL ORIENTATION IN (SOME) HUMAN MALES AS A MODEL[11]

In July 1993 Dean Hamer caused a stir in the behavioral genetics community, as well as a much broader reaction, with the publication of his group's report in *Science* on "A Linkage between DNA Markers on the X-Chromosome and Male Sexual Orientation" (Hamer et al. 1993a). The study has been criticized from both scientific (Risch, Squires-Wheeler, and Keats 1993; Byne 1994) and social points of view (Rose 1995; CRG 2000), but in general it has been well received scientifically as a first step in clarifying the possible genetic influence on human sexual orientation. Hamer has also interrelated his work with Simon LeVay's anatomical investigations of the basis of sexual orientation (LeVay and Hamer 1994). Collaborating with journalist Peter Copeland, Hamer has amplified on his research and its implications in a recent book (Hamer and Copeland 1994). Hamer's work, now replicated and extended by his group, remains controversial.[12] In the November 1995 issue of *Nature Genetics*, Hamer's group, including investigators with special expertise in the analysis of complex trait genetics, reported a replication of their original finding as well as an extension of their results (Hu et al. 1995). Again the Xq28 region was associated with a disposition to male homosexuality, but not to female homosexuality, in subpopulations of the gay and lesbian populations studied.

Recent reports have raised additional questions about replicability of Hamer's group's work. A study by Bailey et al. (1999) published in March 1999 could not confirm X-related linkage with homosexuality, although they also wrote that their new results "do not exclude the possibility of moderate X-linkage for male sexual orientation." A poster presentation by Saunders (1998) and a recent essay in *Science* by Rice et al. (1999) were more critical and interpreted their studies as nonreplications. Hamer responded to both reports in a comment in *Science* that was published along with a rejoinder by Rice and Ebers, also joined by Risch (Hamer et al. 1999). Hamer argues that a meta-analysis of four genetic studies, the two from his group and the Saunders and the Rice and Ebers nonreplications, indicates that "DNA linkage data continues to support a modest but significant role of the Xq28 region in male sexual orientation. Although there is a 0.01% chance that the observed link represents a 'false positive,' there is a greater than 10% chance that the conclusions in the report by Rice et al. represent a 'false negative,' resulting from their use of a small, apparently nonrepresentative subset of families for genotyping. Moreover, their family pedigree data appear to actually support X-chromosome linkage" (Hamer et al. 1999, 803a). In his recent *Science* comment about Saunder's study, Hamer notes that the study *does* replicate his group's reports.

Achieving replications in behavioral genetics is especially difficult. According to Lander and Kruglyak,

Failure to replicate does not necessarily disprove a hypothesis. Linkages will often involve weak effects, which may turn out to be weaker in a second study. Indeed there is a subtle but systematic reason for this: positive linkage results are somewhat biased because they include those weak effects that random fluctuations helped push above threshold [of statistical significance], but exclude slightly stronger effects that random fluctuations happened to push below threshold. Initial positive reports will thus tend to overestimate effects, while subsequent studies will regress to the true value. . . . Replication studies should always state their power to detect the proposed effect with the given sample size. Negative results are meaningful only if the [statistical] power is high. Regrettably, many reports neglect this issue entirely.

When several replication studies are carried out, the results may conflict –

with some studies replicating the original findings and others failing to do so. This may reflect population heterogeneity, diagnostic differences, or simply statistical fluctuation. Careful meta-analysis of *all* studies may be useful to assess whether the overall evidence is convincing. (Lander and Kryglyak 1995, 245)

The issue about *why* nonreplications might have occurred in the Rice et al. study may involve some of these issues. Hamer believes they do, but Rice et al. disagree.

In spite of the failure to replicate and the subsequent empirical questions about X-linkage just described, Hamer's group's study designs, as reported in the 1993 version and even more so in the Hu et al. (1995) replication, are methodologically and statistically sound. Unresolved questions about these studies, however, have been raised in the literature. Both studies only report a partial association between a behavior pattern (more accurately, an orientation) and a chromosomal region; they do not identify a gene or genes and report a DNA sequence. Thus, not surprisingly, it is not known what protein the gene (if it is just one gene) codes for, where such a protein might act and what it might do, how the gene in homosexual men might differ from heterosexual men, whether it invariably leads to homosexual orientation, or what role this gene might play in women (Pool 1993). Further, Hamer himself notes, citing M.-C. King's additional questions, that we do not know what fraction of all gay men carry an allele in this region that might influence sexual orientation, how many different alleles there might be, and what other factors, including other genes, familial environment, and culture, might affect sexual orientation (Hamer and Copeland 1994, 145). In his brief collaboration with LeVay (LeVay and Hamer 1994), Hamer stated that they see "two broad lines" of evidence pointing to a biological component for male homosexuality. But LeVay and Hamer follow only weakly a "coevolutionary" approach that would carefully match the genetics with neuroanatomy and behavior, as is pursued in studies of *C. elegans* and also by Greenspan and his colleagues in *Drosophila*. LeVay and Hamer do speculate that there may be genetic differences in the way that individual brains respond to circulating androgen in fetal development, but an inquiry to determine if the gene or genes in Hamer's Xq28 region were the gene or genes for the androgen receptor turned out negative: the locus of the androgen receptor was at Xq11 – far from the Xq28 region (Macke et al. 1993).

Thus no "gay gene" has been found, and no molecular or even neurological pathway-circuit explanation for sexual orientation for humans is likely to be forthcoming in the near future. Hamer's group's work is significant, however, even if it frequently tends to be overinterpreted in the scientific literature and in the lay press, because it represents a methodologically well designed genetic study of a complex behavior "trait" – a point to which I return in the next section.

RECENT APPROACHES TO METHODOLOGY

The purpose of this section is to is to review briefly some methodological claims (and caveats) of behavioral geneticists who have been particularly concerned with studies on humans, following on the more specific example discussing male sexual orientation. It is widely acknowledged that advances in the molecular genetics of a variety of *single gene* somatic disorders have become mind-boggling in the past ten years. As is generally known, the gene for Huntington's disease, as well as genes for two types of inherited colon cancer, and two forms of breast cancer, were localized and sequenced in 1994. The genetics of mental disorders, however, has not fared as well, although there have been some important, though controversial, advances in the behavioral realm. The finding that there was a gene associated with schizophrenia on chromosome 5 turned out to be fallacious (Watt and Edwards 1991) and a report identifying a gene for manic-depressive disorder on chromosome 11 had to be withdrawn (Kelsoe et al. 1989). I have already discussed Hamer's findings about homosexual orientation. A point mutation in the structural gene for monoamine oxidase A that was associated with a ill-defined aggression phenotype was reported in *Science* in 1993 (Brunner, Nelen, and Breakefield 1993). Although this work seems to represent methodologically sound science, one can raise some questions about the phenotype of this "defect" which was described as "characterized by borderline mental retardation and a tendency toward aggressive outbursts, often in response to anger, fear, or frustration. . . . Other types of impulsive behavior that occurred in individual cases include arson, attempted rape, and exhibitionism" (579). Another study by DeFries and his group (see Cardon et al. 1994) using a more complex quantitative trait loci (QTL) approach has provided suggestive evidence localizing a gene for a reading disorder to two chromosomal locations. These single gene examples are, however, rare and may well be misleading.

There are several problems with this focus on single gene disorders, particularly in the behavioral-mental disorder realm. The Huntington's disease paradigm of a single dominant 100 percent penetrant gene may be more the exception rather than paradigmatic, for the search for such single genes, as already indicated for bipolar disorder and schizophrenia, has *not* been successful.[13] Investigators in this area have, accordingly, begun to think seriously about multigenic causes for behavioral disorders. In a 1990 essay Robert Plomin argued on rather theoretical grounds that there will be a very limited number of behavioral traits that will segregate as Mendelian single loci. Plomin pointed out that (1) most behaviors are not inherited in a dichotomous (either-or) fashion, (2) most behavioral traits appear to be influenced by many genes, each with small effects, and (3) behavior is substantially influenced by nongenetic factors (1990, 183–184). This nongenetic or environmental factor influence is underscored by data presented in a review article on "The Genetic Basis of Complex Human Behaviors" by Robert Plomin, Michael Owen, and Peter McGuffin (1994), which display probandwise concordances for identical and fraternal twins for several behaviors, including some classical psychiatric disorders, as well as a range of some common medical disorders. Genetics is *not* the entire explanation for these disorders, however; a large environmental component will figure as well in any complete explanation.

In the absence of any results that localize traditional DSM-IV mental disorders or other well-characterized stereotypical (or otherwise) human behavior to specific single genes (with the possible exceptions mentioned earlier), genetic-epidemiological findings from twin, family, and adoption studies represent the soundest evidence for a genetic component. The epidemiological studies, important as they are however, do *not* provide answers to questions such as the mode of the genetic component's transmission, the degree of biological heterogeneity, or the pathophysiological process underlying the behavioral manifestations (cf. Gershon 1990, 373, on manic-depressive disorder). Further studies involving linkage, as well as other types of approaches, are required to elucidate these important dimensions, including the possibility of a biochemical means of alleviating or curing these disorders. We have already seen this approach exploited in a very preliminary way in our account of Hamer's research, although on the latter point I hasten to add that Hamer construes homosexual orientation as a normal variant and not a disorder to be treated or "cured." Using genetically based information to alter sexual orientation, of either homosexual or heterosexual form, is viewed by Hamer et al. as "unethical" (see Hamer et al. 1993a, 326).

In an attempt to zero in on the genes that underlie behavior patterns that thus are expected to be multigenically caused, behavioral geneticists have begun to explore two genetic approaches for dealing with more complex modes of inheritance: quantitative trait loci, or QTL; and allelic association. In their 1994 review article, Plomin and his coauthors suggest that QTL coupled with allelic association strategies can identify the more complex genetic causes of behavior, and can also serve as a "guide" to "molecular genetic research by identifying the most hereditable domains of behavior and the most heritable dimensions and disorders within domains" (1738). QTL methods were developed by Lander and Botstein about ten years ago and first applied to fruit characteristics in the tomato, and then were quickly extended to epilepsy and hypertension in rats. The latter study led to an identification of angiotensin in human hypertension, and more recently QTLs have been used in connection with animal models of alcohol and drug abuse (Crabbe, Belknap, and Buck 1994).

It would take me beyond the space I have to try to explain these two strategies in any detail. Suffice it to say that QTL methodology required both the development of high density marker genetic maps, as well as advances in the statistical methods, so that QTLs, each with a small effect but collectively having a major effect, could be detected against background variation due to other loci and the environment. Allelic association studies are not per se associated with familial inheritance patterns. As Eric Lander and Nicholas Schork (1994) note in their superb review article on the "Genetic Dissection of Complex Traits," association studies are essentially case-control studies comparing unrelated affected and unaffected individuals in a population. As Lander and Schork put it, "an allele A at a gene of interest is said to be associated with a trait if it occurs at a significantly higher frequency among affected compared with control individuals" (2041). These types of studies have been used in nonbehavioral domains, in particular to associate variations in the HLA gene complex with specific autoimmune diseases. For example, the HLA-B27 allele occurs in 90 percent of ankylosing spondylitis patients, but in only 9 percent of the general population. Association studies, not well controlled by other information, however, can lead to erroneous conclusions. In one of their many cautionary comments concerning methodology in human genetics, Lander and Schork (1994) note that "positive associations can arise as an artifact of population admixture. In a mixed population, *any trait* present at higher frequency in an ethnic group will show positive association with *any allele* that also happens to be common in that group" (2041; my emphasis). The implications of this are, I think, obvious. Plomin and his coauthors, however, as I understand them, view allelic association as a complementary strategy that can be used together with QTL to detect such loci when QTLs have quite small effects (1994, 1738).

A MODEL OF GENETIC EXPLANATION

The purpose of this section is to review some generalizations and themes, drawn from the three extended examples and the more general methodological sections considered earlier, and to introduce a framework for a genetic explanation of behavior. I can only sketch an account of explanation that I think best fits this area, but believe that enough can be said about my approach to make it intelligible.

I have come to think of the types of models with which I began the chapter, such as the Popper-Hempel-Oppenheim or deductive nomological model and other statistical explanation models such as the causal-mechanical explanation, as representing a fairly narrow focus. This notion of a narrow focus applies both to the logical elements of such models as well as to the explicitly appealed-to explanatory elements, such as laws, initial conditions, or statistical models (such as a binomial model). The narrow focus is also replicated in the simple types of examples that are usually provided as illustrative of such models. A number of philosophers who have reflected on such models of explanation have found them incomplete and have looked for a broader context within which explanations function. This includes Kuhn's paradigms, Lakatos's research programs, Shapere's domains, Laudan's research traditions, Kitcher's practices, van Fraassen's pragmatic question-oriented analysis of explanation, and Railton's notion of an ideal explanatory text. I see this broader context as best representing a "field" of knowledge within which explanations are developed. To provide an explanation that makes rigorous scientific sense and is not a "toy" example, fairly deep knowledge of a field and frequently many fields of science (and mathematics) have to be implicitly appealed to. Thus I see an explanation in science as having both a focal element (and within that focus perhaps one dominant logic, such as a deductive inference) and a broader field element, within which the explanation is based and within which it is evaluated. I like to think of this approach as a "2F" approach to explanation, representing *the field* and *the focus*.

Behavioral genetics is an interdisciplinary field, and one that pursues its inquiries at many, often simultaneous, levels of aggregation. In the preceding sections we have seen a number of examples of this multidisciplinary approach. The field of behavioral genetics is thus very difficult to define and circumscribe in any simple fashion, and I do not attempt to do it here; instead, I rely on the illustrations already introduced to do this job by example rather than by precept. Developing an explanation within behavioral *genetics,* however, does not make the explanation a *genetic* explanation. Behavioral genetics makes extensive use of environmental variables and measures of influence in its explanations of behavior, as the preceding examples indicate. Readers interested in considering how one might begin to limn the peripheries of the field could consult the textbook by Plomin, DeVries, and McClearn (1990), Plomin et al.'s (1997) update of that text, or the recent introductory version of the Kandel and Schwartz successive text editions by Kandel, Schwartz, and Jessel (1995), interestingly entitled *Essentials of Neural Science and Behavior.* These texts also provide a glimpse at the breadth of the generalizations and model systems that are appealed to as the backdrop of explanations in behavioral genetics. The difficulty of providing any simple "model" of explanation is illustrated in a microcosmic way by the example from work on *C. elegans* pursued by Bargmann and her associates described earlier.

In Bargmann's explanation of the behaviors of *C. elegans* in response to different oderant stimuli, we encounter implicit references to many parts of the neurosciences and molecular genetics. A research article such as the Bargmann et al. 1993 or the Sengupta et al. 1994 essays, both of which appeared in *Cell,* provides explanations of *C. elegans* behaviors. But even in such focused research articles as this, the broader context of the problem(s) is sketched (however briefly) and assumptions are made that the reader is knowledgeable about the organism, neuroscience, and molecular biology. Within this broad tapestry, such a research article quickly focuses on several well-defined questions, and then proceeds to present answers to the questions in terms of the novel (or not so novel) discoveries that are the rationale for the publication of the paper. Within the context of these answers, it is possible to pick out a focus (or foci), and ask what is the nature of the explanation within that focal area. Here we can usefully begin to appeal to the nature of the law(s), mechanism(s), and model(s), as well as to scrutinize the nature of the inference (deductive, statistical) and to ask whether this explanation is causal or unificatory, and, if causal, what type(s) of causal conditions is operative.

The types of questions about genes and behavior in connection with olfaction, and an animal's behavioral response to olfaction (i.e., chemotaxis), are introduced early in the 1993 article by Bargmann et al. in the following way: "The existence of these [odorant receptor] genes opens many questions about the mechanism of odorant recognition and discrimination. How specific is the interaction of odorants and receptors? How many receptors are expressed on a single olfactory neuron? How is the information about odorants transmitted to the brain to generate appropriate behaviors?" (515). The questions are clearly "interlevel," in the sense that they intermingle entities at different levels of aggregation (odorants, genes, neurons, brain, and behaviors). The core explanation in this artlicle is formulated at the neuronal level, and then later partially unpacked in terms of genes that contribute to the intact neurons' behaviors. In a key passage several pages later, these authors write:

Two cell types of the amphid chemosensory organ, the AWA and AWC neurons, were required for chemotaxis to specific subsets of volatile attractants. Animals in which the AWC neurons were killed were severely defective in their responses to the volatile odorants benzaldehyde, butanone, and isoamyl alcohol. . . . However, these animals responded normally to other volatile odorants and to water-soluble attractants such as Cl$^-$ ions. Animals in which the AWA neurons were killed had defective responses to the volatile odorants diacetyl and pyrazine but normal responses to other volatile odorants and Cl$^-$ ions. . . . Some attractants appeared to be sensed by both AWA and AWC. Killing AWA and AWC neurons together in the same animal reduced the responses to isoamyl alcohol and 2,4,5-trimethylthiazole more than killing either cell alone. . . . Chemotaxis to Cl$^-$ remained intact in these animals. Thus, AWA and AWC neurons are required for responses to distinct, but partly overlapping, subsets of volatile odorants. (Bargmann et al. 1993, 518)

Thus the AWA and AWC neurons are identified as entities that are "required" (i.e., "necessary") for the chemotactic behavioral response that is the focus of this inquiry. Not yet present is a "model" that would incorporate these neurons into a circuit, and trace additional interneuronal connections that would yield behavioral responses via the triggering of motor neurons and their effects on muscle contraction. Nevertheless, this chapter makes progress toward an explanation of behavior by identifying entities that are "necessary in the circumstances" for specified patterns of behavior. The explanation is thus both causal and incomplete, pointing toward several possible directions of additional research that would begin to fill in the gaps in the explanatory pattern, and move the explanation toward a sufficiency explanation – subject, as such things always are, to background ceteris paribus conditions and provisos.

As noted in the text earlier, some half-dozen *odr* genes have been identified, and one of these, *odr-7*, was the subject of some in-depth analysis by this research group. The *odr-7* analysis considers the *genetic* contribution to the AWA down to the sequence level of detail and, in addition, proposes in a still speculative manner a functional role for the protein product of that sequence. This dimension of the explanation of behavior fills in at a level of considerable detail one element in a causal chain of events between the oderants in the organisms' environment and the organisms behavioral responses. When fully confirmed, such a receptor protein would also function as an entity that is "necessary in the circumstances" for the behavioral response of the organism. Thus, from a general epistemological point of view, the molecular analysis deepens, but does not add a different type of logical condition to, the account provided at the neuronal level.

A general account of focal explanation can, I think, be provided that identifies some of the common features encountered in such explanations. A focal account is necessarily incomplete, however, and requires (usually implicit) reference to provisos and ceteris paribus conditions, as well as a backdrop ideal explanatory text, to make the explanation an adequate one. Here I provide a general framework for a model of explanation that I have discussed in more depth elsewhere (1993, esp. ch. 6; also see ch. 9). This model of explanation has *two main focal components*, each of which I believe relates to themes introduced in our examples and methodological discussion, and is backgrounded by an additional *two more field-oriented components*. I sketch in a largely summary fashion each of these four components of a genetic explanation (or GE) and relate the scientific materials covered earlier to these four components. I think that the discussion and the way that the model is introduced does not require that the reader commit to this type of model, although I think it is a generally reasonable one.

Before developing the details of the conditions for a genetic explanation, I provide an intuitive overview of the concept. The general idea is that an explanation is genetic *if and only* if the event (property, trait, behavior of the empirically investigated system[s]) is a causal consequence of the processes characterized by an instantiated "genetic model." The genetic model can be a traditional Mendelian single-locus model, or it can be a complex molecular model where the "genetic" content is specified in terms of DNA and/or RNA. Typically the gene or DNA [RNA] has been "inherited," although a few exceptions, such as the acquired "genetic" basis of cancer, might be appropriately construed as conforming to this account of a genetic explanation. These two notions of a model and a causal consequence are captured in items GE1 and GE2. This general intuition is badly oversimplified, however, because (1) few explained properties, traits, or behaviors are *solely* explained in terms of the action of genetic processes (excluding environmental or learning variables); (2) the causal connection needs to be liberally interpreted to allow for "probabilistic causation," as well as more traditional "deterministic causation"; and (3) the genetic model needs to be situated in a "field" in terms of which various provisos and evaluative dimensions are located. The broader field embeddedness dimensions are captured in a weak way in items GE3 and GE4.

The Model System (GE1)

The model system or MS is a somewhat idealized system that can be quite simple, as when one introduces a single-locus Mendelian model of a dominant-recessive gene pair, say as a Punnett-square representation, and then uses that model to explain the inheritance of Huntington's disease. Alternatively, the MS can be more complex, as in an explanation of volatile odorant detection using specific mutants and neuron types in *C. elegans,* or a Kandel cartoon depicting presynaptic sensitization in *Aplysia.* There is no formal limit on the degree of complexity of an MS, although these are always idealized to a certain extent. One can provisionally think of three increasing levels of detail and complexity for an MS from (1) a simple Mendelian single-locus model; (2) a more complex QTL analysis for multigenic traits, along with a further unpackable "environmental" source variable; and (3) a connectionist system allowing for multigenetic causation, pleiotropy, and multiple neuronal functions in generating behavior. MS can also be interlevel in the sense of levels of integration; in fact, most FAPs seem to be interlevel. A "genetic" explanation can be formulated at any of these levels of detail; *general assumptions* are embedded in the model system that describe the system under study and which are believed to generalize to other like systems, though the generalization may be strain-limited and species-limited, though possibly even broader. (We saw an example of trans-species generalization in connection with the use of the larger neurons of *Ascaris* to explain *C. elegans* neuronal behavior earlier.) The particular systems being explained – if this is a laboratory experiment, for example – are identified with the MS, in the sense that the particular system is an *instance* of the MS. This is a kind of model-theoretic or semantic theoretic approach to explanation (see Suppe 1989 and Giere 1984 for additional examples). For an explanation to be genetic, the generalizations of the model need to use genetic vocabulary in a way that appeals to the genes as (part of) the explanation of the event to be explained (the explanandum) – in our case, a behavior (pattern). The question of the completeness of the explanation or the amount of "variance" in the event(s) to be explained is perhaps better addressed in the next component, and also in components GE3 and GE4.

The Causal Component (GE2)

Explanations in genetics including behavioral genetics are typically causal, and the action words appearing in a molecular level model system such as found in *C. elegans* involve such causal properties as "binds to," "induces," "phosphorylates,"and "blocks," as well as statements that a process "causes" or "results in" some event. In human genetics the causal language is vaguer, often using such terms as "influences" or "predisposes," as well as notions like "results in" and "determines." Causation can be analyzed in depth in terms of its logic, epistemology, and metaphysics (see Schaffner 1993, ch. 6, for one approach). The notion of causation that appears to be appealed to is that of a condition or entity that is "necessary in the circumstances" (see my analysis of Bargmann's experiment on laser ablation in Schaffner 1998). *Suitably hedged by provisos,* such a condition will in addition be "sufficient in the circumstances" and will thus be a causally sufficient explanation. The provisos may, however, be so strong as to point rather to incomplete or gappy explanations, in which only one of several suspected factors has been identified. Causation can have two distinguishable subtypes involving (a) deterministic causation and (b) probabilistic causation. The latter subtype concept of causation has generated a large literature in philosophy of science. One can illustrate the relation between the deterministic and the probabilistic aspects, even at the molecular level, using the examples of neurotransmitter release at the neuromuscular junction or of color-coat determination in genetics (see Schaffer 1993, ch. 6, for details). For our purposes in this chapter, any complete explanation of a behavior will have to include both genetic and environmental causal factors and, more likely than not, a learning history as well (not to mention the influence of the genetics on the environment and back again – the so called nature of nurture, to use Plomin's term). With multiple chains of causation, virtually no explanation of behavior will be *purely genetic,* and this is in accord with the consensus view discussed earlier. One might, by qualifying and circumscribing, be able to claim that Huntington's disease involved a purely genetic explanation, but this would be a rare (and, I think, misleading) example.

I think that deductive logic can suffice to tie the model system and its instances in cases where we have deterministic causation; in cases where there is probabilistic causation and one is not dealing with "infinite classes," a type of inductive logic is required, probably a Bayesian logic of support.

Although appeals to causation are frequently (and typically) involved in explanation, some MSs might involve generalizations that are statements of coexistence, association, or "correlations" (i.e., indicating quantities that vary together, acting as markers, but perhaps as causes). In these cases we seem to get some explanation via a unification of phenomena, without any clear evidence of causation. Something similar may be at work where a direction of causation is assumed, but the gaps are too large to think clearly of any causal mechanisms. This seems to be the case in Hamer's study and also in the HLA-B27 allele association with the rheumatological disease ankylosing spondylitis. (In Hall's work with *Drosophila*, however, by contrast, notice if gaps exist, how attempts are made by him, albeit speculatively, to fill in those gaps.) In such cases where we have interesting genetic associations it may be reasonable to appeal to another weaker aspect of "explanation" and accept as scientifically legitimate MSs that "unify" domains of phenomena, here genetics and behavior. I continue to remain uncomfortable with this weak sense, because I believe that true explanation must be causal, so in general I view these explanations as provisional surrogates for deeper causal explanations, to be found in the future as a result of scientific progress (excluding, probably, quantum mechanical explanations involving a kind of noncausal stochastic bedrock).

The Comparative Evaluational Inductive Component (GE3)

Most models of explanation that have been introduced into the philosophy of science have an explicit or implicit evaluational component, either a truth requirement, as in the classic Hempel-Oppenheim paper, or something weaker involving evidential support, as in van Fraassen's (1980) model. In conformity with the personal probability perspective I favor (Schaffner 1993, ch. 5) any inductive support (whether of the type described in GE2 or more general inductive support for an MS) appearing in explanations can be viewed in Bayesian terms. The notion of support is also construed as "doubly comparative," in van Fraassen's

(1980) sense: for a given explanation one compares (often implicitly) an explanandum with its contrast class in terms of the support provided, and one also (perhaps only implicitly) compares rival explanations for that explanandum. This, we shall see, introduces evaluational components into explanations, a point similar to Sober's (1984, ch. 5). In behavioral genetics, this evaluational component can take the form of (almost) crucial experiments involving mutants, as in Sengupta et al. (1994) regarding the role of the *odr-7* gene's effect on the AWA neurons, or it can be of the statistical significance sort, as encountered in Hamer's work.

The Ideal Explanatory Text Background Component (GE4)

As emphasized several times already, explanations that are actually given in behavioral genetics, however complex, are partial aspects of a complete story, and those aspects focused upon are frequently affected by pragmatic interests. Following Railton (1980), I accept the notion of an "ideal" explanatory text background from which specific explanations are selected. I provide hints in my discussion of *C. elegans* how vast the information available about this animal is. Recall too, that the explanatory backdrop for accounting for an aspect of this animal's behavior will also rely on the even vaster scaffolding of molecular and cellular biology, only a small portion of which will be used in any given explanation. It is here that the "field" aspects of an explanation appear most strongly.

Components of an explanation as sketched in GE1–GE4 represent a first pass at what we might mean when we provide a genetic explanation. In my view, the "genetic" aspects are in the content, provided by the models and theories of genetics, ultimately drawing on the DNA/RNA material of heredity, but requiring, for a complete explanation, the rest of molecular biology. To reiterate another theme, even this type of "ideal explanatory" account will have aspects unaccounted for unless it refers to what goes beyond itself – the environmental component.

NOTES

1. The terms "selectional," "developmental," and "adaptational" are taken from Sober 1984 and from Brandon 1990.
2. See Schaffner 1972 on nineteenth-century aether theories for specific examples.
3. The parallels between mechanical explanations in the nineteenth century

and genetic explanations today could be pursued further to examine the differences between being able to explain all *interactions* mechanically (genetically) and taking a mechanical particle (or a gene) as the *fundamental unit,* a distinction suggested to me by Robert Wachbroit and David Wasserman. Because I use the parallel mainly methodologically, however, I defer such an inquiry until another paper.

4. Kupferman stresses that not only do genetics and environment always interact to produce behavior, but in addition there is "no sharp distinction between learned and innate behaviors. There is instead a continuous gradation" (1992, 989). This section relies extensively on Kupferman's review as a means of illustrating an account found in a classical neuroscience text.

5. *C. elegans* sequencing is being conducted as a joint project at Washington University, St. Louis, and the Sanger Centre (United Kingdom), and was essentially completed in 1998; see the December 1998 issue of *Science* for a series of articles related to the completion of this 97Mb nucleotide sequence.

6. Brenner published his letter to Max Perutz of June 5, 1963, in which he develops this belief in his foreword to Wood's (1988) reference volume on the nematode. In that same foreword Brenner also includes portions of his October 1963 proposal to the Medical Research Council laying out the reasons why the nematode (although at this point it was *C. elegans's* cousin, *C. briggsiae,* that was mentioned) would be a model organism for these studies. Brenner has indicated that though several people have told him they planned to write a history of *C. elegans* and its community of researchers, none to his knowledge has yet been done (S. Brenner, personal communication, March 1995).

7. Bargmann quotes figures from Durbin (1987): "For any synapse between two neurons in any one animal, there was a 75% chance that a similar synapse would be found in the second animal . . . [and] if two neurons were connected by more than two synapses, the chances they would be interconnected in the other animal increased greatly (92% identity)" (Bargmann 1993, 49).

8. I refer to these general principles relating genes and behavior as "rules" because they are like spelling rules: they hold quite generally but there are exceptions that do not disprove the general rule.

9. I consider one type of exception involving an almost "one gene, one behavior type" association in Schaffner (1998).

10. This technique is based on the fact that in *Drosophila* the sex of a cell is based on its sex chromosomes: a double X is a female cell and a single X is a male cell. It is possible to make X chromosomes unstable, such that during mitosis, a clone of the female parent cell becomes a male cell. This can occur at several points in *Drosophila* development and, when it takes place in late stages, can leave small islands of male cells in an otherwise female sea. These mixed sex (at the cellular level) organisms are called "gynandromorphs." See Lawrence 1992, 10, 82–87.

11. Parts of this section are drawn from my 1999 essay. Here this example is used to illustrate explanatory themes, whereas in my 1999 essay it is employed more methodologically. See my essay for further connections be-

tween Hamer's group's work and Lander and Schork's (1994) "fourfold way" of approaching complex trait genetics.

12. See the June 30, 1995, issue of *Science* (p. 1841) for some questions by the Office of Research Integrity about the 1993 study, now apparently resolved (see *Science* 275 (February 28, 1997: 1251). Replication studies of Hamer's result continue, however, to generate controversy.

13. Even the Huntington's disease example now admits of some variation, and a close reading of Lander and Schork's (1994) article suggests that even other traditional single-gene disorders such as sickle cell anemia are in reality complex trait diseases.

REFERENCES

Avery, L., Bargmann, C., and Horvitz, H. R. (1993). The *Caenorhabditis elegans* *unc-31* gene affects multiple nervous system-controlled functions. *Genetics* 134: 455–464.

Bailey J. M., Pillard, R. C., Dawood, K., Miller, M. B., Farrer, L. A., Trivedi, S., Murphy, R. L. (1999). A family history study of male sexual orientation using three independent samples. *Behavioral Genetics* 29 (2): 79–86.

Bargmann, C. (1993). Genetic and cellular analysis of behavior in C. *elegans*. *Annual Review of Neuroscience* 16: 47–51.

Bargmann, C., Hartwieg, E., and Horvitz, H. R. (1993). Odorant-selective genes and neurons mediate olfaction in C. *elegans*. *Cell* 74: 515–527.

Bargmann, C. and Mori, I. (1997). Chemotaxis and thermotaxis. In C. *elegans II*, ed. D. L. Riddle, T. Blumenthal, B. J. Meyer, and J. R. Priess, 717–737. Cold Spring Harbor, N. Y.: Cold Spring Harbor Press.

Benzer, S. (1971). From gene to behavior. *JAMA* 218: 1015–1022.

Brandon, R. (1990). *Adaptation and environment*. Princeton: Princeton University Press.

Brenner, S. (1974). The genetics of *Caenorhabditis elegans*. *Genetics* 77: 71–94.

Brunner, H., Nelen, M., Breakefield, X., et al. (1993). Abnormal behavior associated with a point mutation in the structural gene for monoamine oxidase A. *Science* 262: 578–580.

Byne, W. (1994). The biological evidence challenged. *Scientific American* 270: 50–55.

C. *elegans* Sequencing Consortium. (1998). Genome sequence of the nematode C. *elegans*: A platform for investigating biology. *Science* 282: 2012–2018.

Cardon, L. R., Smith, S. D., Fulker, D. W., Kimberling, W. J., Pennington, B. F., and DeFries, J. C. (1994). Quantitative trait locus for reading disability on chromosome 6. *Science* 266: 276–279.

Chalfie, M., and White, J. (1988). The nervous system. In Woods 1988, 337–391.

Churchland P., and Sejnowski, T. (1992). *The computational brain*. Cambridge, Mass.: MIT Press.

Crabbe, J., Belknap, J., and Buck, K. (1994). Genetic animal models of alcohol and drug abuse. *Science* 264: 1715–1723.

CRG (Council for Responsible Genetics). (2000). Do genes determine whether

Salmon, W. (1984). *Scientific explanation and the causal structure of the world.* Princeton, N. J.: Princeton University Press.

Saunders, A. R. (1998). Poster presentation 149 at the annual meeting of the American Psychiatric Association, Toronto.

Schaffner, K. F. (1972). *Nineteenth-century aether theories.* Oxford: Pergamon Press.

——— (1980). Theory structure in the biomedical sciences. *Journal of Medicine and Philosophy* 5: 57–97.

——— (1993). *Discovery and explanation in biology and medicine.* Chicago: University of Chicago Press.

——— (1998). Genes, behavior, and developmental emergentism: One process, indivisible? *Philosophy of Science* 65 (June): 209–252.

——— (1999). Reduction and determinism in human genetics: Lessons from simple organisms. In *Proceedings of conference on Controlling our destinies: Historical, philosophical, social, and ethical perspectives on the human genome project,* ed. P. Sloan. Notre Dame, Ind.: University of Notre Dame Press.

Sengupta, P., Colbert, H., and Bargmann, C. (1994). The *C. elegans* gene *odr-7* encodes an olfactory-specific member of the nuclear receptor superfamily. *Cell* 79: 971–980.

Sober, E. (1984) *The nature of selection.* Cambridge, Mass.: MIT Press.

Suppe, F. (1989). *The semantic conception of scientific theories and scientific realism.* Urbana: University of Illinois Press.

Suppes, P. (1967). What is a scientific theory? In *Philosophy of science today,* ed. S. Morgenbesser, 55–67. New York: Basic Books.

Thomas, J. H. (1994). The mind of a worm. *Science* 264: 1698–1699.

van Fraassen, B. (1980). *The scientific image.* New York: Oxford University Press.

von Ehrenstein, G., and Schierenberg, E. (1980). Cell lineages and development of *Caenorhabditis elegans* and other nematodes. In Zuckerman 1980, 1–71.

Watt, D. C., Edwards, J. H. (1991). Doubt about evidence for a schizophrenia gene on chromosome 5. *Psychol. Medicine* 21 (2): 279–285.

Wimsatt, W. (1976). Reductionism, levels of organization, and the mind-body problem. In *Consciousness and the brain,* ed. G. Globus, G. Maxwell, and I. Savodnik, 205–267. New York: Plenum Press.

Wood, W. (Ed.) (1988). *The nematode* Caenorhabditis elegans. Cold Spring Harbor, N.Y.: Cold Spring Harbor Press.

Zuckerman, B. (Ed.) (1980). *Nematodes as biological models:* Vol. 1: *Behavioral and developmental models.* New York: Academic Press.

Chapter 5

On the Explanatory Limits
of Behavioral Genetics

KENNETH A. TAYLOR

A fair number of studies in behavioral genetics purport to show that at least some kinds of "criminality" are highly heritable.[1] On first encounter, such claims are bound to seem provocative. They conjure up the frightening specter of malevolent authorities using genetic screening to determine who the likely criminals are from birth. On a more theoretical level, such results appear to challenge assumptions about the causal determinants of criminal behavior. It is widely assumed, on both the left and the right, that criminals are largely made, though made in different ways, and not born. The left assumes that criminals are made by poverty, deprivation, abuse, neglect, ignorance and the like, whereas the right assumes that criminals are made by insufficient social sanctions against criminal conduct. Of course, even if criminality is partly heritable, it does not follow that such "environmental" factors play no role in "making" criminals. But if criminality is largely heritable, then it just may be that such factors play less of a role than one might pretheoretically have imagined.

In fact, though, it is easy to overestimate both the sociopolitical consequences and the theoretical significance of the supposed heritability of criminality. Consider briefly the sociopolitical consequences of such results. Just for the sake of argument, suppose that we one day discover just what gene or assemblage of genes is responsible for an increased liability to criminality. And suppose that a test that detects the presence or absence of that gene or assemblage of genes is developed. Would the advent of such genetic screening be good news or bad? Surely, the only answer we can *now* give to that question is, That depends. It depends, for example, on what we come to know about the way in which genes and "environment" interact to "produce" criminal behavior, whether the relevant environmental variables turn out to be systematically ma-

117

nipulable by us, whether we discover drugs that can counteract or compensate for the effects of the relevant gene or assemblage of genes. It will also depend on social, political, and economic forces. In particular, it will depend on whether such forces conspire to bring it about that the application of such knowledge to the lives of real people is embedded in a network of institutions and practices that guarantee that each person is treated with dignity and respect, independently of his or her initial genetic endowments.

In the current get-tough-on-crime climate that is, to be sure, far from a guaranteed outcome. But it is also far from guaranteed that we will fail to develop such institutions and practices. The real point is this. Just given the fact that a liability to criminality is to some degree heritable, nothing whatsoever follows, at least not directly, about the best shape of our practices and institutions. This is not to deny that, in the fullness of time, advances in our knowledge and in our technical prowess to intervene in the lives of real people may require us to face complex social and political questions. Largely because of the apparently rather impoverished state of our knowledge of how genes and environment interact and our consequent lack of any principled basis for systematic intervention into the lives of real people, we are at present in no position to anticipate in detail the shape of such issues as may await us.

But we are, I think, presently well positioned to evaluate fruitfully some of the more theoretical challenges and issues to which the behavioral genetics of criminality gives rise. In this chapter, I try to confront some of those challenges head on. My main aim is to say just what, if anything, heritability statistics have to teach us about the *explanation* of crime, criminals, and their behavior. I argue that there is, in fact, very little that heritability statistics explain about crime, criminals, and criminal behavior. This is not to deny that there is a perfectly respectable sense of "explanation" in which behavioral geneticists are entitled to their claim that, say, 50 percent of the phenotypic variance among humans with respect to at least certain types of criminal behavior is "explained" by genotypic variation, at least over a range of certain environments. But even granting that result, many questions remain open, questions that the statistical methods of behavioral genetics seem to me largely powerless to answer. I argue that the inability of research in behavioral genetics to answer these questions greatly diminishes the explanatory import of results about the heritability of criminality.

These points, however, are not intended as a dismissal of the very enterprise of behavioral genetics. My aim is merely to set such work in its proper place – to be very clear about what such work can and cannot teach us about the explanation of human behavior. I make three distinct but related points, none of them terribly controversial or particularly revealing about the explanatory limits of behavioral genetics. Taken collectively, however, they show that behavioral genetics is a strikingly limited explanatory enterprise. First, heritability results tell us nothing about the mechanisms by which genes contribute whatever they contribute to determining criminality. Second, heritability results tell us nothing about what I will call the psychodynamics of individual criminal behavior. Third, the discussion of the heritability of criminality has taken insufficient account of the fact that crime is what in some circles would be called a socially constructed category. These three facts taken together should lead us to expect very little about crime, criminals, or criminal behavior to be illuminated by work in statistical behavioral genetics.

Heritability results per se tell us nothing about the mechanisms by which genes and environment cooperate to determine phenotypic traits. Heritability is a straightforward statistical measure of the ratio of genotypic variation to total variance in phenotype in a population, relative to a range of environments.[2] Just to fix ideas, imagine a variable phenotype, a known set of rearing environments $e_1 \ldots e_n$, and a known set of genotypes $g_1 \ldots g_m$. We can plot on an $n \times m$ matrix how (average) phenotype varies as a function of rearing environment and genotype. In each cell p_{ij} of the matrix, we enter a "score" for that genotype in that environment:

	e_1	e_2	e_3	e_4
g_1	p_{11}	p_{21}	p_{31}	p_{41}
g_2	p_{12}	p_{22}	p_{32}	p_{42}
g_3	p_{13}	p_{23}	p_{33}	p_{43}
g_4	p_{14}	p_{24}	p_{34}	p_{44}

This represents the average phenotypic score for all the members of the population who share the relevant genotype and the relevant rearing environment. The following averages are relevant to the eventual calculation of heritability. For each genotype, gi we calculate $p_{gi}(e_1 \ldots e_n)$, the average phenotypic score for organisms that share g_i but not necessarily the same rearing environment. These averages are used to calculate the contribution of environmental variation to total phenotypic variance. Similarly, for each rearing environment ej, we calculate p_{ej} $(g_1 \ldots g_m)$, the average phenotypic score for organisms that share e_j, but not necessarily the same genotype. These averages are ultimately used to calculate the contribution of genotypic variation to total phenotypic variance. We need to take one more average before we are ready to calculate total phenotypic variance – the average phenotype in the entire population of organisms. Call this quantity A, the overall phenotypic average. We calculate the total phenotypic variance (in a population relative to a range of environments) by calculating how the various p_{ij} spread about the overall phenotypic average. For convenience, we assume that each of the p_{ij}'s represents the same number of organisms. First we calculate for each p_{ij} the difference between it and A, square that difference, and then take the average of all the squared differences:

$$V_{p(EP)} = S(p_{ij} - A)^2/nm.$$

We now determine the contribution of environment and genotype to total variance, via the following formulas. For the contribution of environmental variation, we set:

$$V_{e(EP)} = \Sigma(p_{gi}(e_1 \ldots e_n) - A)^2/m$$

For the contribution of the genotypic variation, we set:

$$V_{g(EP)} = \Sigma(p_{ei}(g_1 \ldots g_m) - A)^2/n.$$

The total phenotypic variance is the sum of the variance due to environmental variation, the variance due to genotypic variation, and the variance due to gene-environment interactions. That is,

$$V_{p(EP)} = V_{e(EP)} + V_{g(EP)} + V_{G \times E}.$$

If we assume an absence of measurement error, $V_{G \times E}$ can be thought of as just the difference between $V_{p(EP)}$ and $V_{e(EP)} + V_{g(EP)}$. When $V_{G \times E}$ = 0, the data are said to be additive. Now we are finally in a position to define the quantity we are really after – h^2 – heritability of the phenotype. h^2 is defined as follows:

$$h^2 = V_{g(EP)}/V_{p(EP)}.$$

h^2 is the portion of the total phenotypic variance that is due to genetic variance.[3]

   In the preceding paragraph, we assumed a known set of genotypes, and a known set of environments and known phenotypic variation.

More typically, researchers are confronted with a range of phenotypes, whose genetic basis is largely unknown, with genotypic variance unknown, with it being hard to say *which* environmental parameters matter, let alone *how* those parameters matter. This is where twin and adoption studies come in. These are methods for estimating the heritability of a trait in a population by considering heritability for very special subpopulations: (a) identical (MZ) twins reared apart; (b) adopted children, their biological parents, and their adoptive parents and siblings; (c) MZ twins reared together and fraternal (DZ) twins reared together. Because, for example, MZ twins are genetically exactly similar, any phenotypic variance between a pair of MZ twins reared apart must be due to environmental variance (or to gene-environment interaction). Consequently, if on average MZ twins reared apart exhibit a high degree of phenotypic similarity for a certain trait, then given certain further assumptions, that argues for a high degree of heritability for that trait in the population at large. One such further assumption is that rearing environments for MZ twins reared apart vary from one another as much as do the rearing environments for two randomly selected members of the population at large.

It would be an interesting project to explore the exact logic behind studies of MZ twins raised apart, the somewhat different logic of studies of MZ versus DZ differences, and the still different logic of adoption studies to show in more detail just what they purport to establish and how they purport to establish it.[4] That is not necessary for our present purposes. We have done enough to make clear what heritability is a measure of. All of these are merely different paradigms for estimating heritability in a population at large by considering special subpopulations.

One can establish that a trait has a high degree of heritability without establishing anything at all about *what* gene or assemblage of genes is responsible for the genetic contribution to overall phenotypic variance – though this is not to deny that more powerful methodologies are available for exploring the genetic details.[5] Indeed, as researchers generally acknowledge, it could turn out, consistent with a trait's being highly heritable, that a vast array of genes are jointly responsible for making whatever contribution genetic variance makes to phenotypic variance, with each individual gene making only the slightest contribution on its own. Further, heritability statistics tell us nothing about *how* genes and environments interact to produce phenotypic traits. Heritability is just a measure of how much of the phenotypic variance is due to genotypic variance; it is not a measure of how the genetic contribution to phenotypic variance is implemented. But *how* either the genotypic contribution or the environmental contribution to phenotypic variance is implemented can matter a great deal. For practical reasons, it is clear that unless we understand the *how*, we are unlikely to have any principled grounds for making systematic interventions in the lives of real people.

More relevant in the present context is the bearing of the *how* on our theoretical-explanatory interests and aims. First, recall that heritability is defined for a population and over a range of environments. A trait that has this or that degree of heritability relative to a range of given environments $e_1 \ldots e_n$ can have a quite different degree of heritability relative to a broader range of environments $e_1 \ldots e_n, e_{n+1} \ldots e_{n+m}$. We may think of the $e_{n+1} \ldots e_{n+m}$ as "new" environments involving, perhaps, systematic intervention in the developmental process. But they need not be things of this sort. The new $e_{n+1} \ldots e_{n+m}$ may be new environmental possibilities that result from the adoption of new social arrangements of varying sorts. Typically neither unaided nature nor the human social world provides us with an exhaustive range of environmental possibilities. So it is at least logically possible and may in particular cases be empirically possible that by, as it were, reconditionalizing phenotypic variance on a new range of environmental possibilities, a trait highly heritable relative to the old range of environments has a substantially lowered heritability relative to the new range of environments. For this reason it would be a mistake to infer any sort of genetic determinism from the fact that a trait is highly heritable. Indeed, that genetic determinism does not follow from high heritability can be seen even without allowing the possibility of "new" environmental interventions. The crucial point, again, is that heritability is only a measure of the proportion of phenotypic variance due to genotypic variation.

If a trait is highly heritable then, on average, we get more phenotypic variation by varying the genotype than by varying the environment. But that is consistent with *some* environments making a great deal of difference, at least for some of the genes, and maybe even for all of the genes. So even for a highly heritable trait, it might be possible to make a great deal of difference by changing the rearing environment in just the right way. Consequently, even if, on average, altering the environment matters little, we might still be able to affect phenotype massively by the right kind of alteration in the environment. For example, even with respect to a supposedly highly heritable trait like criminality (or IQ), it may be possible to massively affect a subpopulation that is phenotypically worse off, just by altering its environment in the right way.

Whether such logical possibilities are, in particular cases, genuine empirical possibilities is a straightforward empirical question that can be answered only on a case by case basis. But it is clear in advance that the methods of behavioral genetics alone will generally be insufficient to yield answers to such questions. To answer such questions, we need to know more about the mechanisms by which genes and environment interact to produce phenotypic traits. When we know enough about such mechanisms, we will be able to explain why this or that change in the environment has greater or lesser effect on phenotype, given this or that genotype. And we will be able to explain why given this or that environment, this or that genotype gives rise to this or that phenotype. We may even be able to *deduce* the likely effects of some previously untried intervention from facts about the developmental mechanism and the nature of the proposed intervention.

All such explanations and deductions require us to know more than the merely de facto statistical averages in which behavior genetics deals. We must uncover the structures of *counterfactual* dependence which interconnect genotype, phenotype, and environment. Such is the stuff of which bona fide explanations are made. When we know enough about such structures of counterfactual dependence, then the de facto statistical averages in which behavioral genetics deals may well seem to be largely beside the point, both theoretically and practically. Who would care that on average genetic variation is more responsible than environment for phenotypic variation in, say, IQ or criminality if we knew just the right way to manipulate the environment to insure that fewer people ended up phenotypically worse off? Of what explanatory relevance would such merely de facto statistical averages then be?

Even when a trait is highly heritable, environment can still make a very big difference, partly because heritability has to do with the average contribution of environmental versus genotypic variation and partly because heritability is defined relative to a fixed range of environments. I now stress the other side of the coin. Even when a large portion of the variance among phenotypic traits in a population is due to environment, it does not follow that genes make no or little difference. Consider phenotypic variation in initial native language among the population of humans. Take the set of humanly possible (native) languages $[L_1 \ldots L_N \ldots]$. Within the human population, whether a child acquires initial native fluency in L_i or L_j is entirely dependent on her initial linguistic environment – although, because of the phenomenon of creolization, the rule is not as simple as: acquire the language spoken by one's parents. So *all* of the variance (setting aside people with language deficits of various kinds) among the human population in initial native fluency is due solely to environmental variation. But it does not follow at all that the environment alone is causally responsible for our acquisition of native fluency. Take a broader population of monkeys and humans, and expose them both to the same range of primary linguistic data. The humans will acquire native fluency; the monkeys will not. What explains the difference between humans and monkeys? Well, one very plausible hypothesis is that the shared human capacity to acquire one of the humanly possible languages has a substantial genetic basis, a genetic basis present in all humans and in no monkeys.

If all humans share the genetic material that underlies our capacity to acquire a language, then none of the phenotypic *variance* among humans will be explained by appeal to genotypic variation among humans. So the heritability of initial native fluency will be exactly zero. But there is surely an at least partly genetic *explanation* of our shared capacity to acquire language at all. Moreover, although logically possible languages are free for the thinking up, humanly possible languages are very tightly constrained, much more tightly constrained than considerations of logical possibility alone would demand.[6] An appealing thought is that, somehow or other, the constraints, whatever they are, are themselves a consequence, either directly or indirectly, of our shared genetic endowments. But even if this is so, it seems very unlikely that we will ever be able to, as it were, "read" those constraints directly off the genetic code or even deduce them anew from facts about our shared genetic endowments. There are, in fact, no better methods for uncovering the constraints than those that are currently deployed in the various branches of linguistics – especially linguistic theory, developmental linguistics, comparative linguistics, and mathematical learning theory. This is not to deny that, once we know the constraints on humanly possible languages, we may set ourselves the further task of discovering where those constraints come from. And at that stage of inquiry, research into the genetic basis of our language capacity may well prove illuminating. But the relevant point for our current purposes is that it is difficult to imagine that the methods of *behavioral* genetics would occupy center stage in any such inquiry.

The case of the human language capacity shows that we may need to do a substantial amount of high-level cognitive science before research in genetics becomes particularly relevant to our explanatory aims. When we do get to the stage of inquiry at which genetic research becomes relevant and illuminating, it is unlikely to be behavioral genetics that interests us. A similar point can be made about research into the genetic basis of criminal behavior. Until we have a much better understanding of both the psychodynamics of criminal behavior and the social construction of crime, research into the genetic basis of criminal behavior is premature at best and positively obfuscating at worst. The problem is that until we understand both the psychodynamics of criminal behavior and the social construction of crime better, it is not clear what we are asking for when we ask about the genetic basis of "criminality."

The move from talk of crime and criminals to talk of "criminality" in itself has the feel of a move from the realm of the criminal justice system and its social and political surroundings to the realm of psychopathology. Such talk suggests some distinctive syndrome of abnormal or deviant behavior, rooted, perhaps, in some distinctive pathological condition, with an etiology waiting to be uncovered by future biological, neuroscience, and/or cognitive science research. Unless we are meant to take such talk purely behavioralistically, it suggests, that is, that "the criminal mind" is a broken mind, the basic cognitive or affective architecture of which may deviate from the architecture of the cognitively and affectively normal mind in some yet to be discovered way. Once we have "pathologized" criminality as such in this way – that is, once we have elevated criminal behavior to the level of a distinctive pathology, with a potentially distinctive but still unknown etiology – it is tempting to suppose that that unknown etiology may ultimately be genetic. Indeed, the more independent reasons we have for thinking that criminality involves a distinctive pathological condition of the mind's *fundamental* cognitive of affective architecture, the more irresistible the genetic hypothesis becomes.

Conversely, the more studies that establish the heritability of criminality, the more tempting becomes the hypothesis that "criminality" is a malady of the fundamental cognitive or affective architecture. If criminality is "in the genes," it is tempting to conclude that the psychodynamic differences, whatever they are, between criminals and noncriminals do not arise from differences in those aspects of our cognitive and affective makeups such as are shaped by experience. Such differences would seem to arise, rather, from differences in genetically encoded and relatively fixed aspects of the mind's basic cognitive or affective architecture. Data appear to lend some credence to this expectation. For example, some studies have found that while only 7 percent of the general population exhibits antisocial personality disorder as many as 45 percent of a collection of convicted violent felons exhibit that disorder (Maughan 1993). Other studies link habitual impulsive aggression (which often characterizes antisocial personality disorder) with reduced central nervous system serotonin turnover (Volavka, Martell, and Convit 1992; Virkkunen and Linnoila 1993). Evidence even suggests that deficient serotonin metabolism may be rooted in a genetic mutation (Brunner, et al 1993).

Although there is a rising trend to pathologize crime and criminal behavior, there are also countervailing tendencies. The workaday conception of criminals and crime enshrined in many of our legal practices and institutions clearly presupposes that at least some criminals are more or less normal cognizers. That conception presupposes that, like the rest of us, criminals sometimes act for what they take to be good reasons, that they adjust means to ends in more or less rationally coherent ways, that they are generally willing to run downside risks of capture, conviction, and punishment only when they believe that the rewards of criminal activity outweigh those downside risks. The very practice of openly and negatively socially sanctioning certain actions as criminal and attaching certain punishments to them already implicitly presupposes the rationality of those whom such sanctions are supposed to deter. We intend that the criminal justice system be such that criminals will be arrested, swiftly tried, and severely punished upon conviction. We intend the certainty, swiftness, and severity of the system to be understood by all. We intend the fact of that certainty, swiftness, and severity to enter into whatever risk-reward calculations prospective criminals perform. Indeed, we take the presumed fact that transgressors are capable of such calculations to be further justifications for the very punishments that we mete out upon conviction.

It is an open empirical question just what proportion of criminals have minds that are cognitively and/or affectively abnormal. But even in advance of further inquiry, it is a fair bet that the class of "criminal minds" is very heterogeneous indeed. Some criminals may suffer from correctable cognitive or affective deficits of one sort or another – poor impulse control due to abnormal serotonin levels, for example. Others may suffer from quite severe and fundamental pathologies of some not yet fully understood nature. Still other criminals may be perfectly normal cognizers who find themselves in quite extraordinary circumstances. If the class of criminal minds does turn out to be heterogeneous in the extreme, then any urge to pathologize criminality *as such* is deeply mistaken. For the class of criminal minds will not, in that case, constitute a uniform kind either from the perspective of biology, neuroscience, or psychology. To say this much is not to deny the importance of biological, neuroscientific, or cognitive scientific research into the etiology of whatever pathologies are disproportionately prevalent among those who happen to commit crime. It is to suggest, however, that researchers investigating the biological or neuronal bases of crime may come to find themselves investigating an increasingly fragmented class of biological abnormalities.

Such research should not be confused with research into the biological or neuroscientific or cognitive scientific roots of criminality per se. Indeed, if the class of criminal minds turns out to be heterogeneous enough, there will be no serious explanatory work for the very idea of criminality as such to do in the biology or the neuroscience or cognitive science of criminal behavior. But again this thoroughly empirical issue must await the outcome of future research. To deny to the very idea of criminality any legitimate explanatory role in biology, neuroscience, or cognitive science is not to deny to that idea its place in our social and political discourse. Moreover, given the role of that idea in shaping our practices, institutions, and relations with one another, it is clear that the social sciences must take notice of that idea. But that is an entirely different matter. For empty ideas with no basis in reality may shape our practices, institutions, and relations with one another just as profoundly as ideas that have a basis in reality do.

Even if we grant the extreme heterogeneity of the etiology of criminal behavior, we might, if we take a behaviorist turn, continue to talk of what I have called criminality *as such*. And now we would intend by this term not a single syndrome of deviant behavior with a uniform etiology, but a diverse array of behaviors, with diverse etiologies. This purely behavioral construal of criminality is entirely consistent with studies that establish that criminality is highly heritable. This fact should, I think, give us pause. If criminality is construed purely behaviorally, then there is no one thing, at least not from the cognitive or neuronal perspective, of which we have established the heritability when we have established the heritability of criminality. And this implies that results establishing the heritability of criminality lend no credence whatsoever to the view that criminality as such involves a distinctive pathology of mind's genetically encoded cognitive or affective architecture.

This last claim deserves elaboration and defense, for which I shall tell a few "just-so" stories. These stories are designed to show how it is possible for criminality to be highly heritable without involving any distinctive pathology of the mind's genetically encoded cognitive or affective architecture.

Imagine a social climate in which distressed people with a range of psychopathologies receive little care and support. Some of the pathologies have a substantial genetic component, and people with such pathologies are disproportionately represented among the criminal classes. Imagine that they are led to lives of crime not because their genes are directly causally responsible for their criminal behavior but at least partly as a consequence of inadequate social arrangements for their care and treatment. That is, social arrangements are such that relatively independently of facts about the available rearing environments, people with the relevant pathologies do not receive the kind of care and support that might enable them to lead productive or, at any rate, nondestructive lives. Suppose, for example, that people with the relevant pathologies are generally unable to hold down jobs, are too difficult for their families to deal with, and are unable to benefit from any but the most specialized kind of educational regime. Imagine also that the rest of us are unwilling to pay the substantial taxes it would take to treat them or to provide income support for them or to educate them via the costly specialized methodologies they require. Suppose that as a consequence of these social realities, disproportionately many such people end up in desperate situations. Suppose that their desperation leads to feelings of deep alienation from the broader society, that they do not feel themselves to have any particular stake in its perpetuation and do not see why its norms should apply to them. Perhaps in their desperation and isolation they take to performing criminal acts, both petty and violent. Not being very skilled practitioners of the criminal arts, they go in and out of prison. In prison their alienation is only increased. Once out of prison they commit still more desperate acts, as if caught in an endless cycle.

Supposing the original pathologies to have a high degree of heritability, then such standing social arrangements can conspire to make their criminality heritable in the precise technical sense – this is an instance of so-called reactive heritability. But there is no principled justification for thinking that their heritable liability to criminal behavior is itself an additional, genetically based pathology, different from the pathologies that we antecedently identified. If we were to treat their original pathology and/or ameliorate their environment, there would be nothing left to treat them for. Nonetheless, if something like my just-so story turned out to be true, the urge to pathologize criminals and their behaviors, to see criminals as suffering from a peculiar pathological condition that causes them to be criminals, would, I suspect, be strong. But the act of pathologizing such behavior in the kind of case just imagined would be an essentially political act, not an act in service of any legitimate scientific or explanatory interest or aim.

Consider a second just-so story about reactive heritability. This time consider a perfectly benign, clearly nonpathological trait like skin color. Suppose that something like the story we told above about pathological people is true of dark-skinned people, in a society populated with light- and dark-skinned people. That is, suppose that relatively independently of rearing environments, standing social arrangements make it more likely that dark-skinned people will end up alienated and desperate, with no particular stake in preserving the standing social order, with no particular allegiance to standing social norms. Suppose also that because of their alienation and desperation, dark-skinned people come to be disproportionately represented among the criminal classes. Because skin color is highly heritable (over the population at large), if the distribution of light- and dark-skinned people among the criminal and the noncriminal classes is of just the right sort, it will turn out that criminality is highly heritable too. But whereas this is true in the precise technical sense of "heritable," it is highly misleading as an account of the etiology of human behavior. The problem is that in the just-so story I have just outlined environment is implicated in the causation of the relevant behavior in a way that is simply transparent to the methods of behavioral genetics.

The urge to pathologize crime, criminals, and criminal behavior is widespread. And one can justly fear that research in behavioral genetics establishing the heritability of criminality will appear to lend the backing of science to that urge. But I think it is clear that such research, on its own, does nothing of the sort. For nothing in the behavioral genetics research is inconsistent with the hypothesis that the cognitive and affective strategies implicated in a great deal of criminal behavior are not in principle recognizably different from the garden variety cognitive and affective strategies we all deploy over very large segments of our lives. Again, it just may be that many criminals engage in means-end reasoning much like the rest of us, assess costs and benefits much like the rest of us, and exhibit the same sort of emotional responses to the same sort of situations as do the rest of us.

To say this is not to say that criminals in no way differ from the rest of us. Evidently they do. Many of them are willing to employ what we regard as repugnant means to achieve their questionable ends. Some are willing to run what seem to us exorbitant risks, for what seem to us marginal benefits. And at least some criminals seem clearly to differ from the rest of us affectively. In America's violent netherworld of gangs, guns, and drugs, what appear to outsiders to be the smallest of slights can provoke a murderous rage. We rightly abhor the violence. Its prevalence is clearly a sign that something has gone badly wrong somewhere. And it is tempting to say that what has gone wrong lies, as it were, in the affective and cognitive makeup of the perpetrators of the violence. Again, I do not doubt that many criminals do suffer from some affective or cognitive disorder. But we should not hastily conclude that violent behavior per se, even when we rightly reject the violence as unwarranted, is a sign that the perpetrator suffers from some affective or cognitive disorder. It just may be that the differences between criminals and noncriminals with respect to choice of means, assessment of risks, and affective responses is the straightforward outcome of the application of perfectly intact cognitive and affective capacities and strategies to quite extraordinary inputs in quite extraordinary circumstances.

Consider briefly the criminal's assessment of risks and reward. During prohibition, bootleggers manufactured and sold constitutionally prohibited beverages, despite the downside risks of being either murdered at the hands of their competitors or being caught, convicted, and imprisoned by federal authorities. Why? Surely at least part of the answer is that they performed (or can be rationally reconstructed as having performed) straightforward utility calculations to the effect that the upside monetary rewards to be gained from bootlegging outweighed the downside risk of being murdered or imprisoned. For at least some drug dealers a very similar assessment of risk and reward is at work. Risk-reward calculations are, of course, relative to the options perceived to be available to the actor. A robust young male who perceives his options to be restricted to either gang membership and drug dealing or receiving a second-rate education at a dysfunctional school will assess risks and rewards quite differently from a robust young male who believes his options to include education at first-rate schools and colleges.

Poor inner-city drug addicts commit a disproportionate number of property crimes. Drug addiction is itself rightly viewed as a kind of pathology. But the connection between drug addiction and crime is clearly highly contingent. Wealthy addicts in places like Hollywood, driven by the very same overriding desire to put as much of the desired drug into their system, can afford to procure their drugs with their own resources. They too have the option of stealing for their cocaine money, but very few of them do because that path would bring little, if any, additional reward and would carry substantially greater risks. My point is that although the inner-city addict typically pursues a strikingly different path to the procurement of cocaine from that pursued by the wealthy Hollywood addict, the means-end and risk-reward calculations of the inner-city and Hollywood addicts do not appear to differ in any deep or principled way. All that differs are the circumstances in which they carry out those calculations and the options with which they are and take themselves to be confronted.

The point here is not to excuse, justify or rationalize criminal behavior. My aim is only to dampen the temptation to pathologize criminality per se. Disproportionately criminals may well have broken brains. But for many others it is not evident that, on average, they reason about means, ends, risks, and rewards in any way that is fundamentally different from the way the rest of us reason about them. Indeed, I suspect that if we look carefully at the cognitive dynamics of the "criminal mind," we will often find not a particularly extraordinary or abnormal mind, but just an ordinary mind embedded in extraordinary circumstances.

Philosophical just-so stories of the sort that I have been telling cannot settle the open empirical question whether "the criminal mind" is a pathological mind or merely a cognitively normal mind embedded in extraordinary circumstances. Such just-so stories are useful only for addressing "how possibly" questions; they do nothing to settle the empirical facts of the matter. But neither, I think, will research in behavioral genetics settle such questions. For the potential significance of behavioral genetics for our understanding of the "criminal mind" is highly dependent on the exact outcome of *independent* research into the psychodynamics of the mind. If such independent psychodynamic research were to show that the minds of criminals are, by and large, cognitively normal, then results showing that criminality is highly heritable would strongly indicate that the heritability in question is merely reactive heritability. And such results would lend no support to the view that criminal behavior has a peculiar etiology, distinct from the etiology of "normal" human behavior. On the other hand, if such independent research were to show that the criminal minds are cognitively abnormal, then, and only then, might the heritability of criminality give us reason to think that the criminal mind suffers some genetically encoded defect in its fundamental architecture. Even if that is how it turns out, we will need to go beyond research in behavioral genetics to achieve understanding of the nature of any such defect.

I have been addressing the pathologizing of criminality as such from a theoretical-explanatory perspective heavily influenced by cognitive science. I close by adopting a more sociopolitical perspective. To that end, I tell one last just-so story, introducing the concept of what I call a no-way-up social hierarchy. Those who find themselves at the bottom of such hierarchies may have little reason to feel much allegiance to the prevailing social norms and may often feel themselves to have good reasons to transgress those norms criminally. The misrepresentation of crime as individual pathology, when married to the combined coercive powers of the dominant social group, may itself serve as an instrument by which the advantages of haves over have-nots are magnified and reproduced.

My story requires some stage setting. I begin by imagining a society divided into a group of resource-rich haves and a group of resource-poor have-nots. Suppose that haves and have-nots play a complex game called resource procurement. In resource procurement, players try to increase their resources. A player may increase her resources in at least the following ways. She may engage in "original" acts of acquisition of some resource not already acquired by another. She may "add value" to an already acquired resource. Or she may bargain with other players. The first stage of the game, called the lottery, is antecedent to any bargaining and any exercise of labor. The lottery distributes different kinds of goods: material possessions, initial skills, and labor capacity, even initial genetic endowments. Some resources – for example, genetic endowments – are distributed *only* through the lottery and cannot be redistributed or augmented either through bargaining or the exercise of labor in original acts of acquisition. Other resources assigned by lottery may be redistributed through bargaining; it may also be possible to add value to them through labor.

Haves are the winners of the antecedent lottery; have-nots are the losers of the lottery. No player knows in advance of the lottery whether he will end up a have or a have-not. Every player knows that initial positions in the game are to be determined by lottery, but players do not know how the lottery will distribute resources – whether roughly equally or with great inequality. The game makes no provisions for re-playing the lottery except by explicit agreement of the players. Such agreements are possible only as the outcome of rational bargaining – that is, only as a consequence of further play of the game. The lottery deter-mines for each player an initial bargaining position. Very roughly, a player's bargaining position is a function of his initial stock of resources, including his initial stock of material resources, the productivity of his labor capacity, his access to the means of production, and possibly even his initial genetic endowments to the extent that they distinguish his skills, talents, and abilities from the skills, talents, and abilities of other players. As a rough first approximation, a no-way-up hierarchy is the outcome of a game of resource procurement played from a situation of great inequality in the original distribution of resources.

After the lottery has produced an unequal distribution of initial re-sources, there will be three broad types of bargaining situations: bar-gains involving only haves, bargains involving only have-nots, and mixed bargains involving one or more haves and one or more have-nots. We assume that both bargains among haves and bargains among have-nots involve situations of rough equality. On the hand, mixed bargains will involve situations of great inequality. For simplicity's sake, we as-sume that each agent has perfect information about the initial bargain-ing positions of every other agent. Such information plays a crucial role in determining what bargains agents rationally ought to strike.

In mixed bargains, have-nots will be disadvantaged relative to haves. Suppose, to take a somewhat whimsical example, there are elephants lurking about in the woods and that elephants are hard to capture but valuable to possess. Imagine a bargaining situation of the following sort between a have and a have-not. Suppose that agent H, a have, has an 80 percent chance of acquiring an elephant on his own – perhaps because he has, as a consequence of the lottery, very good tools and is very skilled in use of them – while agent HN, a have-not, has only a 10 per-cent chance of acquiring a elephant on his own. Assume that one ele-phant represents two units of additional value for H and two units of additional value for HN. Suppose that solitary pursuit of elephants con-sumes one unit of value for both H and HN. The net gain from the cap-ture of an elephant is then one unit of value for both H and HN. Acting alone, in solitary pursuit of elephants, H has a positive expectation of .6 units of value. HN, on the other hand, has a negative expectation of .4 units of value from solitary pursuit of elephants.

But now suppose that if H and HN were to cooperate equally in the pursuit of elephants, they would have a 90 percent chance of capturing an elephant, at an expenditure of .5 units of value each. Now consider a pair of possible bargains between H and HN. In bargain one, H and HN agree to consume .5 units of value each in pursuit of the elephant and agree to a 50–50 split of the elephant upon capture. Both H's and HN's expectation from such a bargain would be .4 units of value. Clearly, though HN would better his expectation by such a bargain, H would worsen his expectation. So bargain 1 is not rationally strikeable between H and HN. Contrast bargain 1, with bargain 2. In bargain 2, H and HN agree to an 85-15 split of any captured elephant and to an equal expenditure of resources in pursuit of the elephant. H's expectation from bargain 2 is .76 units of value; HN's expectation is .09 units of value. Because bargain 2 improves upon both H's expectation from solitary pursuit of elephants and HN's expectation from solitary pursuit, bargain 2 is rationally strikeable between H and HN. It is significant that H's expectation from bargain 2 is significantly higher than HN's. On the other hand, the difference between HN's expectation from bargain 2 and HN's expectation from solitary pursuit is much greater than the difference between H's expectation from bargain 2 and H's expectation from solitary pursuit. This means that bargain 2 is of greater value to HN than it is to H. That is, H's cooperation is more valuable to HN than HN's cooperation is to H. The asymmetrical character of bargains between haves and have-nots is one of the defining features of no-way-up hierarchies. It means that have-nots are highly unlikely to be able to improve their lot merely through rational bargaining with the haves.

More than asymmetrical bargains between haves and have-nots is needed to make for a no-way-up hierarchy. Alternative moves available to the have-nots in resource procurement must also have dim prospects for success. In particular, have-nots must have dim prospects for increasing their resources either through solitary toil or rational bargaining with other have-nots. It is not my goal here to say in detail just how it might come about that have-nots are unlikely to increase their resources through either solitary toil or cooperation with other have-nots. Part of the answer will lie in the initial endowments of the have-nots as determined by lottery. But part of the answer will also depend on standing social arrangements, which will be at least partly responsible for rendering both the solitary labor of the have-nots and bargains among the have-nots relatively unlikely to increase their resources substantially.

The standing social arrangements I have in mind concern labor law, property law, criminal law, and the law of contracts. From the point of view of our game, we may suppose that those arrangements are themselves a product of postlottery bargaining. Given a lottery that produces a situation of great inequality between haves and have-nots, we may assume that the standing social arrangements will reflect this initial radical disequality. That is, we may expect that the standing social arrangements will tend to magnify the initial advantages of the haves and the initial disadvantages of the have-nots.

Consider briefly bargains among have-nots. As before, assume that a have-not has only a 10 percent chance of capturing an elephant by unaided pursuit. Assume also that cooperation between two have-nots increases their joint chance of capturing an elephant to, say, 30 percent – a result that makes cooperation between have-nots substantially more "productive" than solitary pursuit by either. Now suppose that our two have-nots agree to a 50–50 split of any captured elephant. Each still has negative expectation of .20 units of value from the bargain. This example illustrates two facts that hold in the general case. First, because bargains between have-nots are negotiated from positions of rough equality, they typically lead to a more "equitable" distribution of any payoffs and a more equal distribution of any increased expectation. On the other hand, even though bargains between haves and have-nots tend to perpetuate the antecedent inequality between haves and have-nots, such bargains in the general case are more valuable to a have-not than a competing bargain with another have-not. A have-not who is offered a choice between a bargain with a have that yields him a 90 percent chance at a 15 percent share of an elephant, with only a 10 percent chance of unrewarded effort, and a bargain with another have-not that yields him only a 30 percent chance at 50 percent share of the elephant, with as much as a 70 percent chance of unrewarded effort, betters his expectation over solitary pursuit by accepting the bargain with the have and rejecting the bargain with the have-not.

Much more can be said about the basic structure of no-way-up hierarchies. We have said enough to make it clear that have-nots embedded in a no-way-up hierarchy are very unlikely to increase their resources substantially through either unaided honest toil or rational bargaining. It requires no great leap of logic or imagination to see that have-nots often have good reasons to transgress the rules of the game selectively and secretly. Why toil honestly but in vain in solitary pursuit of difficult-to-capture living elephants, when one might more easily steal elephants that have already been subdued by another? Why bargain at a disadvantage with those who are more adept at capturing elephants, winning through the bargain only the promise of a paltry portion, if one can get away with taking the whole thing for oneself? To be sure, the standing social arrangements and social norms that outlaw and discourage the forceful taking of the property of others are themselves the consequence of presumably good faith postlottery bargaining to which the have-nots themselves are a party. But these bargains too are struck from a situation of inequality. So they too tend to magnify the advantages of the already advantaged.

There is little reason to expect have-nots to feel deep allegiance to such norms and arrangements. Even supposing that such arrangements represent the best available grand bargain between the haves and the have-nots about how they shall conduct their lives together, that shows only how very hard it is to bargain one's way up from the bottom of a no-way-up hierarchy. It would not be quite fair to say that no-way-up hierarchies are the products of the malevolent intentions of the haves. It is merely that given the outcome of the antecedent lottery, the haves find themselves in a position to extract, through rational bargaining, favorable bargains indeed from the have-nots. One might think that an obvious solution is simply to replay the lottery and rig it so that it produces an initial situation of rough equality. But only the losers will want the lottery to be replayed, at least if the initial advantages of the winners are great enough. So there is not likely to be any rationally strikeable bargain between haves and have-nots involving a replay of the lottery.

I close by considering briefly the social construction of crime as pathological in the context of a no-way-up hierarchy. Suppose that the received wisdom among the haves in a no-way-up hierarchy is that criminals have malfunctioning minds, an assumption that is reflected in all penal institutions, as well as in all social, political, and even scientific discourse about crime, its causes, and its remedies. A set of institutions founded on the representation of crime as individual pathology is bound to look very different from a set of institutions that acknowledges the role of social inequality in causing crime. The exact details, of course, depend on a variety of further factors. Concluding that criminality is rooted in some genetically determined pathology of the individual mind would not itself settle the structure of penal institutions. Such a belief might easily coexist with a belief in the underlying dignity of the pathological individual. This pair of beliefs might lead to the adoption of a therapeutic approach to crime and criminals. On the other hand, the representation of crime as individual pathology may also coexist with the view that the criminal is nothing more than a diseased threat to the social order and deserves to be countered by the harshest of means.

So there is no saying, in advance, what penal institutions are bound to be like even given a representation of crime as individual pathology. Nonetheless, it is worth imagining, if only as a cautionary tale, the worst that can happen when misrepresentation is married to the combined coercive power of the haves. At worst, marriage of misrepresentation and coercive power may serve to reinforce and to reproduce an already present no-way-up social hierarchy. First, when the haves misrepresent crime to themselves as individual pathology, they are likely to underestimate the role of social inequality in causing crime. That fact alone may lessen any pressure they may feel to bring into existence more equitable and inclusive social arrangements. Indeed, they might mistakenly, but not entirely unreasonably, convince themselves that bringing into existence such arrangements would do little to alter crime.

Moreover, consider the matter from the point of those who are subjected to penal institutions that are backed by the combined coercive powers of the haves and founded on their shared misrepresentations of the nature of crime. If the have-nots endorse the false presuppositions of that system, they become its victims in the truest sense. For they too then misrepresent themselves as pathological and misunderstand the real sources of their alienation from the prevailing social norms. Even if they recognize the falsity of the presuppositions on which the system is founded, we cannot suppose that they have the power to alter the system forcibly. Nor can we suppose that they have access to those arenas in which the social misrepresentation of crime and criminals may be contested. Because that system, with its false representations, is intended to delegitimize the best opportunities that have-nots may have in a no-way-up hierarchy for significant resource procurement, it is not unreasonable to suppose that a clear-eyed have-not may react to this system as merely another instrument for maintaining the advantages of the haves. But then there is every reason to expect that have-nots regard themselves as having good reasons to resist this system at every available turn. So the misrepresentation of crime as individual pathology, when married to combined coercive powers of the haves, may serve to create even more alienation on the part of the have-nots from the prevailing social norms.

NOTES

1. The literature here is very large and growing. Three recent review articles are Carey 1991: Billings, Beckwith, and Alpert 1992, and DiLalla and Gottesman 1991.
2. For a more complete discussion of the concept of heritability see chapter 2, "Separating Nurture and Nature," of Rowe 1994. For nice analysis of the concept of heritability and the logic of twin studies, see Sober, Chapter 3, in this volume.
3. We have defined here what is called broad-based heritability. Broad-based heritability must be distinguished from so-called narrow-based heritability. Narrow-based heritability measures only additive genetic effects. See Rowe 1994 for a discussion.
4. See Sober, Chapter 3, in this volume, and Rowe 1994 for more detailed discussion.
5. In particular, molecular and developmental genetics employ far more powerful and revealing methodologies. The criticisms I make here of behavioral genetics do not apply directly to molecular and behavioral genetics. My claim here is that the results of behavioral genetics have little explanatory significance on their own and that the methods of behavioral genetics cry out for augmentation by the methods of molecular and developmental genetics. For a discussion of the methods of molecular genetics, especially of so-called linkage analyses, see Plomin, Owen, and McGuffin 1994; and Billings et al. 1992 for a discussion of molecular genetics and so-called linkage analyses. As an example of developmental genetic approach to antisocial behavior, see Gottesman and Goldsmith 1994.
6. See Pinker 1994 for a helpful introductory discussion.

REFERENCES

Billings, P. R., Beckwith, J., and Alper, J. S. (1992). The genetic analysis of human behavior: A new era? *Social Science Medicine* 35 (3): 227–238.

Brunner, H. G., Nelen, M., Breakefield, X. O., Ropers, H. H., and van Oost B. A. (1993). Abnormal behavior associated with a point mutations in the structural gene for monoamine oxidase a. *Science* 262: 578–580.

Carey, G. Genetics and violence. Unpublished manuscript, commissioned for the National Academy of Sciences/National Research Council.

Dilalla, L. F., and Gottesman, I. I. (1991). Biological and genetic contributions to violence – wisdom's untold tale. *Psychological Bulletin* 109 (1): 125–129.

Gottesman, I. I. and Goldsmith, H. H., (1994) Developmental psychopathology of antisocial behavior: Inserting genes into its ontogenesis and epigenesis. In *Threats to optimal development: Integrating biological, psychological, and social risk factors,* ed. C. A. Nelson. Hillsdale, N.J.: Erlbaum.

Hodgins, S. (1993). *Mental disorder and crime.* Newbury Park, Calif.: Sage Publications.

Maughan, B. (1993). Childhood precursors of aggressive offending in personality-disordered adults. In Hodgins 1993.

Pinker, S. (1994). *The language instinct.* New York: William Morrow.

Plomin, R., Owen., M. J., and McGuffin, P. (1994). The genetic basis of complex human behaviors. *Science* 264: 1733–1739.

Raine, a. (1993). *The psychopathology of crime: Criminal behavior as a clinical disorder.* San Diego: Academic Press.

Rowe, D. C. (1994). *The limits of family influence: Genes, experience and behavior.* New York: Guilford Press.

Virkkunnen, M., and Linnoila, M. (1993). Serotonin in personality disorders with habitual violence and impulsivity. In Hodgins 1993.

Volavka, J., Martell, D., and Convit, A. (1992). Psychobiology of the violent offender. *Journal of Forensic Sciences* 7 (1): 237–251.

Chapter 6

Degeneracy, Criminal Behavior, and Looping

IAN HACKING

This chapter was written for a workshop whose stated topic was *criminal behavior*. Crime and criminals have been with us always. The idea of criminal behavior has not. In an unproblematic sense of the words, you can engage in criminal behavior without committing a crime. The burglar assembles the tools of his trade and sets out, mask in hand – criminal behavior, we'd unreflectively say – but he falls into bad company on the way, drinks too much, and becomes too drowsy to burgle the mansion. But, of course, that is not what is meant when talk of crime is replaced by a discussion of criminal behavior. We are supposed to think of a tendency or disposition to behave in a certain way. Crimes, we are to imagine, are committed not just (tautologically) by people who behave in a criminal way but by those with a propensity for criminal behavior. In simple statistical modeling, we find the expression "criminal behavior" meaning no more than criminal acts – offences against the criminal law – of any type: embezzlement, burglary, assault, rape, murder, bank card fraud (e.g., Rowe, Osgood, and Nicewander 1990). In that literature it appears that "acts" could be substituted for "behavior" without change of intended meaning. (My burglar who fell asleep before committing an offence did not engage in criminal behavior.) But in sociological, psychological, and genetic work, the word "behavior" is treated more seriously. Most often violent criminal behavior is in view, and not, for example, bank card fraud. No one has produced a genetic or pharmacological correlation with bank card fraud. Criminal behavior, as studied, tends to be associated with the lower orders of society. It has become not only a class concept but also one driven by fear (only bankers fear embezzlers).

The switch from crime to criminal behavior also flags a change in ownership. Crime belongs, on the one hand, to criminals, their families, their intimates, and their victims, and on the other, to police, judges, lawyers, prison guards, parole officers, prison reformers, philanthropists. Criminal behavior, on the other hand, belongs to criminologists, psychologists, sociologists, and their allies, the statisticians. It is, in turn, competed for by biologists.

In this volume Allan Gibbard argues that "biological" is not a helpful term to introduce into discussions of criminal behavior. What behavior, he asks, is not biological? I disagree with this criticism on two counts. The first point is not that the behavior is biological, but that it is investigated using quite specific bodies of technique and knowledge developed in the biological sciences. Second, specific biological conditions are proposed as indicators or markers of an unusually high predisposition to criminal or related behavior. Occasionally there are suggestions that these are more than markers: we are on the verge of identifying the causal mechanisms or underlying structures, in biochemical or molecular terms, that enable us to understand criminal behavior. Kenneth Schaffner, also in this volume, has provided a useful philosophical analysis of potential explanations of familiar biological phenomena in terms of genetic structure. There is a strong tendency, amply discussed by Rutter (1996), to hope that we will identify a criminal propensity and associated conditions at the biochemical or molecular level. Then we would succeed in biologizing criminal behavior in an altogether specific way, Gibbard's worries about the universality of biology notwithstanding.

Several distinct kinds of workers hope to biologize criminal behavior. The neurologists propose chemical or electrical or physiological predictors of, or even explanations of, criminal behavior. The pharmacologists study the relation between chemicals and criminal behavior and propose the use of drugs to curb it. The behavioral geneticists correlate anomalous genetic material with criminal behavior, or behavior often associated with crime. But not every connection between crime and biology is an instance of biologizing. For example evolutionary psychology, discussed by Gibbard, and briefly by myself, may include some biological reflections on crime, but it does not attempt to biologize criminal behavior. And there can be other types of intermediate position. Thus Powledge (1996) argues that genetics is not sufficiently predictive to be of much use in crime prevention or control but that other biological information may be more useful.

The present volume aims at a wider understanding of current fundamental research on criminal behavior than can be had from the fine details of genetic, pharmacological, and statistical studies. To this end I aim at enlarging the discussion in two dimensions. First, I expand the *time frame,* and recall that present-day research on criminal behavior is firmly planted in a tradition as old as criminology itself. Of course no past tradition can vie with tomorrow's pharmacological and genetic technologies. Yet we are here partly to discuss ethical questions. The structure, projects, and moral aims and ethical dilemmas of criminology have been remarkably constant for far more than a century. In the second part of this chapter I turn to *frameworks of classification,* speaking about far more kinds of people than criminals, violent offenders, and whatever. Certain very general observations about how we classify human beings should be brought to bear on the specific case of criminals.

THE TIME FRAME

The Degeneracy Research Program

Suicides and criminals were among the first two types of deviants to be systematically studied. The social sciences began by counting and classifying crimes, criminals, types of crime, rates of conviction, and recidivism about the same time that they started on suicides – or, to be more exact, such countings inaugurated social science itself. Like crime, there have been suicides forever, but *the suicide* was not thought of as a kind of person, with various subkinds of self-destructiveness, until early in the nineteenth century (e.g., Esquirol 1821). These remarks seem impressionistic, but it is easy to operationalize them. Something is thought of as a "scientific" kind of person when experts begin to propose laws about the kind. Laws about crime and suicide became plentiful in the 1820s, to the extent that they created a veritable paradox of free will (Porter 1986). If the suicide rate was a fixture, subject to annual rhythm and geographical variation, how then could suicide be an act of free will? As for crime, we read plenty of statements like this: "Each year sees the same number of crimes of the same degree reproduced in the same regions, each class of crimes has its own particular distribution by sex, by age, by season. . . . we are forced to recognize that *the facts of the moral order* are subject, *like those of the physical order,* to invariable laws" (Guerry 1833).

Suicides and criminals are kinds of people – not just any kind but kinds about which the new social sciences tried to discover laws as invariable as those of the physical order. They are examples of what I later call human kinds. The statisticians had then, as now, primarily a service-industry role, because the competing experts for the ownership of classifications such as criminals and suicides were the lawyers and doctors, creating a new discipline, legal medicine, which was intimately connected with psychiatry, and which generated a new and well-paid avocation, the expert medical witness in courts of law.

The social sciences came into being in order to help people: deviants needed help. The social sciences also fed on fear. Deviants were frightening for two reasons. One is personal. People are scared not only of criminals, but also of the mad, who in their aberrations simply frighten us, at a gut level. Suicide in real life fills anyone who is involved with a sort of incredulous despair. But deviants are also frightening for social reasons. They mock or challenge custom and convention. They seem to display on their faces what is awry in society. In nineteenth-century France, troubled by a relative decline in both power and birthrate, there was a special fascination with certain types of deviants, for it seemed that these degenerates lowered the capital of the nation, and, because they did not breed, also contributed to the low birthrate (Nye 1984). By the end of the century middle-class English women and men (especially socialists and feminists) thought that their problem was the excessive fecundity of the weaker and stupider members of society.

Criminals, suicides, prostitutes, vagrants, the mad were thought of as degenerates. Degeneracy is not a purely descriptive term. Not only does it have moral implications, but there is also the meaning of defect or decline. From the first it was conjectured that degeneracy is inherited, and this connection between inheritance and social deviance has persisted to this day. It leads me to talk of the *degeneracy research program,* in the sense of the philosopher of science Imre Lakatos (1970). Lakatos did not mean what we ordinarily call a research program, with its proposals, budgets, staff, goals, and a time span of five years or so. Indeed one is tempted to keep his British spelling, "research programme," to be reminded of what a different animal he had in mind. His research programs can last a century or more, and are characterized not by investigators with proposals, but by a sequence of theories, T(1), T(2), ... T(k), each of which replaces its predecessors. The Ts share a *hard core* of basic and unquestioned hypotheses; successive Ts result from previous Ts by modifying the *protective belt* of auxiliary hypotheses.

For those unfamiliar with Lakatos, I should add that he saw his work as a stage in improving Karl Popper's model of conjectures, refutations, and fallibility; he thought he presented a rational counterbalance to the more socially oriented picture of science due to Thomas Kuhn. It is one of Lakatos's theses that research programs do not in general exist in isolation, and that all competing research programs are "wallowing in a sea of anomalies" or counterexamples. Competing programs have fundamentally different ways of dealing with problems, for each has a distinct set of auxiliary hypotheses that can be modified. Only a few of today's philosophers of science regard Lakatos's methodology of scientific research programs as a viable account of science-in-general. But even its critics can use it as a helpful model of bits of science. I find that criminology can be usefully characterized, in part, by certain long-standing and competing research programs.

The hard core of the degeneracy program has two parts. First, there is a group of interconnected types of deviancy that are profoundly antisocial, possibly threatening society itself; these types of deviancy are forms of degeneracy or innate defect in individuals. Second, degeneracy (as the word implies) is intergenerational; it is inherited, or when it is a novel defect, it is the result of a throwback or mutation.

The protective belt of the degeneracy program allows an immense amount of flexibility around this core. For example, it allows a fairly free choice of the types of deviancy that constitute degeneracy. At different times different types of deviant behavior are particularly feared. There is an adjustable degeneracy portfolio, of kinds of deviancy that constitute an inheritable degeneracy. I am here making use of the risk portfolio idea of the anthropologist Mary Douglas (Douglas and Wildavsky 1982). That is a set of contingencies or kinds of event to be feared, which characterizes a social group at a time. One year it is nuclear winter, the next year it is global warming, both real dangers, but weighted enormously differently in the portfolio of concerns that can vary annually. In a similar way I speak of the degeneracy program as fear-driven; the deviancies singled out form a subset of those which, at a time, are deemed to be social problems. The expression "social problems" itself has become a slightly ironic label for the metastudy of what gets counted as a social problem, so that a recent presidential address to the Society for the Study of Social Problems, published in its journal *Social Problems*, discusses the way in which ideas such as a genetics of homosexuality or violence acquire catchphrases such as "gay gene" or "mean gene" (Conrad 1997).

A protective belt of auxiliary hypotheses, surrounding a Lakatosian research program, permits diverse opinions about both the lines and the mechanisms of inheritance. Some versions of the program favored direct lineage: alcoholics beget alcoholics, male criminals beget male criminals, child abusers beget child abusers. Others favored what I shall call allotropic degeneracy (the word "allotrope" is taken from physical chemistry, meaning any of two or more forms in which a chemical element may exist; carbon, for example, may exist as coal, diamond, or the bucky balls named after Buckminster Fuller).

Thus it has been urged that alcoholism appears in one generation, crime in the next, epilepsy in a third, all allotropic forms of the same degeneracy. One classic bastion of the degeneracy program was the great Paris neurologist Jean-Martin Charcot (1825–93), with his famous ward, and his public performances. His degeneracy portfolio included the usual roster of criminals and suicides, but degeneracy, in Charcot's eyes, was highly allotropic (he and his peers said protean). Hysteria, epilepsy, and hysteroepilepsy were among the degeneracies that were inherited. It was very widely held, at the time, that alcoholism in the parent might appear as epilepsy in one child, and hysteria in another child, with the next generation being full of criminals. The degeneracy portfolio in Charcot's school was very adaptable. Thus, when vagrancy became a pressing social problem in the late 1880s, it was immediately medicalized as *fugue* or *ambulatory automatism,* which was deemed to be one of the allotropic forms of degeneracy (Hacking 1998). Likewise when Charcot's student Gilles de La Tourette identified what is now called Tourette's syndrome, made famous in our day by Oliver Sachs, it was taken to be a specific type of degeneracy, which usually had both alcoholism and epilepsy in its pedigree.

Charcot's *grande hystérie* ceased to be a diagnosis soon after his death, and hysteria in general ceased to be a diagnosis after about 1910. Epilepsy has now been withdrawn from the degeneracy portfolio, in part due to radically new diagnostic tools, but also in part due to quite deliberate destigmatizing of the various types of epilepsy. Because my observations may seem arcane, I should remark that the degeneracy program is in very good health today. From a survey article in *Science* (Mann 1994, 1687): "More recently a group of researchers [in San Antonio and UCLA] announced in 1990 the discovery of a possible link between alcoholism and an allele of D2 dopamine receptor gene (DRD2) ... afterwards, Coming's team at City of Hope reported associations between DRD2 and autism, drug abuse, attention deficit disorder, pathological gambling, and Tourette's syndrome, as well as alcoholism. Because other groups have failed to replicate these findings, investigators are proceeding with caution." (Cf. Holden 1994. The City of Hope research is in Comings 1994 and Comings et al. 1994.) This is a classic instance of the allotropic version of the degeneracy research program in action. Why should the investigators have looked to see whether these kinds of behavior or syndrome were connected to an oddity about a gene? Because these complaints are high on the early 1990s roster of risks and social problems. Notice also how the degeneracy portfolio may be modified on the edges. Thus vagrancy was "in" in 1890, and pathological gambling was "in" in 1990. Another instance of the degeneracy program is the linking of aggressive and sometimes violent behavior, plus "arson, attempted rape and exhibitionism" to a rare mutation on the X chromosome shared by members of a large Dutch family most of whom are mildly retarded (Brunner, Nelen, Breakefield, 1993).

Although I shall mention a number of failures of the degeneracy research program, my invocation of Lakatos was deliberate in more than one way. He insisted that research programs could go on and on, despite repeated failure. In his terminology, a good research program is progressive (and a bad one is degenerating, but to avoid confusion I'll avoid that word). A program could be empirically or conceptually progressive. There is no doubt that the degeneracy program has been conceptually progressive, constantly introducing new concepts and techniques for analyzing the supposed inheritability of types of degeneracy. Also, of course, it has taken full advantage of the flexibility of the protective belt that defines the class of deviancies to be studied.

The degeneracy program has been extraordinarily fertile from long before Charcot, right up to the City of Hope. Its hypotheses have usually not panned out, but it regularly adds new hypotheses and new directions of research. Lakatos gave a number of important historical examples of programs that simply failed, over and over again, for a century, until they found their true goal, and succeeded. One, for example, ended in a permanent triumph, the discovery of isotopes of the same element, chemically identical, physically distinct. The degeneracy program could in principle experience the same good fortune. It could at this very moment be about to triumph. That is, of course, why so many people are running scared. What if next year at this time companies are about to market a test for the suicide or the violence gene? Or a simple blood test (rather than a lumbar puncture to extract cerebrospinal fluid) that will tell if your serotonin is regularly depleted, causing you to be impetuous, violent, pyromaniac or dangerously belligerent?

Criminology

The name of Cesare Lombroso (1839–1909) is nowadays invoked as a grim reminder of gruesome error, or as a bad joke. It should be remembered that criminal anthropology – the Italian school of Lombroso – was above all a reform movement, part of one of history's almost innumerable movements for prison amelioration. According to Lombroso himself, constructing his own history, it all began in 1870 when he did a post-mortem on a famous bandit: "At the sight of the skull I seemed to see all of a sudden lighted up as a vast plain under a flaming sky, the problem of the nature of the criminal – an atavistic being who reproduces in his person the ferocious instincts of primitive humanity and the inferior animals" (Lombroso [1906] 1972). That revolts us. We usually forget that Lombroso's primary thought was that men with such skulls must not be punished for their crimes. Even if they were guilty of monstrous murders, they should not be executed. I suspect that each new wave in the biologization of crime suffers a similar fate. Although the aim is profoundly philanthropic, it is soon perceived by many well-meaning people as both oppressive and reactionary. Right now advocates of a pharmacological approach to crime are appalled when they are accused of being racist, mind-destroying, and worse. This pattern has endured well over a century and will not stop now.

Charcot's version of the degeneracy program was clinical, although his clinical reports would always try to include a genealogy in a case history. Lombroso was among the first to go beyond the clinic and try to obtain physiological markers of criminal degeneracy (or atavism). He used skull shape as an indicator of, among other things, criminal propensity. Yet he did not stop at skull shape; we have typical criminal hair, eyes, eyebrows, and palates, no wisdom teeth. Physiognomy would enable us to tell murderers from thieves, embezzlers from bandits. The criminal became not just someone who committed crimes, but a "kind" of person with a biological predisposition and physiological markers, to the extent that Lombroso could declare of a young man, "He may not be a legal criminal, but he is a criminal anthropologically." The project of producing specific biological markers is exactly as old as criminology – for "criminology" was the name given in English, for criminal anthropology.

Intelligence

There has been a regular play, within the degeneracy research program, between biological and statistical methods. The enthusiasm for crime as degeneracy began with statistics, but the full hereditarian doctrine emerged in clinical practice. The criminal anthropologists turned to biological markers. But were they reliable? Lombroso wanted a thorough statistical comparison of a hundred criminals and a hundred normals, all conducted by an impartial international jury. That project fell through, and it was left to the British, in the full flush of the new statistical technologies flowing from University College, London, to conduct such studies. It is now common wisdom, at least in the English-speaking world, that the Italian doctrine of specific physiological stigmata was refuted by Charles Goring's official report, *The English Convict* (1913). Goring observed that the heads of about one thousand Oxford undergraduates differed less from as many criminal heads, than the heads of a roughly equal number of Cambridge undergraduates did from the Oxford ones. The bodies of 802 convicts were relentlessly compared with those of 108 soldiers in a battalion of Royal Engineers, and there was no significant difference except in stature. In Lakatosian terminology, Goring refuted the auxiliary hypothesis, that hereditary traits for degeneracy had physiological stigmata.

Goring's study was nonetheless a major contribution to the degeneracy program. His auxiliary propositions were four in number. (1) A statistical deduction: the only significant difference between convicts and the population at large is that convicts are less healthy (weaker and smaller) and that convicts are less intelligent. (2) A statistical hypothesis: there is a significant correlation between parental and filial criminality. (3) A theoretical construct: every adult has a criminal diathesis (viz. susceptibility to committing crimes). (4) A causal conjecture: this degree of susceptibility is normally distributed in the population, and is strictly heritable, in the broad sense so clearly explained by Elliot Sober in this volume.

The assumption of normalcy, the bell-shaped curve, the Gaussian distribution, is enormously powerful as a mathematical tool for analyzing the tendency to criminal behavior. Where did it come from? The idea of inherited diatheses for diseases such as pulmonary tuberculosis was current when Goring wrote. The pioneering statistician Karl Pearson had argued that susceptibility to TB was normally distributed. Pearson (1900) also maintained that moral and mental qualities are normally distributed. Moreover (it appeared from the data of the day), there are many similarities between TB and criminality. Oldest male children are most susceptible. Because the tendency to TB is constitutional, "We would, accordingly, be inclined to attribute the increased tendency of elder members to be criminally convicted to their possessing, in some way, an increased intensity of constitutional criminal taint" (Goring 1913, 280).

Talk of a criminal diathesis – a propensity to commit crimes, which different people may have in different degrees – must seem both preposterous and antiquated. It is not antiquated. It is alive and well in the latent-trait model of Rowe et al. (1990): "a relatively stable and generally latent propensity to engage in crime" (237). Our 1990 authors are at one with Goring in the axiom basic to their mathematical analysis: "Our first assumption is that individuals' positions on the latent trait of propensity toward crime are normally distributed" (245).

More central features of Goring's version of the degeneracy program are equally with us today. He stated that a weak constitution and mental defect are heritable and are concomitants of crime, and that a disposition to crime is heritable. The most famous recent exponent of this doctrine has been Richard Herrnstein (in Wilson and Herrnstein, 1985; Herrnstein and Murray, 1994). The modern version deletes physical health altogether, even retrospectively: "The British physician Charles Goring mentioned a lack of intelligence as one of the distinguishing traits of the prison population" (Herrnstein and Murray 1994, 241). It was rhetorically important for the authors to drop constitutional weakness and small stature, which has been rapidly altered by diet in a couple of generations. These two authors, ideologically committed to the nonalterability of intelligence by social measures, do not want to remind us that the other of Goring's predictors of criminality was altered by social measures.

Antisocial Personality Disorder

Shortly after Goring's death in the great influenza epidemic of 1919, his statistical mentor, Karl Pearson, honored him by pursuing his work, and discussing intelligence and the convict. Pearson admitted that final judgment must await further research being conducted at those great "laboratories," the American penitentiaries. But tentatively he proposed, on the basis of Goring's data, that a large part of the population of convicts, so far as concerns strictly intellectual ability measured by Binet-Simon mental age, is *not* sensibly different from the population at large. Low (intellectual) intelligence is not intrinsically connected with criminality. What Pearson called social intelligence is what matters. The convict population consists less of mental defectives (intellectually understood) than of social ineffectives, "who feel no social or moral responsibility." Pearson was convinced that this trait is heritable. As Pearson put it, it is "obscure" to say that a tendency to suicide is inherited, when it is melancholia that is inherited. It is likewise obscure to say that a tendency to crime is inherited when it is an antisocial personality trait that is inherited. What Pearson called his "working hypothesis" is that antisocial personality is a trait passed down from generation to generation, conducive to the commission of crimes in certain environments (Pearson 1919, xv).

Pearson's version of the degeneracy program is strikingly different from Goring's. Its modern descendants propose that various psychiatric disorders may be heritable and conducive to crime. Nowadays an investigator must use the coding and diagnostic criteria of the current edition of the *Diagnostic and Statistical Manual of Mental Disorders* (DSM-IV) of the American Psychiatric Association. Thus Pearson's version of the degeneracy program is now couched in terms of, for example, antisocial personality disorder. The most recent *Manual* defines APD as follows:

A. There is a pervasive pattern of disregard for and violation of the rights of others occurring since age 15 years, as indicated by three (or more) of the following:
   (1) failure to conform to norms with respect to lawful behaviors as indicated by repeatedly performing acts that are grounds for arrest;
   (2) deceitfulness, as indicated by repeated lying, use of aliases, or conning others for personal profit or pleasure;
   (3) impulsivity or failure to plan ahead;
   (4) irritability and aggressiveness, as indicated by repeated physical fights or assaults;
   (5) disregard for the safety of self or others;
   (6) consistent irresponsibility, as indicated by repeated failure to sustain consistent work behavior or honor financial obligations;
   (7) lack of remorse, as indicated by being indifferent to or rationalizing having hurt, mistreated, or stolen from another.
B. The individual is at least age 18 years.
C. There is evidence of conduct disorder [another "menu" disorder, applying to persons 15 years or younger, and involving at least three listed kinds of bad behavior].
D. The occurrence of antisocial behavior is not exclusively during the course of Schizophrenia or a Manic Episode. (American Psychiatric Association 1994, 649–50).

Aside from niceties, such as the required age of eighteen, Pearson would have thought that the DSM-IV criteria captured much of the behavior of his "social ineffectives." I return to antisocial personality disorder when discussing impulsiveness and serotonin depletion as in Virkunnen, Rawlings, Tokola, et al. (1994).

We need not extend our prehistory beyond Goring and Pearson to perceive the structure of the degeneracy research program. My examples from publications of the past couple of years suffice to show that the program is not only alive and well, but preserves, to an astonishing degree, the verbal formulations of days so long ago that we all thought that they had disappeared. The hard core is not only intact, and also the forms of current competing auxiliary hypotheses were in place early in this century. Thus, for Goring read the iconoclastic but terribly influential Herrnstein, Wilson, and Murray. For Pearson read mainline psychopharmacological-genetic-statistical investigations of the antisocial personality disorder. Recent pharmacology and genetics have, of course, given radically new life to the program. Biological stigmata of degeneracy are back, not as Lombrosian physiognomy but as serotonin depletion or genetic markers.

It is unkind of me to recall the old word degeneracy. No contemporary American investigator would use that word in a publication. I use it deliberately because I am interested in the way in which certain kinds of people are retained in the hard core of the research program. They are the deviants who are, at a given time, taken to pose a threat to society. One point is that the targets of this research program are chosen by social values and social fears. A second point is that the degeneracy program has always been philanthropic in character, intended to find ways to help degenerates, to improve them – or at least to make them less of a nuisance. Only in one respect is the situation of the degeneracy program significantly different from earlier days. It comes from the cachet of genes. With the media obsessed by genes, every purported correlation is immediately broadcast. That is true for "medical" conditions but, if anything, is more true for "moral" conditions such as suicide, addictions, violence. This obsession creates policy problems for issues that were less pressing in the past.

Rival Programs

According to Lakatos's model, research programs usually exist in competition with rivals. The chief rival to the degeneracy program has long been the *societal* program. The hard core of this program states that crime is a product of social conditions, and the frequency and character of crimes in a population can be changed by changing social conditions. As Adolphe Quetelet, the great founder of social statistics wrote in 1832, after discovering what he thought were profound regularities in criminal behavior, "Society prepares the crimes." The competition between the two programs has been aphorized as nature versus nurture. That famous dyad is catchy but too coarse-grained. Thus an investigation could focus either on the nurture of individuals or of a community within a society. Much American inquiry after World War II focused on individuals; only quite recently has the societal program reemerged in the United States with some force (Hagan 1993). Being both empirically and conceptually progressive, it has unearthed new phenomena and introduced fruitful new concepts. But both the degeneracy program and the societal program are silent in the face of one of the few agreed-upon criminological facts of the past two decades. Crime rates have increased enormously. The societal program can explain this piecemeal – for example, correlations between delinquency rates and placement of new public housing (Bursik 1989). In public perception such achievements in prediction, explanation, and recommendations for the future inevitably seem trivial.

Because I have chosen to present criminology in a framework of Lakatosian research programs, I should add evolutionary psychology to the list of programs. In the present volume Allan Gibbard states his vision of this relatively new spur on the evolutionary mountain. The project is to explain and even predict relatively universal aspects of human behavior in terms of evolutionary adaptation. Some of our nasty characteristics are particularly singled out, our proneness to beat our wives, abuse our children, and murder our brothers. Compared with all other programs of research on criminal behavior, evolutionary psychology has one cardinal virtue. It addresses the only known cross-cultural fact about violence. Most violence is committed by young men between the ages of, say, sixteen and thirty. This fact, plus other specifics about violence, are explained by the mating habits of early humans (Daly and Wilson 1988; 1994).

Evolutionary psychology can be grafted to both the degeneracy and the societal program. It could admit a normal distribution of propensity to crime, or a set of genetic or chemical markers. It could also urge that the transformation of an adaptive trait into actual behavior requires the right social or personal conditions. I admire evolutionary psychology because it addresses the only known universal fact about violence, namely that most violence is perpetrated by young men. After that, I find its explanatory and predictive power less impressive than Gibbard does.

FRAMEWORKS OF CLASSIFICATION

Human Kinds

I have used the phrase "human kind" to refer to kinds of people or their behavior subject to scientific study. To begin with a logical nicety, I intend to be deliberately ambiguous between the concept (or "intension") picking out people of that kind, and the class (or "extension") of people picked out. If I refer to the "criminal" as human kind, I am not talking about criminals but about the concept that picks out criminals, or the class picked out by that concept. If I refer to "criminal behavior" as a human kind, I am talking about the concept, class, or, to continue the ambiguity, the system of classification according to which criminal behavior is distinguished from other behavior.

I do not intend this ugly phrase, "human kind," to refer to a definite and clearly bounded class of classifications. I mean to indicate *kinds* of people, their behavior, their condition, kinds of action, kinds of temperament or tendency, kinds of emotion, kinds of experience. I use the term "human kinds" to emphasize kinds – the systems of classification – rather than people and their feelings.

But I don't mean any kinds of people. I choose the label "human kinds" for its inhumane ring, and mean the kinds that are studied in the marginal, insecure, but enormously powerful human and social sciences, which of course include criminology. By human kinds I mean kinds about which we would like to have systematic, general, and accurate knowledge; classifications that could be used to formulate general truths about people; generalizations sufficiently strong that they seem like laws about people, their actions, or their sentiments. We want laws precise enough to predict what individuals or groups of individuals will do, or how they will respond to attempts to help them or to modify their behavior. The model is that of the natural sciences. One event brings about another, although the causal laws may be only probabilistic laws of tendency. Or, as Kenneth Schaffner reminds us in his contribution to the present volume, we might even strive for underlying structures that characterize human kinds, the "inner constitution" (Locke's phrase) that characterizes a kind of person, in the way in which an inner constitution can characterize gold or a nematode.

The term "human kind" is patterned after "natural kind," and so, for philosophers, I have to make some disclaimers. I am very mealy-mouthed about natural kinds. Some have argued that there are very few natural kinds (maybe none, if there turn out to be no ultimate classifications in physics). I take the other route: I see nature as incredibly rich in kinds. Some seem fairly cosmic: quarks. Others, such as genes, are probably terrestrial. Others are mundane: mud, the common cold, headlands, sunsets. On my view, the common cold is as natural a kind as cystic fibrosis, and sunsets are as real as quarks. Lawlike regularities about mud are more widely known than ones about quarks – known to youths who play football, parents who do the family laundry, and to mud engineers on oil rig sites. The regularities about mud do not have profound consequences for theoreticians. That does not make mud any the less a natural kind of stuff. Thus my undemanding idea of a human kind does not invoke any serious philosophical theory about natural kinds – not at all. When I speak of human kinds I mean: (a) kinds that primarily sort people, their actions, and their behavior; (b) kinds that are in the first instance studied in the human or social sciences; (c) kinds about which we hope to have knowledge of the sort that we gain in the natural sciences, for example, of lawlike regularities, underlying structures, and powerful predictive or explanatory theories.

Looping Effects

Although I attach no strict theory to the idea of natural kinds, I think that there is one important difference between natural and human kinds. When we recognize a natural kind, and learn some laws about it, we often interfere with it, using those very laws to guide us. The same is true with human kinds. But simply naming a natural kind makes no difference to it at all. Human beings, in contrast, often get to know when they or their behavior is classified in a certain way. The sheer classification may influence their behavior and their attitudes, as well as that of their neighbors. Labeling theory suggested this long ago. But also the class and its properties may change in the light of being classified, creating a feedback effect, or what I call a looping effect, to make clear that this is a two-way process that can go on and on (Hacking 1995).

Classifying always involves knowledge of, or belief in, regularities about items of a class. As Nelson Goodman pointed out long ago, philosophical issues about lawlike regularities, dispositional properties, and projectible classifications go hand in hand. I am interested in the relations between people, on the one hand, and their classification into a kind (plus associated regularities), on the other. There is a whole gamut of relationships. I have argued at length that "child abuse" and "abuser" have displayed very strong looping effects since 1962 (Hacking 1991). Childhood autism is an example in which the people classified cannot (by definition) strictly comprehend how they are classified. Yet the autistic and the classifiers, the knowers (psychologists, speech therapists, even philosophers) have been extraordinarily malleable since "this" disorder was defined in 1943 by Leo Kanner.

Most of our classifications of people – expressed by half the nouns and adjectives in the dictionary, perhaps – have little to do with science. Call them *common* kinds: fat, loveable, rude, tired, witty, fast, . . . athlete, lover, misogynist, brute, . . . on and on. I shall say that a common kind becomes a human kind when it becomes an object of scientific scrutiny, when investigators propose lawlike regularities about the kind, or causal structures underlying it.

Finally, I say that a kind, common or human, is fully *biologized* when previous criteria are superseded by biological criteria and associated regularities or structures. These criteria can be chemical, electrical, physiological, molecular, or whatever. A happy simplistic vision would be this: when a common or human kind is biologized, we have made a discovery; we have come upon a natural kind.

The Criminal as a Human Kind

Ever since the advent of criminology, the criminal has been regarded as a kind of human being. This is shown at once by our willingness to speak of "the criminal," as if that were almost a species of humanity, in the way that the whale is said to be a species of mammal. The criminal justice system would like there to be very specific subkinds of criminal. In the early stages of classification, different kinds of criminals are characterized by profiles. Thus in the *Uniform Crime Reports* for 1965 we find a table headed "Profile of Known Repeaters by Type of Crime." We are informed that "repeating the same crime was highest for the narcotic offender 53 percent, the burglar 48 percent, the gambler 47 percent, and the bogus check offender 40 percent" (Federal Bureau of Investigation 1965, 29). The narcotic offender, the burglar, the gambler, the bogus check offender – these have become kinds of people. Profiling has gone to remarkable lengths today, so that the serial killer has become a kind of person (Jenkins 1994). I would say that burglar was a common kind which the FBI wanted to make into an object of scientific knowledge: a human kind. But serial killer was never, I think, a merely common kind; it came into being as a new classification, in company with a conjectured profile. Jack the Ripper was a serial killer, but the serial killer, as a kind of person, labeled "serial killer," emerges only about 1970.

To repeat a general admonition: I am not saying that there were no serial killers before 1970; only that this classification, this kind as kind, emerged about that time. It is well known that people have conjectured that there has been a looping effect for serial killers. It is suggested that, as the classification and its attributes were disseminated, some individuals decided to become serial killers. Others deliberately decided to act in ways inconsistent with established profiles, and so on. I know of no evidence for such claims, but such looping effects are at least easy to imagine.

Obviously advocates of a biological approach to crime, from Lombroso to the present, hope for some feedback effect, even if it is only self-knowledge. "I am born with a disposition to crime (or violence or whatever) so I will take such steps as I can, medication, therapy, self-help groups, stay away from bad company, as will help keep my disposition at bay." If the criminally disposed acted with self-knowledge like that, then their behavior would change, and what was true of the class would change, so we would require new knowledge of the class. I think all that is just sanctimonious fantasy, but no matter; I wanted only to illustrate the most naive of imagined feedback effects. It probably begs a couple of questions. (Is there a class of persons disposed to criminal behavior under suitable circumstances? Do members of that class want help?)

Kenneth Taylor (Chapter 5, in this volume) has pointed to a quite different sort of looping effect. He asks us to think of young men in a "no way up" situation. He imagines that the degeneracy research program achieves some measure of success. It becomes bruited about that young criminals are genetically (or chemically) disposed to crime. Then the urge to crime could be positively reinforced. "So I'm a criminal by birthright! Nothing can restrain me now." The correlations between some genetic or chemical indicator might then be overconfirmed. If this were to happen, it would be a classic looping effect. First there is a kind, the violence-prone, crime-prone youth. Then comes knowledge about that kind, that violence from such people is made probable by some features intrinsic to them. This knowledge comes into general circulation. The known-about know about it. This class of individuals becomes more reckless than before. This in turn becomes new knowledge about the kind.

I am myself dubious about both kinds of looping effect, the amelioration imagined by philanthropically inclined geneticists, and the far more plausible but very gloomy deterioration imagined by Taylor. I certainly don't think that either kind of effect can be predicted. This is partly because the very criminals who most interest the biologizers of crime inhabit their own network of subcultures. What happens there will determine what happens to the kinds, rather than the knowledge of experts. The experts are totally despised in the subcultures. People are not influenced much by a knowledge for which they have indifference or contempt. But before proceeding to closer scrutiny, it would be good, briefly to escape the dreary diet of deviants, of suicides, criminals, the mad, and serial killers. I'd like an upbeat example.

Exhaustion

One thing that is getting better, everywhere in the world, is athletic performance. Now consider a familiar human experience after intense exertion: exhaustion. In the old days a swimmer knew when she had pushed herself to the limit. Not any more. No one ever doubted that the thoroughly human experience, "feeling physically exhausted after a burst of intense exercise," was also thoroughly physiological. But now we have a measure of this, namely lactose levels. The athlete no longer owns her body. The trainer does. It does not matter much what the seriously competitive swimmer feels. Her lactose levels tell when she is exhausted, or has not, no matter what she says, pushed herself to the limit. For a rather brief period in the lives of relatively few people, a common kind of human experience is replaced by a measurement made on urine. The athlete surrenders her autonomy over her body to the trainer. No longer, while she is in training, does she have what the philosophers call privileged access. We now know what exhaustion after exertion really is. This discovery has helped make a measurable effect on world records, as well as lesser individual performances.

One trouble about this new reality is that when the swimmer quits, she has to reclaim her experience. Sports medicine has created a new specialty: the psychologist who helps former athletes by returning them to their bodies and their experiences. This example hints at a remarkable aspect of the biologization of human kinds. We are uncomfortable over a loss of humanity that seems to ensue. We never, or seldom, feel this for natural kinds. It is, of course, the familiar progress of natural kinds that we develop new criteria, new expertise, and new experts. We don't mind, much, that gold passes from the phenomenological to the assayed, to the Au molecule. But consider a biologization that has been more in the public eye than intense exertion and lactose levels.

Depression hovers somewhere between a feeling, a state, an illness, and a biochemical condition. Depression, the illness, used to be diagnosed in terms of self-report, of feelings, and of behavior, of affect and indifference, loss of interest in everything, or worse. But now, in some quarters, it is responsiveness to certain medications that matters. The numerous drugs, of which Prozac is only the most famous, which counter serotonin depletion, are now used. If they don't work, one initial guess at depression may be replaced by another, such as bipolar disorder, and lithium may be tried. This transition, from human to a biological (pharmacological?) kind, is of course accelerated by finance. Health management organizations strongly prefer drugs to open-ended therapy. Outcome-oriented treatment begins with a questionnaire for the patient and another for the reporting physician. These two reports are transformed by computer, using a large data bank, into a prescription. Because most drugs in the serotonin-active family don't work for a while, there is a waiting period, then new questionnaires, possibly a new dosage or prescription – a process pursued until the questionnaire outcome enters the normal range, as judged above all by the data bank.

The analogy between the swimmer and the depressed person is obvious. The experience of depression is no longer the defining characteristic. Even the depressed behavior is not definitive. What the patient says of herself does matter, but her feelings have no longer anything like a privileged status. Of course, even before the advent of extensive drug therapy, this loss of autonomy was to some extent true. A patient might have to learn that her personal mix of unhappiness, melancholy, inertia, and blackness was a familiar syndrome called depression. But although the language might be learned, from a psychiatrist or a self-help manual, the melancholic still owned her depression. But now, in pharmacology, what she "has" is determined by what drugs do to her. Issues raised in the antipharmacology movement, well represented by Peter Breggin's *Talking Back to Prozac*, are, in the end, about this loss of autonomy. The criticisms go on at the level of chemicals. For example, serotonin is nonspecific, everywhere in the brain, and changes in levels take seconds to have some effect; how then can Prozac-type drugs be specific, and also why do they take a month or more to have any effect? But those, we feel, are important technicalities. What is lost is autonomy – something that by definition is not lost as we find out more about a natural kind. It is not an accident that Breggin is the leading opponent of biological criminology, in both its pharmacological and genetic manifestations (Breggin and Breggin 1994).

From Violence to Impulse

I have spoken of three types of "kinds" (of people): common, human, and biologized. Let us turn these categories to crime, where we expect a fourth type of kind, legal kinds, classifications codified in law. Criminal behavior is "deviant." We want to stop it. What's wrong with crime? There are consequences: people lose property and/or get hurt. But there is something else. The fact that some crime is violent is what scares many people the most. The criminal behavior that is most scary, to middle-class scientists at least, is violent behavior. I suggested at the outset that the focus of criminology on criminal behavior was in part a focus on what we are most frightened of: violence. So I now wish to trace the trajectory of one distinguished body of research on violence. Markku Linnoila and Matti Virkunnen and their colleagues have produced a major sequence of papers starting with Linnoila et al. (1983). Their original finding was that "low cerebrospinal fluid 5-hydroxyindoleacetic acid concentration differentiates impulsive from non-impulsive violent behavior." 5-HIAA is produced when serotonin is broken down by the enzyme monoamine oxidase, so a relatively low concentration is taken to indicate a low serotonin level. The analysis is based on a lumbar tap of cerebrospinal fluid (CSF). Because serotonin depletion is "inferred" rather than "observed" many technical reports are now expressed in terms of CSF 5-HIAA, which is, nevertheless, code for serotonin depletion.

The 1983 Finnish study by Linnoila et al. built upon an American one that had correlated aggression in military cadets with (inferred) lower levels of serotonin. Cadets with the highest aggression scores were more impulsive, and high scores correlated with (inferred) serotonin depletion (Brown et al. 1979; F. K. Goodwin, later to be so fatefully involved in the Federal Violence Initiative, was a coauthor of this paper and the Finnish one just cited). A recent contribution to this work is Virkunnen et al. (1994), which shows that "in the present sample, a low CSF 5-HIAA concentration was primarily associated with impulsivity and high CSF testosterone concentration, with aggressiveness or interpersonal violence." The sample consisted of alcoholic violent offenders and fire setters. In what follows I call this the 1994 article, meaning the 1994 contribution to a body of research begun before 1983.

We are in effect concerned with four categories of kinds: (1) common kinds: violent, fire setter, impulsive; (2) legal kinds: in this instance, legally refined versions of common kinds – that is, the offenders were convicted according to legal definitions, but we would ordinarily say that if their conviction were just, they had indeed been violent or set fires; (3) certain psychiatric kinds of the sort I call human kinds: antisocial personality disorder, or intermittent explosive disorder, as defined in DSM-III-R; and (4) persons with low CSF 5-HIAA under controlled experimental conditions.

You might think that the focus of interest should be the connections between (3) and (4). Have the neurological chemists successfully correlated a clinical description (3) with a biological description (4)? I am as much interested in the connections between (1) and (3) – and thence to (4). The point of the exercise, in the end, is to come to grips with, and limit or prevent, violence itself, as commonly understood. The 1994 article appeared in *Archives of General Psychiatry*, which will not allow the use of common kinds. Only DSM classifications will do. We do not go directly from violence to serotonin. We need a scientifically defined human kind in between. Virkunnen and Linnoila state that the DSM-III-R diagnoses

characterized relatively often by impulsive violence and behavior are the following:

(1) Antisocial personality disorder (APD).
(2) Conduct disorder (solitary aggressive type in particular), a precursor of antisocial personality disorder in adolescence. [As we have seen, early conduct disorder is now incorporated into the DSM-IV definition of APD; it applies only to individuals under the age of fifteen. It may in fact be diagnosed before adolescence.]
(3) Borderline personality disorder.
(4) Intermittent explosive disorder.
(5) Pyromania in "impulse control disorders not elsewhere classified.'" (Virkunnen and Linnoila 1993: 227)

Intermittent explosive disorder is defined, in DSM-III-R (and, in essentials, in DSM-III) by these conditions:

A. Several discrete episodes of loss of control of aggressive impulses resulting in serious assaultive acts or destruction of property.
B. The degree of aggressiveness expressed during the episodes is grossly out of proportion to any precipitating psychosocial stressors.

C. There are no signs of generalized impulsiveness or aggressiveness between the episodes.

D. The episodes of loss of control do not occur during the course of a psychotic disorder, Organic Personality Syndrome, Antisocial or Borderline Personality Disorder, Conduct Disorder, or intoxication with a psychoactive substance. (American Psychiatric Association 1987: 322).

In DSM-IV, condition C is deleted. I have already given the DSM-IV definition of antisocial personality disorder. It is brief, compared to the DSM-III-R definition, which had to be used for Virkunnen et al. (1994). The criteria run on for two pages. Clause B lists twelve ways to be a bad child up to age fifteen and demands three of them (these are now taken over by conduct disorder). C lists ten ways to be bad after age fifteen, of which four must be displayed in order to be an antisocial personality; these in turn are broken down into as many as six options, of which only one need be manifest.

The DSM-III-R definition is a mess, but it paints a clear picture of an antisocial or at least antiauthoritarian person. APD people are a pain in the neck. If they stretch things, they do go to jail. As boys they play truant, run away from home, lie, beat up people and animals; when they are a little older they drink and drive, lie, pick fights, don't pay their bills, don't care properly for their own children, bully, and rape.

We have a somewhat complicated dialectic here. The conclusion of the 1994 article speaks of two clinical entities, IED and APD, and even conjectures a chemical distinction between them. Did the investigators actually apply the clinical criteria? On the contrary, when they speak of impulsive behavior, they are referring to something more like the common kind. A criminal act that harms a person is defined as impulsive if the convict did not know the person harmed. A man who starts a fire for insurance is not performing an impulsive act, whereas one who does it for no rational reason, and with almost no planning, is impulsive. This is common, not clinical, impulsivity. Various psychological tests were administered to the jailed alcoholics who were studied, but the looseness of fit between the criteria actually applied and the conclusion of the article is remarkable. That is, a chemical difference between APD alcoholics and IED alcoholics is conjectured and is suggested as a difference between these two clinical entities. But the two classes were not segregated according to clinical but rather common experience. This is not a criticism of the article. I am making an observation about the dialectic of human kinds.

What we are beginning to see here is the making of a biologized human kind, in which there is a complex mix of biological criteria, common criteria, and clinical criteria. The clinical criteria are in constant flux. Thus in 1987, the DSM demanded that in IED "there are no signs of generalized impulsiveness or aggressiveness between the episodes." (To confirm my assertion that the 1994 article did not apply DSM-III-R clinical criteria, there is no indication that the investigators even considered this criterion.) This criterion has been deleted from DSM-IV. Now, what is the kind of behavior that is being caused by a biological condition, of which low CSF 5-HIAA is an indicator? It is to be a biologized category. Would not this be great, if we finally got the "right" categories? The situation is not so clear.

Recalcitrant Kinds

Should we expect there to be a looping effect from biologizing a human kind such as antisocial personality disorder, or a common kind such as "impulsive"? Recall that by a looping effect, I mean a cycle composed of (a) an effect on persons classified by a kind K, with associated law-like regularities L, which is caused simply by the classification and knowledge of the classification, and (b) the necessity to modify the criteria for the kind K, or the knowledge embodied in L, because of the effect (a).

I am inclined to think that the kinds that interest victim, witness, and criminal are the common kinds, commonly understood, and not the human kinds or the biologized kinds. Common kinds will, for a long time, trump the new kinds. By that I mean that whenever it matters, we use the common and not the biologized category. The dynamics of looping are not working at all.

Elsewhere I have proposed a taxonomy of human kinds, including inaccessible, administrative, self-ascriptive, and others (Hacking 1995). I would add a new category of recalcitrant human kinds, kinds that are not picked up by anyone. The present generation of research on the criminal, the violent offender, and the like will resemble generations of past research on recidivism. Due to our high technology of reidentification, we can detect recidivists with great ease. Yet no knowledge about recidivists has stuck. One reason for this may be simple. Criminology exists in an entirely different culture from its object. Create human kinds as it will, the only effect on criminals is on the degree of punishment meted out. We can understand the current popular American predicament as a cry of frustration. Increased punishment and increased incarceration do not decrease the rate of crime. But the categories in terms of which we try to help or comprehend crime are futile. So Americans by a large majority want to incarcerate and kill criminals because it is the only way in which they can bring power to bear on criminals. I suggest that the dialectic in the dynamics of human kinds of criminals is the exact reverse of many other human kinds.

A Reality Check?

At any time, the distinct Western democracies have pretty much the same portfolio of social problems, modified by history and local circumstances. This is not an accident, because social problems get passed along, in the following sense: what counts as a problem, what is recognized as a problem, gets passed from region to region. For half a century after 1945, most social problems originated (in this sense) in the United States. For example, child abuse originated in Colorado about 1961, quickly spread across the United States and to the other English-speaking countries. It was taken up, much more slowly, for example, in Italy and later still in France.

Crime is and always has been a shared social problem. Criminology is shared knowledge. For example, Scandinavia is a prime source for certain types of statistics, such as twins and crime. I have referred to Finnish studies of prisoners that quite possibly would not be allowed in American jails. A Dutch report of a troublesome family (Brunner, Nelen, Zandvoort, et al., 1993) is used in the American state of Georgia by the attorneys of the murderer Stephen Mobley as part of his last-ditch plea for a stay of execution. Criminology knows no borders. The United States is nevertheless a distinct society. This comes out in many ways connected with crime itself – in the practice of executing murderers, or in the proportion of young men kept in jail. In respects like that, there is nothing much like America anywhere else in the world. But let us turn to a more pleasant contrast. In a Gallup poll, it is very hard to pose a question to which almost everyone gives the same answer. Gallup asked Americans about a number of possible causes of crime. Is this critical, important, or not important as a cause of crime? The results were unusually uniform by region, age, income, political affiliation, race, education. On exactly two of these causes there was effectively complete agreement. First came drugs, with over 60 percent thinking this to be a critical cause, and 98 percent thinking it important. Next, with 99 percent agreement (greater than 99 percent in many subsets), and with over 50 percent thinking it a critical cause, was "lack of moral training in the home" (Bureau of Justice 1993, 172).

This phenomenon leads to two observations. First, although I can cite no data, I very much doubt that you could find such moral unanimity anywhere else in the West. Second, genes and serotonin be damned. The entire American public thinks moral education in the home is what counts. The much-vaunted and much-feared work on the connection between crime and genetics may be, at the decisive level of public belief, completely irrelevant. Or is it the other way around? Is the body politic shielding itself from oppressive biologizing by the constant refrain of restoring family values? Here we have a question indeed. Has Gallup directed us to a reality check? Or is this rather a fantasy check?

REFERENCES

American Psychiatric Association. (1980). *Diagnostic and statistical manual of mental disorders*. 3rd ed. Washington, D.C.: American Psychiatric Association.
   (1987). *Diagnostic and statistical manual of mental disorders*. 3rd ed. rev. Washington, D.C.: American Psychiatric Association.
   (1994). *Diagnostic and statistical manual of mental disorders*. 4th ed. Washington, D.C.: American Psychiatric Association.
Barinaga, M. (1994). From fruit flies, rats, mice: Evidence of genetic influence. *Science* 264: 1690–1693.
Breggin, P. R., and Breggin, G. R. (1994). Genetics and crime: Violence and the danger of psychiatric social control. *21st Century Afro Review* 1, 1.
Brown, G. L., et al. (1979). Aggression in humans correlates with cerebrospinal fluid metabolites. *Psychiatric Research* 1: 131–139.
Brunner, H. G., Nelen, M., Breakefield, X. O., et al. (1993). Abnormal behavior associated with a point mutation in the structural gene for monoamine oxidase A. *Science* 262: 578–580.
Brunner, H. G., Nelen, M. R., Zandvoort, P., et al. (1993). X-linked borderline mental retardation with prominent behavioral disturbance: Phenotype, genetic localization, and evidence for disturbed monoamine metabolism. *American Journal of Human Genetics* 52: 1032–1039.
Bureau of Justice. (1993). *Sourcebook of criminal justice statistics*. Washington, D. C.: U.S. National Criminal Justice Information and Statistics Service.
Bursik. (1989). Political decisionmaking and ecological models of delinquency: Conflict and consensus. In *Theoretical integration in the study of deviance and crime,* ed. S. Messner, M. Krohn, and A. E. Liska, 105–177. Albany: State University of New York Press.
Comings, D. E. (1994). Genetic factors in substance abuse based on studies of Tourette Syndrome and ADHD probands and relatives. II. Alcohol abuse. *Drug and Alcohol Dependence* 35: 17–24.

Comings, D. E., Muhleman, D., Ahn, C., Gysin, R., and Flanagan, S. D. (1994). The dopamine D2 receptor gene: A genetic risk factor in substance abuse. *Drug and Alcohol Dependence* 34: 175–180.

Conrad, P. (1997). Public eyes and private genes: Historical frames, news constructions, and social problems. *Social Problems* 44: 139–154.

Daly, M., and Wilson, M. (1988). *Homicide.* New York: A. de Gruyter.

(1994). Evolutionary psychology of male violence. In *Male violence,* ed. John Archer, 253–288. London: Routledge.

Douglas, M., and Wildavsky, A. (1982). *Risk and culture: An essay on the selection of technological and environmental dangers.* Berkeley: University of California Press.

Esquirol, J.-E.-D. (1821). Suicide. *Dictionnaire de sciences médicales* 53: 213.

Federal Bureau of Investigation. (1965). *Uniform crime reports for the United States.* Washington, D.C.: U.S. Department of Justice.

Goring, C. (1913). *The English convict: A statistical study.* London: His Majesty's Stationery Office. Reprinted in Goring 1972.

(1972) *The English convict: A statistical study, including the schedule of measurements and general anthropological data.* Montclair, N.J.: Patterson Smith. Reprint of the 1913 edition, including Pearson 1919.

Guerry, A.-M. (1833). *Essai sur la statistique morale de la France.* [Paris].

Hacking, I. (1991). The making and molding of child abuse. *Critical Inquiry* 17: 253–288.

(1995) The looping effects of human kinds. In *Causal cognition: A multidisciplinary approach,* ed. D. Sperber, D. Premack, and A. J. Premack. 351–383. Oxford: Clarendon Press.

(1998). *Mad travelers: Reflections on the reality of transient mental illnesses.* Charlottesville: University Press of Virginia.

Hagan, J. (1993). Introduction: Crime in social and legal context. *Law and Society Review* 27: 255–260. (Introduction to a symposium, Crime, class and community – an emerging paradigm.)

Herrnstein, R. J., and Murray, C. (1994). *The bell curve: Intelligence and class structure in American life.* New York: Free Press.

Holden, C. (1994). A cautionary genetic tale: The sobering story of D2. *Science* 264: 1696–1697.

Jenkins, P. (1994). *Using murder: The social construction of serial homicide.* New York: A. de Gruyter.

Lakatos, I. (1970). Falsification and the methodology of scientific research programs. In *Criticism and the growth of knowledge,* ed. I. Lakatos and A. Musgrave, 91–195. Cambridge: Cambridge University Press.

Linnoilla, M., et al. (1983). Low cerebrospinal fluid 5-hydroxindoleacetic acid concentration differentiates impulsive from nonimpulsive violent behavior. *Life Sciences* 33: 2609–2614.

Lombroso, C. ([1906] 1972). Introduction to G. Lombroso-Ferrero, *Criminal man according to the classification of Cesare Lombroso.* Montclair, N.J.: Patterson Smith. (From his speech to the Congress of Criminology, Turin, 1906, printed in the *Proceedings.*)

Mann, C. C. (1994). Behavioral genetics in transition. *Science* 264: 1686–1697.

Nye, R. A. (1984). *Crime, madness and politics in modern France: The medical concept of national decline.* Princeton, N.J.: Princeton University Press.

Pearson, K. (1900). The inheritance of mental and moral characters in man. *Biometrika* 3: 142.

(1919). Introduction: Charles Goring and his contributions to criminology. In Goring 1919: ix–xvi. Reprinted in Goring 1972: xv–xx.

Porter, T. M. (1986). *The rise of statistical thinking, 1820–1900.* Princeton, N. J.: Princeton University Press.

Powledge, T. (1996). Genetics and the control of crime. *Bioscience* 46: 7–10.

Quetelet, A. (1832). Recherches statistiques sur le royaume de Pays-Bas. *Nouveaux mémoires de l'Académie royale des sciences et belles-lettres de Bruxelles* 7: 20.

Rowe, D. C., Osgood, D., and Nicewander, W. A. (1990). A latent trait approach to unifying criminal careers. *Criminology* 28: 237–280.

Rutter, M. (1996). *Genetics of criminal and anti-social behavior.* New York: Wiley.

Virkkunen, M., and Linnoila, M. (1993). Serotonin in personality disorders with habitual violence and impulsivity. In *Mental disorder and crime,* ed. S. Hodgins. Newbury Park, Calif.: Sage.

Virkkunen, M., Rawlings, R., Tokola, R., et al. (1994). CSF biochemistries, glucose metabolism, and diurnal activity rhythms in alcoholic, violent offenders, fire setters and healthy volunteers. *Archives of General Psychiatry* 51: 20–27.

Wilson, J. Q., and Herrnstein, R. (1985). *Crime and human nature.* New York: Simon and Schuster.

Chapter 7

Genetic Plans, Genetic Differences, and Violence: Some Chief Possibilities

ALLAN GIBBARD

"Heritability" is not inheritance. The distinction between the two is one of the first things to learn for anyone who encounters the discipline of genetics. A normal human being inherits two-leggedness, the crucial characteristic of having two legs. This characteristic, though, might have low heritability in the human population.

"Heritability" is a technical term used by geneticists. Heritability is a property not of an individual but of a population: a single characteristic of an individual might have high heritability in one population the individual is in and low heritability in another. Heritability concerns *differences:* it is a matter of the degree to which, in a given population, differences in a characteristic trace back to differences in individuals' genotypes. Let an entire population be two-legged, and the heritability of two-leggedness in that population is simply undefined in that population; there are no differences in two-leggedness to be accounted for. In another human population, suppose, everyone is born with two legs, but occasionally someone loses a leg in an accident. If the accidents have no correlation with genotype, then in that population the heritability of two- leggedness is zero. In another population, imagine, blue-eyed children are fed to lions, but some of them survive, maimed. If eye color is inherited and this grim ritual is the predominant cause of anyone's lacking a leg in that population, then non-two-leggedness, in that population, has substantial heritability. Individuals inherit, but only populations have heritabilities (Tooby and Cosmides 1990, 37–39; Sober, Chapter 3, in this volume).

Geneticists' terminology, then, can easily be misleading. One fine review of findings in genetics is entitled "The Genetic Basis of Complex Human Behaviors" (Plomin, Owen, and McGuffin 1994). This title suggests a broad topic of immense importance: what do human genes have

169

to do with complex human behaviors? The question discussed in the article, though, is much narrower than this. The topic is heritability in the technical sense: what average causal effects, in the environments of a human population, do *differences between* the genotypes of different people have on differences in their complex behaviors? It is important for readers not to think that answers to this question tell us, by themselves, what genes have to do with behavior. The answers may bear on this broader question, but we must be careful in our reasonings about what the bearing might be.

This chapter concerns violence, genes, and scientific method. It asks about the tie between two different but related classes of questions. The first class concerns violence and genetic inheritance: what do human genes have to do with human violence? The second class concerns violence and heritability: to what degree is violence of various kinds "heritable" in various human populations? The two kinds of questions are distinct but perhaps not entirely unrelated. While asking what connections might obtain between them, I also examine the ways in which the two kinds of questions are independent. Even if in certain populations, violence of some kinds turns out to be "heritable" to some detectable extent, what could that tell us about the causes of violence and what questions would it leave open?

Geneticists work with differences, and certainly with violence many of our chief concerns are matters of difference. What features make the difference between high rates of homicide and violent injury in one society and low rates in another? Why do some people act to kill or injure others, and others lead peaceable lives from birth to old age? How can public policy make a difference to the incidence of violence? Answers to these questions have much to do with human genes. They may have *something* to do with the genetics of current human populations. But even though genetics studies differences, what genetic findings might have to do with differences in violence is an involved and somewhat tenuous matter.

Two-leggedness is genetically inherited, but the "heritability" of two-leggedness in a population is of little interest. The characteristic is inherited, in that the normal human genome includes an intricate "plan" for the development of two legs. The plan has a "purpose": efficient locomotion that leaves the hands and arms free to do other things. The *heritability* of two-leggedness, though, is beside the point: two-leggedness, as an adaptation, long ago came to fixation in all human populations. Should we then expect the same pattern with propensities to violence? The human genome includes a complex of "plans" for the brain. Among these will be "plans" for whatever psychic mechanisms make the difference between violence and nonviolence in one situation or another. Could we hope to understand the psychology of violence without searching for clues as to what these inherited plans might be? If, though, these "plans" too came long ago to fixation in all human populations, then the *heritability of* violence in a current population might have little to do with violence and its causes. And if it did have a bearing, teasing out what that bearing might be would be far from straightforward and obvious.

I begin with general questions about genetic "plans" for behavior and then turn to violence in particular. Both the general theory I sketch and its application to violence have been studied and expounded at length in first-rate scientific studies, some of which are highly readable. I pass rather quickly, then, over this vast terrain, asking specifically about heritability. What might heritability have to do with violence? What questions about genes and violence would even the clearest findings about heritability, if they could be attained, leave open?

GENES AND THE HUMAN PSYCHE

In recent years, a new picture of the human mind has been emerging, although no doubt only among a small minority of those who study the human mind professionally. Mental activity, the hypothesis is, consists of the actions and interactions of a set of *domain-specific mechanisms*.[1] The reasoning behind the talk of domain specificity (Symons 1979; Tooby and Cosmides 1992) can be described as follows.

The genetic plan that a human infant comes with, we should all now be convinced, is the product of billions of years of natural selection, with some crucial late modifications cobbled together within the last several thousands of generations. The pressures of natural selection themselves, of course, are mindless, but we know that the results can look intricately planned: Paley rhapsodized on the intricacy of the human eye as a proof of divine planning, and it was this appearance or intricate planning that Darwin explained – or explained away – when he realized that it could have resulted from an extremely long history of small variations, along with a natural analog to selective breeding.

Now what, if this is right, will the seeming plan for a species be like – for the human species, to take an instance? With all animals, the grand plan must include provisions for behavior. Having a marvelous heart,

lungs, vision, digestion, and legs for running do not promote one's reproduction unless one does the right things: unless one does things that
turn out to work, in the long haul, to pass one's genes on into distant
generations. To think about the results of natural selection, then, we
have to think, among other things, about how a genetic plan can best
ensure behavior that promotes reproduction.

An organism faces an intractable complexity of reproductive problems. No general mental mechanism, no general "learning mechanism,"
could handle such a complexity at all adequately. What could emerge
from selection pressures on genetic plans for human brains is not a plan
for some general learning mechanism, but for a host of mechanisms to
deal with specific reproductive challenges. This argument has often
been made with respect to the study of "cognition" (Nisbett and Ross
1980). It applies, though, not only to "cognitive" representations of the
world but to behavior: what we do isn't likely to be the outcome of some
general mechanism of "will" or "conditioning." Nor can we expect to
find a general "reproductive prospects calculator." Behavior, we must
expect, emerges from the interplay of mechanisms, each specialized to
cope with some specific kind of historically important reproductive
challenge.

People do not, then, always or even on average, act to maximize their
"reproductive success." Natural selection explains the shaping of psychic mechanisms, not of actions directly. Our genetically "planned" psychic mechanisms have been highly refined by natural selection to enhance reproduction in various ways, in comparison with the
alternatives that mutation made available over thousands of generations. Our hunting and gathering ancestors, though, did not always
make the choice best calculated to enhance their reproduction because
no feasible mechanisms could handle the calculations. And our own environment is not theirs, so that mechanisms that promoted reproduction
among our ancestors may hinder it with us.

A broad research program that stresses genetic plans for behavior is
called by its practitioners "evolutionary psychology." Researchers such as
Donald Symons, David Buss, Leda Cosmides and John Tooby, and Martin Daly and Margot Wilson have offered remarkable analyses of such
areas as human sexuality, male and female, and of human violence of
various kinds in various circumstances – supported by a wide range of
data (Symons 1979; Daly and Wilson 1988, Tooby and Cosmides 1990,
1992; Buss 1994).[2]

Evolutionary psychologists reject any view that we as people spend our psychic energies incessantly plotting how to reproduce. Sometimes people do, but often we end up reproducing just by "doing what comes naturally" in various ways. Few organisms in history have thought about how to achieve the greatest representation of their genes in distant generations. When evolutionary psychologists speak of "planning" for reproduction, they normally have the as-if planning of the genes in mind. This as-if planning accounts, among other things, for the psychic mechanisms whose workings constitute planning in the literal sense.

A deep mistake we must avoid in thinking about genes and psyche concerns what a "biological" theory of human nature must be like. Man and woman as painted in biology, it is all too easy to think, must be rigid and impervious to environment. To whatever degree genes turn out to matter, it is widely thought, what we do turns out to be *determined* by our genes. Thinking this way seems to allow just two alternatives: to embrace a pernicious "genetic determinism," or valiantly to reject any claim that genes could matter for human thought and action.

The same error, strangely, can be made by enthusiasts for scrutinizing the role of genes in behavior. For each kind of behavior we want to explain, it is tempting to think, we can find the right point on a spectrum from genes to environment: there exists a degree, from 0 to 100 percent, that the behavior is "genetic." At the extremes, a kind of behavior may be "biologically" determined, or it may be "environmentally" or "culturally" determined. Or it may be to some degree the one and to some degree the other. A "biological" explanation will appeal to the genes alone, whereas if an "environmental" explanation is correct then genes are irrelevant.

These are prime fallacies in thinking about genes and behavior. "Environmental" explanations must be biological and genetic, once they are fully spelled out. All learning depends on genetically coded devices for learning.

Biology tells us that genes matter immensely, but nothing in biology says that genes are *all* that matters – far from it. Much of the as-if planning represented by the genes concerns how to cope with *differences* in kinds of circumstances that the genes might encounter. The genetic plan for a human being will be full of contingency plans: full of schemes that in effect say "If *A* then do *X*, whereas if *B* then do *Y*." Failing to realize this, people often take it that a "biological" explanation of why two people are different in what they do must attribute the difference to a difference in their genes. They think that an "environmental" or a cultural explanation of a difference is not "biological." All this is confused. Given a difference in how two people act, it is perfectly biological to say something like this: the two people's genetic plans are the same in relevant respects. They've encountered, though, different cues as to their circumstances. The cues the two have encountered differ in ways for which the single genetic plan they share makes provision. The plan they share is to respond one way given the one set of cues and another way given the other. The cues in question may be immediate ones, or they may be cues that came years ago in childhood and have affected the development of psychic mechanisms or the setting of parameters for them.

A thoroughly biological explanation for a difference between two organisms can be as "environmentalist" as you might want. But evolutionary ways of thinking do impose a discipline on thinking about responses to environment. An evolutionarily coherent account of an organism's responsiveness to environment can take any of a number of forms, and in the last two sections of this chapter I consider some chief possibilities. One especially important kind of account explains responsiveness as an adaptation. To do this, one must first suppose that there are mechanisms that respond differently to those differences in environment that are claimed to make the difference. One then explains why the plan to respond one way to the one set of environmental cues and another way to the other set would have tended to enhance reproduction in the environments of our ancestors. Some naive "environmentalist" accounts cannot plausibly be elaborated to fit this pattern. Other patterns are available: one may explain the responsiveness as not itself having a biological function, but as a by-product of other adaptations, or the working of adaptations in an environment outside the range of environments for which they were "designed." Some explanations, though, cannot be fit into any evolutionarily coherent pattern, and these we should find implausible. (Or: we should find them implausible unless the evidence for them is strong, in which case we should be puzzled and investigate further.) But one crucial thing to note is this: accounts of responsiveness to environment as an adaptation can attribute crucial roles to the environment.

"SOCIOBIOLOGY" DEBATES

Evolutionary approaches to human behavior have sparked polemical fireworks. Some of these have touched on genuine and serious methodological questions. The evolutionary study of behavior is young, and any pioneering work in a complex field is bound to include confusions. In a way, to be sure, the field dates back to Darwin, who made penetrating suggestions about behavior, emotions, and the like, mixed in with what now strike us as dubious tales and Lamarckian speculations. Current serious work on the subject, though, stems from the 1960s, with W. Hamilton's model of kin selection and G. Williams's attack on naive group-selection arguments (Hamilton 1964; Williams 1966). I myself am impressed by the argumentative rigor of the best recent work on genes and human behavior. This work offers sketches of aspects of the human psyche that fit common experience in important ways, and that fit better with empirical data than do many alternative treatments that are not evolutionarily informed. They integrate a view of current humanity with what we know of the natural processes that shaped human nature and made the human newborn so different in potentiality from a brick, a new-hatched lobster, or a newborn chimp.

One broad charge that has been leveled against evolutionary treatments of the human psyche is that it is naively "adaptationist." Charges of "adaptationism" have been leveled eloquently by leading biologists, but as blanket charges they must be viewed with caution.[3] Not all features of an organism are adaptations. If one hypothesizes that a given feature constitutes an adaptation, the hypothesis needs support, and the supporting argument should appeal to both cogency and evidence. On the other hand, adaptation does play a special role in the living world. When, as with the human eye, there is clear evidence of intricate "special design" for a job, with an elaborate and detailed suitability of the organ for a function that promotes reproduction, we know that the only possible kind of explanation for all this must be a vast history of natural selection.[4]

What, then, of behavior? Obviously, not everything that humans or other organisms do is adaptive, in the biologists' sense. But evolution must have paid great attention to behavioral mechanisms, for behavior so much affects reproduction. With behavior, matters will be more subtle than with eyes or the shapes of limbs. We have to be on the lookout for what the intricate "special design" of behavior might have been and expect that often false hypotheses will look attractive. Still, the job has to be to examine all the clues we can as to what the "special design" of human behavior might have been. Whether such a design would have worked is an important part of the case for or against a claim as to how we are "designed," but empirical support is crucial.

Why, one might ask, should we appeal to selective histories in a prehistoric past that we can know little about? Why not stick to what we can observe now, and just study what current human beings are like? The philosophy of science of the past few decades should make the answer clear. The field has been dominated by arguments that no mechanical "scientific method" leads, in a straight, Baconian way, from the observation of nature to scientific understanding. Scientific understanding requires somehow getting to the point where to some degree, the right kinds of observations are made and the right kinds of considerations are brought to bear. What, then, are the chances of this in the study of human beings if we ignore what we know about the processes that made the species what it is? Hypotheses must be tied to current evidence, true enough, and much work in evolutionary psychology is devoted to empirical testing and elaboration. It is far better, though, to work with hypotheses suggested by a correct view of what shaped the human species than to work with hypotheses that just seem appealing.

Science is far more than the collection of data. Research takes place against a background of preconceptions. In the worst case, preconceptions can guide the outcome of research with considerable freedom from empirical data. Even, though, when investigators use methods that detect when nature is not cooperating – well-controlled experiments, cautious statistical analysis of nonexperimental data, and a lively canvassing of what the possibilities are and how what we observe would respond – preconceptions very much matter. Until one begins to have a picture that allows observation to guide one along paths that lead closer to the truth, one gets perhaps chaos, or perhaps localized theories that do not integrate into a wider picture.

In thinking about genetics and violence, then, we do best if we have a set of empirical preconceptions that promises to mesh with the ways things are, guiding us to observations that can refine and correct those preconceptions. We need to start with a picture that incorporates what we already know and suggests what the major possibilities are. Much writing on this subject does proceed against a background of awareness of a wide range of possibilities for how genes might connect with violence; it certainly is not my aim to wipe the slate clean. I do want to suggest ways that a picture of what genes could have to do with violence delineates kinds of possibilities.

Still, can such evolutionary hypotheses be more than mere "just-so stories"? The hypothetico-deductive method in science is to devise hypotheses, confront them with the evidence, and use the result either to reject hypotheses or to refine them for further confrontation with evidence. The legitimate concern, then, can't be that evolutionists traffic in hypotheses. What distinguishes "just-so stories," apart from their sheer implausibility, is the paucity of support for them, their way of being highly unconstrained by evidence and by the demands of coherent theory. Now whether or not this charge applies to some of the work it has been leveled against, we should be clear on this: in the 1970s when this charge gained currency, its target was not the corpus of work in "evolutionary psychology" that I am drawing on, for that work came later. The writings of evolutionary psychologists span the range from speculative hypotheses to work with impressive empirical support. This is as it should be, so long as speculation is recognized for what it is: a science without speculations that suggest paths for investigation must be blind and aimless. I can't imagine, though, how any fair-minded reader could read Symons and Buss on human sexuality, or the joint work of Daly and Wilson on homicide, and think that they were not subjecting their hypotheses to strong tests of theoretical coherence and empirical support.

COMMON SENSE

One alternative to thinking about human beings in evolutionary terms is to rely on common sense, on the "folk psychology" we all use to navigate through daily life. A psychosocial science that starts with common sense need not proceed blindly; it has a basis for generating its initial hypotheses. Does common sense refined by empirical investigation offer an alternative to thinking evolutionarily? Indeed, can evolution tell us anything about human beings that refined common sense cannot?[5]

No quick and adequate response is possible; we must let the best evolutionary treatments speak for themselves and see how they stack up against approaches that ignore evolutionary considerations. But some initial remarks may be useful. I agree that common sense is an indispensable asset in the study of human phenomena. But consider what folk psychology leaves mysterious, and how few resources it has for resolving some of these mysteries.

The term "folk psychology" can be applied in at least two ways: first, there's a philosopher's invention, a schematized "belief-desire" psychology that has people acting to satisfy their desires in light of their beliefs. Explain, then, a murder: the man aimed a loaded pistol at his wife and pulled the trigger. Was that because he wanted his wife dead and believed that doing so would bring it about that she was dead? A marvelous explanation, fine as far as it goes, but it is the man's desire that his wife be dead that we really need to explain. Now perhaps this tru-istic start of an explanation can be supplemented in a belief-desire mode. What we need is an account of what he further desired that he thought her death would bring. Let's try: she has just told him, with bit-ing contempt, that the children he thought were his were by another man.[6] What, then, does he aim to accomplish by her death? Revenge? Why is this now his dominant aim? Desires, we can agree, figure cru-cially in the picture, but the puzzle is to explain the desires, and why one desire prevails over many others.

The real "folk psychology" of common sense countenances a much wider range of factors: emotions of humiliation, rage, and hatred, and moods of despair, and lore as to what will bring them on. Do these ma-terials of real common sense obviate the need for evolutionary thinking? Well in the first place, common sense is often wrong; fairly common-sense assumptions that contribute to racism, for instance, now seem mostly mistaken. But more than that, we need to remember how much common sense leaves unexplained, rightly or wrongly. Why marriage, and why divorce? People fall in love and so they get married. Then they come to hate each other, and they get divorced. Why, though, do the same people at one point love each other and then later hate each other? Survivors of divorce often find the question baffling. She kept doing things that made him furious. Why, though, was he furious at things that a few years earlier he would have found mild annoyances? Or he became morose and listless, and she lost patience with him. Why, though, the depression, and why are some people at some times hero-ically patient and others sometimes not? These phenomena cry out for explanations that even a rich folk psychology cannot provide.

Once, indeed, we think of the mind as involving the interplay of var-ious specific mechanisms designed for the contributions they can make to action, we can far better accommodate psychic life as we know it than we can if, say, we try to classify everything neatly as a belief or as a de-sire. Emotions enter easily into the picture, as do moods and personal-ity traits. We can see emotions as the workings of mechanisms designed to deal with specific features of human life: disgust, for instance, keeps us away from likely sources of disease, guarding us from eating, drink-ing, and breathing dangerous microbes.[7]

Moods, for example, we can think of as longer-term modifiers of ac-tion, responsive over the medium term to cues indicating that some kinds of action are likely, in one's circumstances, to enhance one's re-productive prospects and others to diminish them. The mood scale that ranges from depression to mania, for instance, affects one's reaction to opportunities for possible gain that are fraught with risk. It affects how much one shuts down old enthusiasms and engages in brooding, which we might think of as mental processing that is alert for cues as to what quite different modes of life might be good; we are thus "designed" so that what appears good to us tracks, in the best rough way that evolution could devise, what enhances one's reproductive prospects.

Like things might go for long-range personality types, such as being hard-bitten on the one hand, or soft and sentimental on the other. These might be "designed" responses to cues as to the circumstances one is in and the mode of life that, in those circumstances, best enhances one's reproductive prospects (Tooby and Cosmides 1990, on personality psychology).

WHAT TO ASK ABOUT VIOLENCE

Adaptation is a biological notion it would be crippling to do without. An "adaptation," in the biological sense, is what the genes have provided for, directly or indirectly, as a result of natural selection. To speak of adaptation is to cite a history of differential reproduction in an ancestral population that accounts for organisms' now having the features they do.

Being an adaptation does not mean being good or even tolerable – and it does not mean inevitable. Ancestral history by itself does not tell us what's good or bad, what to accept and what to struggle against. Nor does it, by itself, tell us what we're stuck with: the ancestral environment does not have to be ours, and what human psychic mechanisms were shaped to do in ancestral environments they may not do now, because our current environment may differ.

Human violence is sometimes good and far more often bad. Humanity owes deep gratitude to those who, opposing violence with violence, defeated the Nazi, Fascist, and militarist regimes of World War II, but the praiseworthiness of what they did is not a matter of whether it was biologically normal or adaptive. A man, on the other hand, may tend easily to fall into rages against his mate's children who are not his own, and so to act abusively or even murderously toward them (Daly and Wilson 1988, 82–93). This may constitute an "adaptation" in the biological sense: killing one's mate's child leaves her ready to put her care undivided into one's own children by her. But if this tendency stems from an ancestral history in which men with these psychic propensities had more genetic great-grandchildren on average, that does not justify it or make it tolerable. Nor does it make such atrocities inevitable: many stepfathers take loving, considerate care of their children. This story – the story of child abuse as the work of psychic mechanisms that are biologically adaptive – does suggest that the extent of child abuse will be difficult to alleviate. But we know that already from experience, and a better understanding of the psychic mechanisms in play may tell us more about the conditions under which they will be activated, and more about what might avoid their being triggered.

Once we conceive that violence may often be the work of adaptive psychic mechanisms, how should we conceive the relation between genes and violence? That is a large question on which others have said much and I can say little. Still, some of the possibilities are worth surveying. As I have stressed, a chief moral of recent philosophy of science is that preconceptions very much matter, and the better we can make our preconceptions, the better scientific investigation is likely to proceed.

An appreciation of ways things are likely to be should make us skeptical that, at least in America, a major difference between men who are more or less prone to violence lies in anything that fits any ordinary idea of a "gene for violence." We need to ask what the main possible patterns are that our prior knowledge of matters allows for. Two weak conclusions will be supported by the survey: first, although *genes* very much matter, genetic *differences* among men now alive may not at all be the chief place to look for what makes the difference between more and less violence on the part of men; second, even if current genetic differences between men do have a detectable causal effect on observed violence, we won't know much about what's going on until we begin to identify the kinds of mechanisms at work. Population statistics give evidence concerning mechanisms, but to get far on questions of what the mechanisms may be, one will have to resort to a more eclectic array of evidence.

Martin Daly and Margot Wilson, working jointly, have applied the evolutionary psychologists' program to homicide in a superb group of studies presented most fully in their book *Homicide* (1988). For anyone with significant interest in genes and violence, the book should be required reading; I only summarize their findings here.[8]

Some human psychic mechanisms involved in violence are *adapted* to produce violence in certain kinds of circumstances. Other mechanisms may be adapted to inhibit violence in certain kinds of circumstances. We can expect men, then, to be equipped with a complex of psychic mechanisms adapted to "assessing" when to act violently and when not to. With violence and refraining from violence, after all, the reproductive stakes are immense, and the selection pressures favoring certain kinds of discriminations are intense. Will evolution have refined one's eye and hand, as it so marvelously has, and then leave one with no provision for directing that hand to respond to what the eye transmits in ways that will tend to enhance one's reproductive prospects? Will it make no provision for those very circumstances in which one's reproductive prospects most hinge on what one does? These are evolutionary considerations in advance of most of the evidence: natural selection must provide carefully for discrimination in when to be violent.

In human affairs too, though, nontheoretical evidence that violence matters is dismayingly plentiful. What psychic mechanisms make the difference between violence and nonviolence will be difficult to discover, but common experience suggests a few things. Truistically, rage can lead to violence, and rage is an emotion with its own characteristics. One loose but highly plausible hypothesis is that rage, along with the control of when to be enraged and when not, is the work of psychic mechanisms shaped by natural selection to produce violence in some kinds of circumstances and not in others. Genes that coded for these mechanisms, we can expect, proliferated among our ancestors because the mechanisms produced violence in circumstances where the particular kinds of violence they produced tended to promote the reproduction of one's genes. We have mechanisms that respond to certain cues, cues that tended to go, among our ancestors, with violence of certain kinds promoting reproduction or hindering it.

In some cases, the adaptive causal paths that resulted – the work of interacting mechanisms coadapted to produce reproduction-enhancing violence in determinate circumstances among our ancestors – are indirect. Cosmides and Tooby, for instance, note the finding that abusers of children often were abused themselves as children. They speculate that an adaptation might be in play: we might be adapted to respond to childhood cues of violence by setting our own thresholds for violence lower. These might, say, be thresholds for rage or for violence given rage. Such a mechanism for an early, lifelong setting of thresholds might be adaptive, if the cue of being treated violently as a child was a good cue, on average over a long period of human evolution, that violence tends to pay reproductively in one's society, and the cue of not being treated violently as a child were likewise a good cue that violence did not on average enhance one's reproduction in this society. (The child, of course, need have no such thoughts; the point is the as-if reasoning of the selection pressures that shape the psychic mechanisms that figure in a man's harming a child.)[9]

Such a mechanism, we should stress, is informed speculation, not a hypothesis in which we should put great credence. But its possibility illustrates a number of points. This is not a speculation that those more prone to child abuse differ genetically from those less prone: the story could hold true with no current genetic variation for the trait. The hypothesis that this is an adaptation would mean that there was an ancestral history of variability in the genes that program for the mechanisms involved, but the genes for these mechanisms could by now have long since come to fixation. Second, if this were the full explanation of why some men abuse children and others do not, it would be fully "environmentalist" in the sense that differences in behavior stem from environmental differences and not at all from genetic differences among current individuals. Nevertheless, genes play a crucial role: there can be no coherent environmentalism without genetic adaptations for mechanisms to respond to environmental cues.

Finally, behavioral adaptations are bound to consist in contingency plans. Any interesting kind of behavior tends to promote reproduction in some circumstances and stymie it in others. An adaptive mechanism, then, must produce the behavior given some complexes of cues, and, given others, not produce it. A mechanism may respond to immediate cues, or set thresholds or otherwise adjust mechanisms for responding to later cues. Some adjustments may be lifelong, some long term, and some a matter of passing mood. We can expect highly refined mechanisms for "assessing" when to be violent in various ways and when not. These can include mechanisms of mood and emotion.

DIFFERENCES WITHOUT GENETIC DIFFERENCES

If this picture is on the right track, what kinds of differences might account for the difference between people who commit certain kinds of violence and people who do not? What will the statistical signs of these patterns be? How likely is it that various of the possible patterns will be realized? These broad questions cannot be answered without careful investigation. Although they are questions that a philosopher is not specially equipped to address, I offer some tentative answers.

Because extreme violence is chiefly the work of men and to a lesser degree, women in response to violence in men, I speak of violence in men and not in women. To claim that violence is unisex would be to ignore the realities.

My discussion is organized around the kind of question that geneticists collect statistics to answer. Men differ sharply in their histories of violence: some men kill someone at least once in their lives, whereas others go through their entire adult lives without so much as hitting anyone. Geneticists gather statistical evidence as to whether genetic differences between men in a population have affected, in some way or other, how violently they act. Genes might interact with environment to produce differences in levels of violence, even if none of these differences trace, on average, to genetic differences between men in the population. After exploring causal paths to violence that do not depend on genetic differences among current human beings. I consider ways in which current genetic differences might make a difference.

The chief causes of violence may well involve adaptive mechanisms working pretty much as "designed." Some of these mechanisms will constitute adaptations for violence versus nonviolence – that is, mechanisms the genetic plan for which was shaped by the ways their workings produced violence or not in an array of ancestral circumstances. Others will be mechanisms shaped by more general kinds of selection pressures. The psychic mechanisms whose workings constitute being enraged and attacking out of rage involve, we should expect, specific adaptations for violence in some circumstances and not in others.

A refined array of mechanisms adapted for violence versus nonviolence may also figure in the reactions of men acculturated to a "culture of honor": Nisbett and co-workers have compared whites from the American north and the American south (in roughly the eastern half of the country).[10] Cities and towns in Iowa and Oklahoma, matched demographically, have roughly the same murder rates except in one specific category: murders from confrontations where "honor" is at stake and will be lost if one wimpishly acquiesces in insult (Daly and Wilson 1988, chs. 6 and 10). Differences were marked even among college undergraduates at the University of Michigan: insulted northern young men were little affected by the insult, and uninsulted southerners were much like the northerners in various measures. The southerners, though, if they had been insulted and been unable to respond, showed heightened adrenaline. They were more confrontational: faced with a big man striding confidently past a table obstructing a hallway that the subject had to pass in the other direction, they backed down only in the last second – three feet from the table, on average, compared to nine feet for the other groups. They reported feeling that others would think badly of them.

Now we know little about the mechanisms that produced these reactions, but we can make a few remarks. Growing up in the south, a young man evidently responds to certain kinds of cues by permanently becoming "prickly." We can say, uninformatively, that this consists in some sort of setting or adjustment of psychic mechanisms – including, presumably, ones designed by evolution to respond to cues as to whether proneness to violent response to insult tends to be reproduction-enhancing. Subsequently, in the north and given the same immediate cues as a nonprickly young northern man, the prickly young man responds to a gratuitous insult as a crisis.

Consider now, in contrast, a set of likely mechanisms that could be important for violence, though not specifically adapted to cope with questions of when to be violent. Steele (1992), Boyd and Richerson (1985, ch. 8), and others have studied these mechanisms for identifying ways of life that are reproductively promising in one's own circumstances, given one's own characteristics and potentialities. This can manifest itself as finding an "identity" under which one can value oneself and to whose criteria of success one attaches one's self-esteem. Boyd and Richerson model this quest for an "identity" and the emulation of selected models as a genetic stratagem for discerning what life strategy is likely to pay off in reproduction. Life is too complex to analyze what makes for success, and so the stratagem is this: take someone who shows signs of success, along with signs that one could emulate his strategy with like success. Then do as he does, whether or not you can see how each feature of his life bears on success. This is the "reasoning" of the genes; they code for mechanisms that might manifest themselves in finding certain things "neat" or "worthy," the "way to be," or "the way to be for me."

A vast amount of the difference, then, between those with more or less violence in their histories, or more or less current proneness to violence, may well stem from adaptive mechanisms, working as designed, responding – with some considerable chaos and fortuitousness in the causal paths involved – to cues as to when violence will be reproductively promising and when it will not.

One way this can work is through adjustments of psychic mechanisms that other psychic mechanisms are adapted to make. Some such adjustments are hormonal. Hormonal differences are classed as "biological" by those who speak as if there were an interesting biological-nonbiological dichotomy in explanations of violence. It would be a grave confusion – though easy to fall into if one maintains this false biological-nonbiological dichotomy – to think that hormonal differences between people suggest differences in genes for those hormones. Hormones are biological, true enough, but in the only clear sense I can identify of the term, so is everything that happens in the brain. Hormonal differences among human beings need no more depend on genetic differences than need differences in clothing.

Take the hormone serotonin, for instance. Low levels are implicated in violence in some studies. Serotonin is heavily involved in dominance hierarchy behavior and in depression (Kramer 1993). We all share psychic mechanisms for negotiating dominance hierarchies to reproductive advantage. Evolutionary game theory can model the advantages of sometimes "knowing one's place" in a dominance hierarchy, and sometimes taking risks to rise above it. Cues as to how others treat one and how confrontations with them come out indicate one's place, and one's serotonin levels adjust accordingly to produce the right degree of assertion and the right degree of diffidence. Highly varied serotonin levels, then, can result from the workings of a psychic mechanism that all normal humans share (and that we share, to some degree, even with lobsters).

Violence tied to low levels of serotonin could well be biologically "normal." Normality consists in being born with genetic developmental contingency plans, including plans to develop mechanisms for assessing whether or not to be violent in various ways. These plans may include a plan to lower serotonin levels in some circumstances, and a plan to be sensitive to certain kinds of cues if one's serotonin is low and to respond violently. Such plans could be adaptive, in the sense that in our ancestral population those with plans more like the ones current now among humans tended to reproduce more.

That won't mean that such violence should be tolerated, or that it is in any way inevitable. We know that normal men in some environments live nonviolent lives, apart from bits of minor violence in childhood. We face the problem of how to attain a society where boys and men won't get cues that evoke bad kinds of violence. (The same applies to girls and women, but the problem of violence by women is far less pressing.)

We may also face dilemmas. In a world where bad kinds of violence are not ruled out, we need police and soldiers who are prepared to be violent in certain circumstances. Can we maintain a set of cues that prepares at least some men for the kinds of violence needed to defend a generally nonviolent way of life, without evoking other kinds of violence?

It might be argued that where there is genetic adaptation, there must be heritable differences. Adaptation, after all, is the work of natural selection, and genetic natural selection can work only with heritable differences. Natural selection, though, need not depend on levels of heritability that are large, especially when it comes to matters of life and death like propensities to violence. A selection pressure of 1 percent per generation in favor of an allele will make the allele come to 99.9 percent fixation in roughly 700 generations. Once a gene disappears, alternatives to the gene that has reached fixation might reappear by mutation, affecting the workings of the mechanism the gene codes for. Whether such mutations figure causally in any substantial proportion of current violence, though, is not a question one could answer on the basis of general evolutionary considerations.

I have so far discussed one way in which some men may act violently and others not, though their genotypes differ in no relevant way: psychic mechanisms they share may respond differently to differences in their circumstances. What other possibilities are there for differences in violence without current genetic differences? Normal genes can face novel environments, and can respond to novel environmental features in "unplanned" ways. We then have evolutionary "artifact" as opposed to adaptation.[11] We must be on the lookout to find ways in which such nonadaptive, evolutionarily novel responses figure in violence and its lack.

Deciding whether a response is a case of mechanisms "working as designed" may be difficult. The cues received will be novel in some ways, and also have features in common with cues that figured in the selection pressures that shaped the mechanism. Like things could be said of the response that the mechanisms produce. The question is how to "project" classifications of cues and responses in the ancestral environment to apply to the evolutionarily novel situations of today. (The existence of effective policing that one can call on even though one doesn't oneself participate is presumably evolutionarily novel, for instance.) This difficulty is not peculiar to biology, though; any science projects predicates from one set of cases to another.

In the case of violence, the distinction can probably be made in a rough but satisfactory way, and evolutionary artifact will not loom large. The consequences of violence for reproduction are huge: being violent at the wrong time or in the wrong way can put an end to one's re-productive prospects, as can failing, at the wrong time, to be violent. These selection pressures must have given rise to quite refined contin-gency plans for violence and its alternatives. Enough in current cues and responses bears analogy to things our ancestors encountered – and en-countered frequently enough for genetic contingency plans to make some provision for them – that the mechanisms can be said to be "work-ing as designed" when they produce these responses. (Whether these responses promote reproduction in current conditions is another ques-tion, and probably not a very interesting one for most purposes.)

My discussion of differential responses to environmental differences that are either designed into the human genetic plan, or matters of re-sponse outside the range of circumstances that shaped that genetic de-sign, includes plans for learning, modeling, and the development of emotional styles and long-term moods. But some ways we are as a re-sult of what happens to us are not a matter of genetic design. Bones are not designed to break, and we are not designed to lose blood supply to the brain or get infected with measles. Differences in the traumas, in-juries, and infections to which individuals are subjected make for dif-ferences that are not part of the human genetic contingency plan. These can affect behavior.

These assaults also evoke defensive responses that may have side ef-fects on behavior. Fever, for instance, is an adaptive response that helps fight off infection, but it can also injure. Behavioral mechanisms can be injured as side effects of adaptive responses to infection or trauma. En-vironmental differences make a difference to behavior that is not a mat-ter of learning or any other sort of genetic planning to make such a dif-ference. (If, as recently suggested, schizophrenia is associated with maternal measles, it could be a result either of infectious assault, or a side effect of adaptive responses to infection.)

This last kind of environmentally induced difference among people is an example of the working of pleiotropy – that genes can have effects other than those for which they were "designed" (Gould and Lewontin 1979; Dawkins 1982, 34). With a high-stakes matter like violence, selec-tion pressures are strong and pleiotropy minimized. In responses to trauma and disease, though, the stakes are likewise high and the genetic plan faces design limitations: victory over disease may be a close mat-ter, and mechanisms that make for behavior may be casualties.[12]

Thus, causes of differences in violence and proneness to violence may not, in the technical sense, be "heritable." (Of course, they might be found heritable in studies that were badly designed, or were designed with assumptions that turned out not to be true – say, that the environments of adopted twins are uncorrelated, if that were in fact not so.) Apart from this one feature that they can arise without current genetic differences in the population, they have little in common. Some are matters of "learning" and other adaptive responses to different histories. Some are adventitious. Some are a matter of trauma or infection, or side effects of defenses against trauma or infection. When a difference is nonheritable, the genetic design of the human psyche figures crucially in an adequate account of how the difference arises. Nonheritability by itself, though, indicates almost nothing about the form a correct account takes.

CURRENT GENETIC DIFFERENCES AND BEHAVIOR

Might there be "genes for violence" that differ from one man to another? That, of course, becomes an empirical question, once it is clear what a "gene for violence" is to mean. Methodological discussion cannot settle the question by itself. A finding that violence in a population is "heritable" would not show that there are "genes for violence." In one sense, there are sure to be "genes for violence"; there are genes that figure in genetic "plans" to be violent in some kinds of circumstances and not in others, even if the "heritability" of violence is zero in all current populations. These need not, though, be genes that come in different variants in different men, with some current variants causing a higher level of violence than others. One can have "plans" for violence versus nonviolence in the human genotype without any current "heritability" of violence. What, though, if violence is, to some degree, heritable in a current population. Does that mean that there are "genes for violence" that differ from person to person? Not in any sense that the words suggest. Some possibilities do not fit one's intuitive picture of how genetic difference between men could make differences in violence.

Take first *reactive heritability*. Steele (1992) discusses the *disidentification* of large numbers of "black" schoolboys from what they see as "white" school aspirations. The effects on the subsequent lives of some of these boys are in many cases terrible, and in the American population, if Steele is right in his diagnosis, there are large effects on such measures as IQ test scores (Steele 1992).[13] Suppose, then, we are all equipped with genetically "planned" mechanisms of identification and disidentification, adapted to respond to certain kinds of cues that went, in ancestral populations, with a way of life's being propitious or not for one's reproduction. These mechanisms won't specifically be mechanisms *for* violence or nonviolence. Still, identification with a violent ethos can lead to violence, and identification with a nonviolent ethos can lead away from it. One may identify with a violent way of life, dressing in the style of a thug, walking with the gait of a thug, talking in the accents of a thug, and engaging in thuggish violence. Or one may identify with an ethos that is not in itself violent, but that stymies opportunities for nonviolent success. Other psychic mechanisms that all men share may subsequently respond to special cues that one's way of life now makes likely. Then too, failure to identify with a way of life that can offer rewards may start one on a path to violent criminality by another path.

The story of young Malcolm Little (later Malcolm X) illustrates not a path to violence but a path to criminality. The "first major turning point" of his life, Malcom X tells us, came in eighth grade when a teacher told him, "You've got to be realistic about being a nigger. A lawyer – that's no realistic goal for a nigger" (Malcom X and Haley, 1964: 43). He advised carpentry. Malcolm's response fits Steele's picture: "It just kept treading around in my mind. . . . It was then that I began to change – inside. I drew away from white people. I came to class, and I answered when called upon. It became a physical strain simply to sit in Mr. Ostrowski's class" (44). Previously at the top of his class or close to it, he now searched for other models, moving to Boston and eventually joining the world of numbers runners and burglars. Little's own criminal career was not violent, but his progression illustrates how others might be led to violence. A cue portended that it would be futile to pursue the kind of rewarding, nonviolent model of life that had hitherto directed his efforts. In his agitation, he bitterly disidentified with his old aspirations and sought other models.

We have only hints of how the psychic mechanisms involved in such a story may work, but we can guess that they constitute highly refined adaptations, coded for in the genes. Adaptations of the kind I'm speculating about would constitute the "genes' way" of getting one to select, in the face of a blooming confusion of cues, a way of life with reproductive promise.

Of course in Malcom X's case, a genetic difference from others in his school was much in play. He had genes for a somewhat dark skin, and these genes had a profound indirect effect on his behavior. They weren't, though, "genes for criminality" in any normal sense. In another social setting they would have had nothing to do with a propensity to crime;

indeed, in other social circumstances, a similar complex causal chain might be triggered instead by genes for light skin. In Malcom X's case, the causal path went from distinctive genes for dark skin, to his having dark skin, to others' responding to him as a "nigger" with all that the term connoted in their minds. Others acted, then, in ways that served, in effect, as cues to Malcolm that a life devoted to study was futile. This led to psychic crisis and to the search for a different identity. The cues he faced, then, triggered adaptive mechanisms that all human beings share, but the cues were directed to him because of a genetically coded characteristic – skin color – that not all human beings share.

In the specific environment of present-day America, then, tendencies to an underclass model of life may be "heritable" in a technical, statistical sense, because of this kind of causal path alone. Skin color is heritable in the normal sense, and in our particular social circumstance, skin color tends to cause, along a devious psychosocial causal path, underclass behavior. Genes whose sole direct effect, though, is the production of dark pigment in the skin are not what anyone had in mind in looking for "genes for violence."

Nothing I have said rules out genetic differences with less devious and fortuitous causal paths to being prone to violence. But we should be cautious in developing expectations that such differences exist, or that such differences loom large among the causes of violence in countries like ours with high rates of homicide. A recent *Scientific American* report on research to identify such genes was full of predictions that within ten years, we should find a number of genes for violence, though none have been found yet (Gibbs 1995, 107).[14] Those who make such predictions, we should bear in mind, might be in the grip of a wrong picture. Some evidence does exist that genetic abnormalities can make a difference to violence, often in a restricted range of other circumstances, such as difficult childbirth or social conditions conducive to violence (Bruner et al. 1993; and Brennan, Mednick, and Mednick 1993). My aim is not to rule out such possibilities but to explore their place in a wider range of possibilities.

I spoke earlier of side effects of responses to trauma and infection (and other kinds of parasitization). These responses could produce differences in behavior without current genetic differences, and so in that case would not be "heritable." Recent theories suggest another kind of nondesigned, side-effect degradation of mechanisms of behavior that would be "heritable."

The function of sexual reproduction, the hypothesis is, is to produce genetic (and hence chemical) variety to stymie germs and other parasites. A gene that many of one's neighbors do not share may be adaptive in that it makes one harder to infect, but produces an inferior version of some mechanisms that depend on the gene. In this way genetic variation can have behavioral consequences, without the gene's being, in its entire average effect on balance, maladaptive or abnormal (Tooby and Cosmides 1990).

Another kind of adaptive difference in genes can be the setting of parameters, for hormones and the like. Skin pigment is an obvious example. Survival and vitality are reproduction-enhancing, other things equal, and when there is much sun, dark pigment promotes survival and vitality, whereas with little sun, survival and vitality are promoted by having low levels of skin pigment. One might imagine that the same kind of pattern could hold for kinds of behavior.

Might such effects be substantial, though? There are grounds for thinking not. The pattern requires genetic selection of different characteristics in different environments, as opposed to contingency plans for the same individual's having one set of characteristics if in one kind of environment and another set in another. This requires special conditions. The difference in which characteristic is more advantageous must last, on average, over many generations. Climate can make such stable differences, and accounts for such things as differences in skin color. With violence, relevant differences of life circumstance are likely to be much more volatile. When violence "pays" reproductively and when it does not will depend chiefly on characteristics of one's society and one's position in it. Whether social differences will be stable enough to produce different selection pressures in different lineages over large numbers of generations is doubtful.

It is amid all these other possibilities that we should see the chief focus of much research on the "genetic basis" of violence: that maladaptive mutations may degrade behavioral mechanisms, so that they do not work "as designed." Even here, though, we must distinguish many causal paths by which a degradation could lead to violence. All violence is at least in part the work of adaptive mechanisms; violence requires, for instance, an intention to injure, and the workings of intention are the workings of adaptive mechanisms.

The paths to violence that are "provided for" by the human genetic plan will be extremely complex. That may have something to do with the size of the human brain in which these plans for mechanisms are realized, a size that kills mothers and orphans babies in childbirth, and so has been achieved in the face of great selection pressures against it. Degradations may operate at many points in a complex array of paths.

The degradation caused by a maladaptive allele might be directly a defect of the mechanisms for controlling whether to be violent. But less direct paths are plausible – indeed, perhaps more plausible. Mechanisms for assessing violence may work "as designed" on cues evoked by the malfunctioning of other mechanisms – the responses of others to physical blemish, for instance, or to damaged mechanisms for sociability. Or mechanisms for assessing violence may work "as designed" in many respects, but mechanisms of chemical signaling – through serotonin, testosterone, and the like – may be damaged.

THE UPSHOT?

The aim of this survey of various causal paths that can lead to differences in violence has been in part to impress us with the large range of possibilities. I classified these causal paths by whether they would make violence "heritable," in the technical, statistical sense. The question of heritability, though, turns out to cut across many other distinctions in the ways genes can interact with violence. Well-known statistical methods can, if their background assumptions can be established, discover that certain effects are "heritable" or not. This by itself, though, tells us very little about the etiology of differences in proneness to violence, and very little about what genes have to do with violence. If violence is "heritable" in certain populations, that need not mean that there are "genes for violence," and if violence is nonheritable in current populations, that does not mean that genes are irrelevant to violence.

Suppose, moreover, that a "gene for violence" were found. Suppose, that is to say, that a genetic abnormality in certain men was found to constitute a defect in their genes' "plans" specifically for when to be violent and when not, and that the impact of this abnormality on a man's chances of being violent, in some set of current environments, was substantial. It would not follow that any substantial proportion of the violence in one's society traces to that defect – or, indeed, to genetic defects of all kinds combined. That a condition makes a substantial difference to violence in the rare cases when it is present does not entail that it is the cause of any substantial proportion of the violence in the population. If the condition is rare and violence is far from rare in a group, then by simple arithmetic the condition *cannot* be the cause of the bulk of that violence.[15]

Going from studies in one group to conclusions for another brings more pitfalls. Suppose it were established that in Denmark genetic defects with a direct relevance to violence accounted for a nontrivial proportion of some index of violence in that country.[16] Would that suggest anything about major causes of violence in America? If the index of violence in Denmark is far lower than a like index in America, it would not. Homicide, for instance, figures in a variety of syndromes, with different kinds of causal histories. The proportion of homicides that are of a given pattern varies greatly from country to country, from region to region, and from group to group within a region. Only a small proportion of the homicides in America fit any of the patterns that account for most of the homicides in a country like Denmark, with its far lower homicide rate. If anything about homicides in Denmark transfers to America, it may well apply only to that small proportion of homicides in America that resemble the few homicides in Denmark. Even a predominant cause of rare murders in one country need not be a substantial cause of frequent murders in another (Daly and Wilson 1988, 284–286).

In this chapter, I have not discussed the many difficulties of arriving at well-founded estimates of "heritability" for kinds of violence, or whether announced research findings of "heritability" are likely to be reliable. Instead, I have talked of why current "heritability" may be close to irrelevant to the causes of current violence. Let a syndrome connected to violence be found "heritable" or not, in a well-crafted study whose findings bear credence. Even then, we have no more than a start to learning the etiology of this violence. Either finding – of heritability to some degree or of nonheritability – is consistent with a variety of importantly different causal paths.

Violent death is a serious and tragic problem in the United States, as it is in many other countries. Do genetic studies hold out good promise to help us understand what afflicts us? This question cannot be answered on the basis of the kind of quick survey I have presented: I have laid out a number of ways in which genes might bear on violence, and for a few of those, "heritability" matters. This leaves open the question of whether the patterns for which "heritability" matters account for any great proportion of American violence.

We know that there are big differences in violence between groups. Differences in rates of genetic defects are very unlikely to account for these differences; human groups are genetically highly similar. That leaves open the possibility that within certain groups, genetic defects make a substantial difference to levels of violence. We need to ask, though, whether this possibility obtains for groups with high rates of violence. I am not aware of any indication that it does. It is with these groups that violence is by far the most serious problem, and so whether current heritability matters for violence is chiefly a question of whether it matters in these groups.

Heritability may well be irrelevant to the bulk of violence in such places as America. Understanding American violence may be almost entirely a matter of understanding differences that either are not heritable in populations with high rates of violence, or are heritable there only adventitiously. An important focus in the study of violence should indeed be on genes: we need to understand the provisions the normal human genetic "plan" makes for violence and nonviolence. We need to understand such psychic states as rage, hatred, and humiliation, their causes and their effects. We need to ask what the psychic mechanisms are likely to be that bear on when men will be violent and when not. It is much less clear that the statistics of "heritability" will tell us about any large proportion of the violence in the world or in our own country.

NOTES

1. See, for instance, Hirschfeld and Gelman 1994 for a fine set of articles on domain-specific mechanisms.
2. See Barkow, Cosmides, and Tooby 1992 for a recent collection of papers representing this program. Symons coined the term "Darwinian psychology" for this program, and two of his essays (1987; 1992) supplement the findings in his book (1979).
3. Gould and Lewontin 1979 is the font of many of these charges; see Dennett 1995 for a discussion.
4. Symons starts his book (1979) with a methodological discussion of findings of adaptation.
5. Kitcher at some points challenges a preferred sociobiological explanation by asking whether it tells us more than we could learn from some commonsense lore about such things as "greed" and desires for wealth, power, and prestige (1985, 286–288).
6. Daly and Wilson (1988, ch. 9) treat spousal homicide, and this example is drawn from a man's account of why he killed his wife (212).
7. See Gibbard 1990, ch. 6, and Frank 1988 for considerations of possible functions of some social emotions.
8. An even more accessible treatment of their program appears in the *New Yorker*; see Wright 1995.
9. Tooby and Cosmides (1990; 1992) on personality psychology.
10. Nisbett and Cohen in press. The point of restricting the study to whites is not, of course, that whites and blacks will respond differently, over their lifetimes, to the same sequence of cues – that seems quite unlikely. The point is

that in America, north and south, being regarded as "black" is likely to result in quite different treatment from being regarded as "white," so that the two groups may be presented, in the course of growing up, with relevantly different sequences of cues. See Steele 1992. In the discussion, I do not continue mentioning the restriction to whites.

11. Symons (1979) begins his book by arguing that female orgasm is an evolutionary artifact.

12. See Nesse and Williams 1995 on evolutionary factors in disease.

13. I attempt to put Steele's findings in an evolutionary framework in Gibbard 1993.

14. The grounds offered are, roughly, that small numbers commit most crimes, and those who start youngest are most likely to become chronic offenders.

15. Bruner et al. (1993) report on a Dutch family in which five males who share a point mutation have "a complete and selective deficiency of enzymatic activity of monoamine oxidase A" (578) and are "characterized by borderline mental retardation and a tendency toward aggressive outbursts" (579). The frequency of the syndrome in the population had not been determined, and so we know neither in what proportion of the low rate of Dutch violence such a disorder figures, nor in what proportion of the much higher rate of American violence it figures. The report is consistent with this and similar disorders' accounting for little even of the violence in the Netherlands, much less of the violence in America.

16. Brennan, Mednick, and Mednick (1993, 245–246) conclude from a study of adoptions in Denmark that biological sons of one group – "criminally recidivistic parents . . . who had *also* been hospitalized one or more times for a psychiatric condition" – were roughly three times as likely to have been arrested for violent crimes as were either (i) biological sons of noncriminal parents or (ii) biological sons of recidivistic, nonpsychiatric parents. It appears that eight sons of recidivistic psychiatric parents had been arrested for violent crimes, out of ninety-six men arrested for such crimes in the groups reported. Because fewer than 40 percent of male adoptees in the cohort were in the groups reported, I am not certain of the adopted men in the cohort who were arrested for violence, what percentage were in this special, higher likelihood group, but it appears to be in the range 3 to 8 percent.

REFERENCES

Barkow, Jerome H., Cosmides, L., and Tooby, J. (1992). *The adapted mind: Evolutionary psychology and the generation of culture.* New York: Oxford University Press.

Boyd, R., and Richerson, P. J. (1985). *Culture and the evolutionary process.* Chicago: University of Chicago Press.

Brennan, P., Mednick, B. R., and Mednick, S. A. (1993). Parental psychopathology, congenital factors, and violence. In *Mental disorder and crime,* ed. S. Hodgins, 244–261. Newbury Park, Calif.: Sage.

Bruner, H. G., Nelen, M., Breakefield, X. O., Ropers, H. H., and van Oost, B. A.

(1993). Abnormal behavior associated with a point mutation in the structural gene for monoamine oxidase A. *Science* 262: 578–580.

Buss, D. (1994). *The evolution of desire: Strategies of human mating.* New York: Basic Books.

Daly, M., and Wilson, M. (1988). *Homicide.* New York: Aldine.

Dawkins, R. (1982). *The extended phenotype.* San Francisco: Freeman.

Dennett, D. (1995). *Darwin's dangerous idea: Evolution and the meaning of life.* New York: Simon and Schuster.

Frank, R. H. (1988). *Passions within reason: The strategic role of the emotions.* London: W. W. Norton.

Gibbard, A. (1990). *Wise choices, apt feelings: A theory of normative judgment.* Oxford: Oxford University Press.

(1993). Sociobiology. In *A companion to contemporary political philosophy,* ed. R. E. Gooden and P. Pettit, 597–610. Oxford: Basil Blackwell.

Gibbs, W. W. (1995). Seeking the criminal element. *Scientific American* 272 (3): 100–107.

Gould, S. J., and Lewontin, R. (1979). The spandrels of San Marco and the Panglossian paradigm: A critique of the adaptationist program. *Proceedings of the Royal Society of London* B 205: 581–598. Reprinted in *Conceptual issues in evolutionary biology,* ed. E. Sober, 252–270. (Cambridge, Mass.: MIT Press, 1984).

Hamilton, W. D. (1964). The genetic evolution of social behavior. *Journal of Theoretical Biology* 7: 1–52.

Hirschfeld, L. A., and Gelman, S. A. (Eds). (1994). *Mapping the mind: Domain specificity in cognition and culture.* Cambridge: Cambridge University Press.

Kitcher, P. (1985). *Vaulting ambition: Sociobiology and the quest for human nature.* Cambridge, Mass.: MIT Press.

Kramer, P. D. (1993). *Listening to Prozac.* New York: Viking.

Nesse, R., and Williams, G. (1994). *Why we get sick: The new science of Darwinian medicine.* New York: Random House.

Malcom X and Haley, A. (1964). *The autobiography of Malcom X.* New York: Ballantine, 1992.

Nisbett, R. and Ross, L. (1980). *Human inference: Strategies and shortcomings of social judgment.* Englewood Cliffs, N.J.: Prentice-Hall.

Nisbett, R., and Cohen, D. (In press). *Culture of honor: The psychology of violence in the south.* Boulder, Colo.: Westview Press.

Plomin, R., Owen, M. J., and McGuffin, P. (1994). The genetic basis of complex human behaviors. *Science* 264: 1733–1739.

Steele, C. M. (1992). Race and the schooling of black Americans. *Atlantic* 269 (4): 68–78.

Symons, D. (1979). *The evolution of human sexuality.* Oxford: Oxford University Press.

(1987). If we're all Darwinians, what's the fuss about? In *Sociobiology and psychology: Ideas, issues, and applications,* ed. C. B. Crawford, M. F. Smith, and D. L. Krebs. Hillsdale, N.J.: Erlbaum.

(1992). On the use and misuse of Darwinism in the study of human behavior. In Barkow, Cosmides, and Tooby 1992, 137–169.

Tooby, J., and Cosmides, L. (1990). On the universality of human nature and the uniqueness of the individual. *Journal of Personality* 58: 17–67.

——— (1992). The psychological foundations of culture. In Barkow, Cosmides, and Tooby (1992), 19–136.

Williams, G. C. (1966). *Adaptation and natural selection: A critique of some current evolutionary thought.* Princeton, N.J.: Princeton University Press.

Wright, R. (1995). The biology of violence. *New Yorker*, March 13, 68–77.

PART II

Chapter 8

Crime, Genes, and Responsibility

MARCIA BARON

As ethicists and metaphysicians we have been asked to discuss freedom, responsibility, and desert in connection with the possibility that a genetic marker will be discovered for "criminality" or aggressiveness. Some have thought that if there was a marker, then individuals with the marker could not be held responsible for whatever criminal or aggressive behavior they might engage in. Would we, they worry, be morally justified in punishing a murderer or a rapist if this person's genes predisposed him toward violent behavior?

Before answering this question, we should note a problem in its formulation. So far I have used as if they were interchangeable all of the following: "criminality," "aggressiveness," and "violent behavior." Of course, they are not interchangeable terms, but the problem is that we don't know what a genetic marker, if found, would mark. Toward what would a person with the marker be predisposed? Anger? Or calculated, premeditated violence? Some sort of generic criminality (whatever that might be)? Or a sort of generic aggressiveness – a tendency to try to get one's way by force rather than by asking or negotiating? Or something more specific: a tendency to beat "loved" ones when they seem to be getting too independent? Or something specific and criminal but nonviolent: tax evasion or embezzlement or shoplifting?

The fact that we don't know what a marker, if discovered, would

I would like to thank Ted McNair and Fred Schmitt for helpful discussion of the issues, and Judith Short for secretarial assistance. In revising my paper I have benefited from written comments by P. S. Greenspan, Kit Kinports, Alan Strudler, and David Wasserman, and from the papers written for the April 1995 meeting of the working group on genetic explanation and responsibility ascription.

mark, generally goes unnoticed, and I think it is of considerable importance. Not knowing what a marker would mark, we put some face on it. But we do so without realizing it. And the face we put on it then affects our view of the meaning and importance of research into a link between genes and crime. When someone speaks of crime without giving any indication of what sort of crime is meant, most of us don't think of embezzling or writing bad checks. We picture violent crime of one sort or another. And indeed it seems that what we think we are looking for is a tendency toward violence of one sort or another.[1] (It is striking, then, to read the much touted twin and adoption studies and to discover that to the extent that they support a genetic link to crime, it is primarily just to minor property offenses, not to violent crime.)[2] Having emphasized the importance of bearing in mind that we don't know what a genetic marker, if found, would mark, I sometimes, to avoid cumbersomeness, use the usual terms as if they were interchangeable, despite the fact that if we are not on our toes, doing so may mislead us into supposing that we do know what it would mark.

So, to ask the question again: would we be morally justified in punishing a murderer or a rapist if this person's genes predisposed him toward violent behavior?

The answer is, I think, fairly straightforward. Suppose there were a genetic factor in crime and a test that disclosed whether the individual had the marker. The individual who tested positive would simply be genetically *predisposed* toward such behavior. He or she would, as philosophers sometimes say, be *inclined but not necessitated* to act violently. The person's position would be the same as that of people whose environment – including, especially, their past environment and, in particular, their childhoods – significantly increased the chance that they would become violent offenders.[3] Both the person with the genetic marker and the person whose environmental factors significantly increase the chance that he or she will become a criminal are likely to have a harder time than they otherwise would (and than other people have) leading a life in which they do not act violently. Neither is "caused" to commit violent crimes; it is simply harder for them than it is for others not to commit violent crimes. If we do not regard the individual whose upbringing significantly increased the chance that he would become a violent offender as therefore not responsible for his crimes, I see no basis for thinking that someone with a genetic marker for criminality was not responsible for crimes that he committed.

Why might there be a temptation not to hold someone responsible for a crime if we knew that she was genetically predisposed to commit crimes of this sort? The reason, I think, is that we would judge her to be less free to refrain from committing such crimes than the rest of us are. But if she is less free, so is the person who is "environmentally predisposed" to commit such crimes (and, for that matter, so is the person whose nongenetic medical condition predisposed her to commit such crimes). So again, I don't see that the possibility that a genetic marker for criminality will be discovered and readily tested for raises new problems concerning freedom and responsibility.

The same point holds regarding desert: the fact that the person is genetically predisposed to commit violent crimes should be a mitigating factor just to the extent that the fact that environmental factors dispose the person to do so is a mitigating factor. And so my tentative conclusion to the question of responsibility, desert, and freedom is this: the possibility of genetic markers for aggressiveness or criminality or violent criminality does not pose any new problems regarding responsibility, etc. that are not already posed by environmental (and other familiar) factors.

Now, a complication. I have said that a genetic predisposition to act in certain ways – or to suffer certain behavioral problems – would only incline the person to such behavior, rather than necessitate it. There are different ways of being inclined, however, and some ways may provide reasons for considering the person so minimally free to refrain from the objectionable behavior that to hold her fully responsible for her conduct is not justifiable.[4] If someone has a condition that drastically reduces her willpower, so that no matter how firmly she believes that arson is wrong, and no matter how fervently she wishes never to set fires, and no matter how many precautions she takes – keeping no matches or lighters in her house, for instance – and despite seeking counseling or taking special classes for pyromaniacs, she nonetheless commits arson, this condition surely calls for our sympathy and may be grounds for not holding her fully responsible for her actions. It certainly should be a mitigating factor, a reason to give her a more lenient sentence, if in fact we do hold her responsible and do convict her.

Now, if certain genetic influences were to affect us much more dramatically than any other influences, predisposing us to violent, criminal behavior more, or in a significantly different way than other influences do, this would of course be noteworthy. I don't see that the research points in this direction, however; and in any case, the point remains that the fact that the influence is *genetic* is itself of no importance. Moreover, the primary reason for thinking that in a case such as the one I describe the agent deserves a lighter sentence than usual (and arguably should not be held responsible at all) is that she tried so hard not to commit arson (not that it was so hard for her to refrain from doing so).

Consider the following three examples. Imagine someone – let's call him A – who feels a strong sexual attraction to children. He abhors this desire in himself, tries to extinguish it, tries to avoid situations in which the temptation will be strong and where the opportunity to satisfy the desire will be present. Despite his efforts, he sexually molests children. Now imagine B. What was just said of A is true of B, as well. But unlike A, B has a genetic abnormality which has been correlated with impulsivity. In the case of B we have a bit more of an explanation of why it is that B fails to refrain from molesting children. But I don't see that we have any more reason not to hold B responsible for his acts of child molesting than not to hold A responsible, or any more reason for leniency in sentencing B than for leniency in sentencing A (if we do hold them responsible and do convict them). Finally, consider C, who like B has a genetic abnormality that is correlated with impulsivity and, like both B and A, has sexually molested children. But in the case of C, we understand the mechanism at work and see that the underlying problem is that, as Greenspan puts it, C has inadequate resources for behavioral control. There is "an absence of 'enabling' causes of normal control such as [a] supply of serotonin and other electrotransmitters." (I assume for simplicity that no treatment for the defect was available at the time the defendant committed the offenses.) Here there may be some temptation to think that C deserves greater sympathy, perhaps, but I think that is only because we are more certain that C is really doing his best and just cannot control his untoward impulses.[5] If we fully accept that A and B are doing all they can, is there any reason to hold them more responsible than C? I don't think so.[6]

My more general point is that (with an exception I'll note in a moment) it does not matter what the source of the problem is (the problem being that it is unusually difficult for the person to do what he firmly believes he should do, and fervently wants to do). Whether it is genetic or not does not matter. The one exception is that if the source of the problem is something that he did, we *may* (reasonably) feel less sympathetic toward A than toward B or C. We *may*; it will depend on what he did and why.

Still, some might feel that, although they are not sure why, a gene linked to criminality would render the person *less free* to refrain from committing crimes of the sort to which he is disposed than would environmental factors. Is there reason to believe that a genetic factor limits one's freedom more than environmental factors? Are there (and the answer to this question might possibly be different, though it probably would not be) reasons for treating a genetic predisposition toward committing (certain) crimes as a greater, or more genuinely, mitigating factor than an environmental predisposition? I very much doubt it (in each case); but I can think of reasons why some might think otherwise. Genes sound like the sorts of things that determine us completely. If a trait has a genetic basis, we may erroneously assume that, in some interestingly robust sense, we have to be that way. In a sense it is true: if we are genetically predisposed to x, we're disposed to it. My guess is that many think that any trait with a genetic basis is rather like having blue eyes; it is something determined by one's genes, and one cannot do anything about it. (Of course one can wear tinted contact lenses that alter the way one's eyes *look*. But still one's eyes are blue.) By contrast, we've all been taught that no matter how impoverished one's background, any American can, if he or she only tries hard enough, become wealthy, famous, and maybe even a U.S. president. Thus, environmental factors are seen by many to be more superficial, more surmountable, then genetic factors.

Obviously, one's political leanings will influence how one thinks about these matters. Those who accept the idea that any American can become wealthy, no matter how dire the poverty into which she was born and how grave the injustices that her parents and grandparents suffered, will see environmental factors to "incline" one less than will those who reject it. But this disagreement does not run very deep. For the point remains that genetic predispositions only predispose; they don't necessitate. Thus, while acknowledging political differences, we can all agree that with respect to conduct (as opposed to eye color) environmental factors and genetic factors alike incline but do not necessitate.

There is, however, a somewhat stronger reason why jurors might be more sympathetic to an unfree-to-do-otherwise argument in the case of someone who is genetically predisposed to commit crimes than they are to the same argument used in the case of someone whose childhood environment was marked by violence, sexual abuse, lack of love, dire poverty, hunger, terrible schools, and so on. We have *medical proof* in the first case that the person is genetically predisposed to commit crimes – that is, assuming a scenario in which a marker had been found and where medical tests would show decisively whether the individual had the marker. But there's no "proof," a lawyer might convince jurors, that the murderer with the terrible childhood was *really* predisposed to commit murders. There's something irrefutable in the first case, and not in the second. To put it differently: "Anyone," it might be claimed (with some exaggeration), "can be said to have had so lousy a childhood that it marked him for life; but not everyone can be said to have the genetic marker. There's a test for the genetic marker, and no one will say under oath that the person has been tested and found to have the marker unless it's true."

The difference seems to be one of evidence – of how convincing the claim is – rather than that environmental factors limit one's freedom less than a genetic marker for criminality does. If this is right, there is no reason to think that those with the genetic marker are less responsible for their crimes than those who lack the marker but were disposed by environmental factors over which they had no control – the socioeconomic and familial conditions into which they were born – to commit crimes that they indeed did commit. There is, in the scenario we are imagining, merely more room for doubt about whether the individual who has committed a crime was "environmentally disposed" than there is about whether such a person was genetically disposed. Even this is not altogether clear; however. For the fact that we do not know what the marker marks bears on the issue. If it marks a tendency to unaggravated property offenses, do we say that the marker made someone less able to refrain from the crime of armed robbery that she committed? It may be difficult to judge how much the marker disposes her to. If this is the case, the medical testimony may in fact leave as much room for doubt as the social worker's or the psychologist's testimony that her early childhood marked her for life, leaving her less able than most of us to refrain from committing crimes.[7]

How have defenses claiming that the defendant should not be held responsible for his crime because "he couldn't help it" fared in the courts? Although my treatment of this huge topic has to be somewhat superficial, it is helpful to have some idea of the reactions of judges, juries, and state legislators to such defenses. Their reactions provide some indication as to how a "genetic predisposition" defense might fare.[8]

One defense for which a proven genetic predisposition to violence or impulsivity might be relevant is the insanity defense. The heart of the insanity defense is the M'Naghten test, which requires that "at the time of the committing of the act, the party accused was labouring under such a defect of reason, from disease of the mind, as not to know the nature and quality of the act he was doing; or, if he did know it . . . did not know he was doing what was wrong."[9] By the middle of the twentieth century the test was widely held to define legal insanity too narrowly. The problem, as Circuit Judge Kaufman wrote in *United States v. Freeman*, is that it does not permit the jury to identify those who can "distinguish between good and evil but who cannot control their behavior." To address this defect, the Model Penal Code (MPC) provides (section 4.01) that "a person is not responsible for criminal conduct if at the time of such conduct as a result of mental disease or defect he lacks substantial capacity either to appreciate the criminality [wrongfulness] of his conduct or to conform his conduct to the requirements of law."[10]

The relevance of the MPC provision for our topic is clear. If I assume that what the genetic marker, if discovered, would mark would not be an incapacity (or vastly diminished capacity) to distinguish between good and evil (or some related cognitive incapacity), the genetic marker defense would be relevant to the insanity plea only if the insanity plea includes a volitional component, as does the MPC provision. Someone with a genetic defect that severely reduced his or her capacity for self-control might meet the MPC definition of insanity (though see my subsequent discussion for reasons why this might not be the case) but would not satisfy the M'Naghten test.

How widely accepted is the inclusion of a volitional component? If we were asking this question twenty years ago, in the late 1970's, the answer would have been "very." The MPC definition of insanity was adopted by ten of the eleven federal courts of appeal and more than half of the states in the two decades after its official promulgation in 1962. The trend underwent a reversal after John Hinckley was found not guilty of his attempted assassination of then president Ronald Reagan by reason of insanity.[11] In the ensuing three years, two-thirds of the states took steps to limit the insanity defense. Some narrowed the definition of insanity; others abolished the insanity defense altogether. Very frequently what was jettisoned was the volitional component – crucial if a genetic predisposition to impulsiveness were to be relevant.[12] At present, the M'Naghten test is the majority rule. The trend away from including a volitional component and the fact that the insanity defense is controversial make it somewhat unlikely that a genetic marker defense would succeed under the rubric of an insanity defense.[13]

How extreme must the incapacity, or reduced capacity, be, according to the MPC? The MPC says that to qualify as insane for the purposes of the insanity defense, the defendant must *"lack substantial capacity* either to appreciate the criminality . . . of his conduct or to conform his conduct to the requirements of law"; but what does this amount to? In their comments on the code, the authors explain that by "substantial capacity" they mean "a capacity of some appreciable magnitude when measured by the standard of humanity in general, as opposed to the reduction of capacity to the vagrant and trivial dimensions characteristic of the most severe afflictions of the mind." A "test requiring an utter incapacity for self-control" was rejected, and so was a test that was at that time employed by military courts – namely, would the presence of a policeman at the defendant's elbow have caused him to refrain from committing the offense? If so, then the defendant did not meet the test for volitional incapacity. The MPC explicitly rejects that test as too restrictive an understanding of what a "lack of sufficient capacity . . . to conform" one's conduct to the requirements of the law amounts to. The authors explain: "Whether the defendant could have controlled himself in an imaginary situation of automatic and instantaneous punishment is a very different question from whether he could have refrained from criminal conduct in the situation in which he actually found himself. For defendants who are in fact incapable of conforming their conduct to the law's requirements, the normal deterrent function of criminal punishment is inoperative, and posing the test of responsibility in terms of hypothetical conduct under other circumstances requires speculation that may be of limited relevance to the real issue" (*Model Penal Code and Commentaries* 1985, 171–172).[14]

It is instructive to consider how a different genetic abnormality defense fared in the brief period when the abnormality was (erroneously, as it turned out) thought to be causally linked to criminality. I am referring to the view, popular and highly publicized in the late 1960s and early 1970s, that someone with XYY chromosomes, rather than the usual XX or XY, was innately aggressive and predisposed to be a violent criminal. (It is interesting, and sobering, to note how widespread this belief was.

Genetic links to virtually anything – but especially criminality and intelligence – are widely publicized when some evidence for them emerges; evidence against such links is, by contrast, just not newsworthy. Even as a child, I heard about the XYY syndrome; not until I was pregnant, some twenty years later, and educating myself about chromosomal abnormalities in order to decide whether to opt for genetic testing, did I learn that the link between XYYs and aggression had been debunked.)[15]

Before it was debunked, several courts heard arguments that appealed to the fact that the defendant was an XYY individual. In no instance did the defense succeed. In *Millard v. State*, a 1970 Maryland case, the appellate court rejected the insanity defense, holding that "to simply state that persons having the extra Y-chromosome are prone to aggressiveness, are antisocial, and continually run afoul of the criminal laws, is hardly sufficient to rebut the presumption of sanity and show the requisite lack of 'substantial capacity' under the statute." In 1970 a California appellate court upheld the murder conviction in *People v. Tanner*, arguing (among other things) that "the evidence did not suggest that all XYY individuals are aggressive by nature." Both are arguments we can expect to hear if a genetic marker for criminality, aggressive behavior, or lack of self-control is discovered and if an insanity defense is predicated on evidence that the defendant has the marker. It is a large step from showing that the defendant has a genetic marker to rebutting the presumption of sanity (even in jurisdictions that recognize a volitional, and not only a cognitive, test for legal insanity); and because not all who have the marker will display the problematic tendency in question, the link is unlikely to be seen as sufficiently tight to establish a strong defense.

In one respect, however, a genetic marker defense might prove much stronger than the XYY defense: in the case of the XYY defense, even before the much-publicized studies had been discredited, all that was (supposedly) established was a correlation. If a genetic basis for (certain types of) criminality is found that actually explains the connection between the genetic abnormality and the undesirable behavior, courts might look more positively on such a defense (provided that a huge proportion of those with the abnormality display the behavior). Thus, in *State v. Roberts*, a Washington appellate court held: "There does in fact seem to be an incidence of the XYY defect among institutionalized men. However, the behavioral impact of this chromosomal defect has not been precisely determined. . . . Presently available medical evidence is unable to establish a reasonably certain causal connection between the

XYY defect and criminal conduct." If, as seems likely, a causal story rather than a mere correlation will be provided – for example, a deficit of serotonin leaves the agent with inadequate resources for behavioral control – courts might be more willing to consider the contemplated genetic abnormality as strong evidence in support of an insanity defense than they were in the 1970s, when the abnormality was the extra Y-chromosome.

But here again, the same questions will arise: what proportion of those with the abnormality commit violent crimes? Unless the proportion is very high indeed, why think that this person was unable to control himself, when others with the same genetic marker were not so disabled? And what about this person's other behavior? If we see self-control (of the sort alleged to have been unavailable to the defendant at the time) exercised in other arenas or at other times, why think that his genetic abnormality causes him to lack "substantial capacity to conform his conduct to the requirements of law" (in the language of the MPC)? Courts will want to be convinced that the genetic abnormality does not just make it harder for this individual to refrain from (certain) crime or, more generally, to exercise self-control, but that it makes it so very difficult that there is no basis for holding him responsible for his conduct.

Even if a genetic marker for violence or criminality does not, in the eyes of the courts, negate responsibility, it might nonetheless serve as strong evidence for a partial excuse. Two partial excuses are relevant here: "diminished capacity" and "heat of passion." Unlike the insanity defense, which, if successful, issues in a verdict of "not guilty by reason of insanity," the other defenses are partial defenses, seeking to show not that the defendant is free of culpability, but rather is guilty only of a lesser offense than that with which she or he would otherwise be charged: voluntary manslaughter rather than first- or second-degree murder, in the case of heat of passion, and second-degree murder instead of first-degree murder (and occasionally voluntary manslaughter rather than murder) in the case of diminished capacity.

A diminished-capacity defense aims to show that the defendant has a diminished capacity for premeditation, which usually is required for a conviction of first-degree murder,[16] or for malice, required for first- and second-degree murder under common law.[17] The defense is allowed in very few jurisdictions. A notorious case in which it was successful was that of Dan White, who killed both George Moscone, the mayor of San Francisco, and Harvey Milk, a fellow member of the Board of Supervisors and a gay activist. Public outrage over the verdict – acquittal of first- degree murder, conviction only of voluntary manslaughter – led to an alteration of the California Penal Code, which now provides that "evidence concerning an accused person's intoxication, trauma, mental illness, disease, or defect shall not be admissible to show or negate capacity to form the particular purpose, intent, motive, malice aforethought, knowledge, or other mental state required for the commission of the crime charged" (Saltzburg et. al 1994: 303–304). (In White's trial the defense had introduced evidence to show that because of the severe pressure he was under and because he suffered from depression, he lacked the capacity to harbor malice.)

Although this defense is now permitted in very few states, one can still introduce evidence that the defendant lacked the mental state required for conviction of the particular crime.[18] So although a diminished capacity for self-control, such as might be had by someone with a genetic marker for violence or impulsivity, would be a defense in very few jurisdictions, *if* the incapacity were extreme enough to prove that the defendant could not possibly have premeditated the act, this would be strong evidence that the elements necessary for first-degree murder were lacking. It seems very unlikely that a genetic marker would mark so extreme an incapacity, and if it did, some evidence other than the marker would surely be needed to show the extreme incapacity. It is more plausible that a diminished capacity might be shown, in those few jurisdictions that allow a diminished-capacity defense.

Far more common than the defenses discussed earlier is the heat-of-passion defense, by which murder can be mitigated to voluntary manslaughter if the defendant's reason was, at the time of the act, "disturbed or obscured by passion to an extent which might render ordinary men, of fair average disposition, liable to act rashly or without due deliberation or reflection, and from passion, rather than judgment."[19] The key idea is that some provocation roused the defendant's passion to the extent that he or she lost self-control, and the provocation is such that the reasonable person would also be provoked by it and would probably act rashly.[20] How is it to be determined whether the provocation is sufficient or reasonable? Not just any anger or frustration will do. And are we to take into account that some people "lose it" a lot more easily than others? Is the standard a generic reasonable person, or could it be a reasonable impulsive person – someone who, measured against the backdrop of impulsive persons, would qualify as reasonable? Could the fact that a defendant has a serotonin deficit that makes him much less able to exert self-control than are most others be taken into consideration in judging, not only whether he was provoked, but also whether the provocation was reasonable?[21] To put it differently, can we ask whether a reasonable person, hampered by this defect, would have been provoked to act rashly?[22]

The consensus in legal thought is that we are not to take such defects into consideration. If we were, then a genetic predisposition toward impulsivity might well be relevant in arguing that the defendant acted in the heat of passion. Not that a disposition to impulsivity would by itself turn any homicide that would otherwise be murder into voluntary manslaughter; there has to be something – some event, or possibly a series of events – that provoked the agent. But arguably a genetic predisposition to impulsivity could bear on whether this agent was adequately provoked and on whether a reasonable person – *if* we read that to mean "a reasonable person with this predisposition" – would be similarly provoked. The same event that would not be adequate provocation for a "normal" reasonable person could count as adequate provocation for someone with a genetic predisposition to impulsivity.

There is some room to argue that it should be so understood. In *Maher v. People*, Judge Christiancy wrote: "In determining whether the provocation is sufficient or reasonable, *ordinary human nature*, or the average of men recognized as men of fair average mind and disposition, should be taken as the standard – *unless, indeed, the person whose guilt is in question be shown to have some peculiar weakness of mind or infirmity of temper, not arising from wickedness of heart or cruelty of disposition*" (emphasis added). The part I have highlighted holds out some hope for those – should there be any – who have a proven genetic predisposition to impulsivity. They could be said to have a peculiar weakness of mind or infirmity of temper, and one for which they are not culpable (one not arising from wickedness of heart or cruelty of disposition). But in fact the courts have been very reluctant to take up the approach hinted at in Maher, and insofar as they have taken it up, have imposed a restriction that would undermine any claim that a genetic predisposition to impulsivity should be taken into account in determining whether the provocation was reasonable. Special characteristics of the defendant have been deemed relevant only insofar as the provocation is conceptually tied to that characteristic (and, typically, makes explicit reference to that characteristic).[23] An English ruling, *Regina v. Newell* (1980), held that "there must be some real connection between the nature of the provocation and the particular characteristic of the offender by which it is sought to modify the ordinary man test." More specifically, "the words or conduct must have been exclusively or particularly provocative to the individual because, and only because, of the characteristic."[24]

It seems unlikely, then, that the courts would treat a genetic predisposition to impulsivity as negating responsibility or even mitigating a charge from murder to involuntary manslaughter.

I have argued that there is no good reason to believe that genetic predispositions to criminality undermine responsibility more than environmental predispositions do, and I have also suggested that there would be little basis in the law for treating such a genetic predisposition as excusing – even partially excusing – the defendant. Nonetheless, many people may believe that genetic predispositions to criminality do undermine responsibility more than environmental predispositions, and some of these people may be jurors. In that sense a discovery of a genetic predisposition to crime may raise issues regarding responsibility, freedom, and desert. It will be seen to raise new issues because genes are believed by many to determine us more fully than they do. The supposition that there is something very special – and wonderful – about a genetic explanation is prominent in the popular literature on genes and crime. It is also prominent in works on related topics, such as *The Bell Curve* (Hernstein and Murray 1994), a book that received a great deal of attention a few years ago.

Readers of *The Bell Curve* – and readers (or "auditors") of the many enthusiastic summaries of that work in magazines, newspapers and on radio and television – are left with the impression that a genetic basis for x means that x cannot be altered by improved education, a more supportive environment, or the like.

Recognizing that a genetic marker for some sort of criminal tendency may raise social issues that it would not raise were it not for (induced?) ignorance, we do well to broaden our topic and ask how things would be different if we discovered and had a test for a genetic marker. In particular, what would be gained?

It is common to assume that the discovery would be a great advance. The following view is, I think, not unusual (though more cautious in its optimism than is typical): "(Most) experts in the field doubt there are 'criminal genes' or that genetic markers could be found that would identify potential wrongdoers at an early age. But, on the other hand, amazing genetic breakthroughs occur routinely. These advances can improve the quality of life for all of us."[25] How so? The assumption that such a scientific advance would "improve the quality of life for all of us" needs to be examined.

Our thinking on this topic is no doubt affected by the way we think of other genetic markers. Perhaps we assume that finding a marker would improve our quality of life because finding some other marker (of a sort that it is easier to imagine finding, or that has already been found) would, we think, do so.

Consider the benefits and the costs of finding a genetic marker for a disease, for example, colon cancer. The negative effects are not negligible. Insurance companies may review applicants' (and members') family histories, require those with a worrisome family history to be tested for the marker, and, if they test positive, reject their applications (or terminate their insurance). There is, moreover, the possibility that prospective (or actual) employers will gain access to such information and avoid hiring those who have the genetic marker or perhaps fire (once they reach an age that the company decides is "too risky") those who have the marker for colon cancer. As markers for more and more diseases are discovered, they might be factored into a ranking system and used by insurance companies and some employers to avoid "costly" clients or employees. Other drawbacks (at least from the standpoint of the person found to have the genetic marker) are anxiety and a sense of doom. Moreover, parents' knowledge that their young children have the marker could have a damaging effect on the child. Education might alleviate the last two problems, and legislation forbidding insurance companies and employers from obtaining this information would offset the former problems.[26] Still, none of these problems can be discounted.[27]

On the other hand, there could be significant benefits to having a marker (and a reliable and reasonably inexpensive test for it). Those who learn that they do have the marker can have frequent diagnostic tests to facilitate early detection and thus reduce the likelihood that they'll die from colon cancer. (With or without a marker, we could, of course, all have these tests, but the cost is high and the tests are unpleasant.) And the news that they have the marker should motivate those in need of motivation to redouble their efforts to improve their diets so as to reduce the likelihood that they will get the disease (after all, having the marker does not mean that one is certain to get the disease) or at least postpone the onset of the disease until old age. These are the sorts of measures that many people are prone to avoid but which perhaps they would pursue if they found out that they are genetically prone to the disease. (Admittedly this is optimistic; the notion that genes determine our fates is hard to shake, and many who learn they have the marker may simply resign themselves to what they assume will inevitably be an early death.) Those who learn that they do not have the marker will enjoy the relief afforded by that news (without assuming, of course, that they are guaranteed not to get colon cancer). In addition, they can reasonably opt for less frequent diagnostic tests than their family history of cancer would otherwise have indicated.

Would the discovery of a genetic link to crime and a test to determine who has the marker benefit people in some similar ways? Not likely. It might benefit people – possibly even the people in whom the marker was found – but not, it would seem, in any way analogous to the way in which a marker for cancer or some other sometimes curable diseases would be beneficial. The dissimilarities are striking. The benefit of a marker for a sometimes curable, sometimes fatal disease, a disease whose early detection greatly increases one's chances of survival, is evident – and it is primarily a benefit for the person with the marker. It is hard to see how someone would benefit from knowing that he or she had a genetic marker for criminality or violence. Whatever benefits there are would seem to accrue to people *other* than the person with the marker.

Now there is one scenario in which a test for a marker for criminality would be highly beneficial, probably even for the affected person, and it parallels the way in which a test for a gene linked to cancer is beneficial. As noted earlier, a genetic predisposition to crime would only make a person more likely to commit crimes; it would not ensure that he or she would commit crimes. But suppose that we knew that in certain conditions the person genetically disposed to commit crimes was far more likely to commit them than he or she would be in other conditions. And suppose that these conditions were not simply the same as the conditions in which everyone is far more likely to commit crimes. After all, we have a pretty good idea of the conditions that render it more likely that a person will commit crimes: the person is addicted to an illegal and expensive drug and is not wealthy (and, we might add, the waiting lists for drug treatment centers are very long); and so on. Suppose, rather, that the environmental conditions that triggered those genetically predisposed to commit crimes were something like this: the person had recently consumed four ounces or more of a sports drink or had recently consumed two ounces or more of blue cheese. Imagine that people who were genetically predisposed to criminality needed only to avoid drinking or eating these substances – and, if they did, their predisposition would, in effect, be nullified. They would then be no more likely to commit violent crimes than people without the genetic predisposition. (I am here supposing, for purposes of discussion, that the predisposition is to commit violent crimes.)

If this were the situation, the existence and availability of a test for the marker would be a real benefit to those who test positive for it. For there would be something to do about it, and the something would not be costly (unlike, say, being kept permanently on drugs that have serious side effects). There would be a benefit to others, as well: there would be less violence in the world, if a substantial number (even if it were a substantial minority) of those with the marker abstained (at least most of the time) from the sports drink or blue cheese or whatever. Even here the existence and use of a test for the marker is not without its costs to the individual. Would it be possible for parents to be instructed in proper diet for their child, and convinced of the importance of adhering rigidly to the diet, if they were not told that the child had this genetic abnormality? Would it be ethical not to inform the parents of the abnormality? If the parents are told, the likelihood that the child will be raised with the expectation, or at least fear, that he or she will grow up to be a criminal is worrisomely high. Confidence that all the usual disturbing behaviors young children exhibit (biting other children, kicking their parents, screaming "I'm going to kill you") mean nothing would be hard to maintain. ("I knew this diet wouldn't work; he's clearly a sociopath.") In addition, there is the risk of social stigma, a further cost to the individual with the marker.

What I have described is, of course, an unduly rosy picture of how, on balance, the person with the genetic marker would be affected by the existence and implementation of a test for it. In the more likely scenario, the predisposition would not be nullified by a simple dietary measure. The conditions that triggered the feared behavior would most probably be environmental, and very likely (I would venture to guess) the same sorts of conditions that trigger violent behavior in people without the genetic predisposition. These conditions include others – parents, in particular – not having confidence in the child. The very knowledge that the child has the marker is likely to increase the risk that the child will exhibit the behavior associated with it. It is, in short, extremely difficult to see how the person with the genetic marker is better off if we have a test for the marker; and in this respect a genetic marker for violence (or aggression or criminality) stands in striking contrast to markers for various cancers.

Are there benefits to others in having (and implementing, say, on all newborns) a test for the marker? Unless there is some reasonably simple way to prevent the problem from ever surfacing – dietary measures or, better yet, administration of a harmless drug at infancy – it is unlikely. (Of course, if we were willing to engage in obviously unacceptable strategies for dealing with it, such as forcible isolation of those with the marker from the rest of the population, knowing who has the marker would be useful.) In general, there seems to be little benefit, all told, if the aim is to implement large-scale testing to find out who has the marker. The risks seem clearly to outweigh the benefits. This is not to say, however, that genetic research concerning criminality does not yield benefits that outweigh its risks; on that point I am agnostic. (The risks, however, are great, and one would like to see a concerted effort to address the risks by correcting common misconceptions – for example, the belief that a genetic predisposition to X means that one is virtually guaranteed to do X, and the related assumptions that environmental conditions such as schooling would have no bearing on whether one ever engages in the undesirable conduct to which one is genetically predisposed.) Genetic research may well be of value in a number of ways. For instance, studies of twins reared apart might yield information on environmental protective factors that help to explain why one twin commits crimes whereas the other genetically identical twin does not (Raine 1993, 50). If it could be known that the twins had the genetic marker for criminality (or whatever the marker marks), that would make studies of twins reared apart all the more useful. We could compare pairs with the marker with those without it, and if there were enough pairs in which one committed crimes and the other did not, we could see if the environmentally protective factors differed between those who have and those who lack the marker. (Notice, though, that this research is tricky: if the parents knew or suspected that their children had the genetic marker, worries about their children would, in addition to – and by way of – harming the children and their relationships with them, throw off the study.)[28]

A discovery of a marker might be helpful within that research. Tracking those with a marker, as well as a control group, to look for environmentally protective factors that explain why some with the marker commit crimes and others do not might even lead to a discovery of what can be done to avoid "triggering" the predisposition to violence or criminality. But, in general, having a test for a marker does not seem to me to be likely to improve either the lives of those found to have the marker or the welfare of society as a whole. It seems most likely that the conditions that trigger the feared behavior in the genetically predisposed person are not of a sort that are very easily corrected (e.g., by diet). It is only a hunch, but my hunch is that the conditions that trigger the undesirable behavior in the genetically predisposed person would turn out to be no different from those that trigger it in everyone else. This is one reason why, in my view, research to find a genetic marker should not be a very high priority.

Would it not be more to the point to focus on altering the social conditions that trigger such behavior? Why so much interest in figuring out ways to detect which people are the "problems" – which people are flawed by an internal defect? If we deemed it morally appropriate to isolate the "problem people" – to send them away to an internment camp, perhaps, or require them to undergo medical treatment designed to tame them – there would be some point to the search for a marker. Or again, if we thought it a good social policy to screen fetuses routinely for the marker and urge pregnant women whose fetuses tested positive to have abortions, then again, there would be a real point to the test. But unless we favor such policies, and unless the triggering conditions turn out to be something like consumption of more than four ounces of a sports drink, altering (some of) the environmental conditions that make the commission of crimes more likely (no matter what one's genes are) seems more to the point.[29]

NOTES

1. Failure to distinguish the various undesirable tendencies that we suspect may have a genetic basis undermines the research into the correlation between "criminality" and the genetic abnormality. This is strikingly evident in the embarrassing history of the supposed "XYY syndrome." As Saulitis (1979) chonicles, in many studies the mere fact that an XYY male was an inmate in a mental or penal institution was taken as evidence, without regard to the circumstances that led the individuals to be locked up, that having XYY chromosomes predisposed one to violence (275; see also pp. 277, 278). No one bothered, in those studies, to distinguish violence from property crimes from mental illness involving neither violence nor commission of crimes. As it turned out (once better studies were performed), the only correlation between XYY chromosomes and criminality was that XYY criminals were remarkably numerous among those convicted of petty property offenses. See n. 13.
2. See table 4 in Raine 1993 for a summary of the adoption studies. The Swedish and Danish adoption studies showed heritability for petty property offenses. The summaries in the chapter are useful; the claims made for

heritability, however, often seem much more confident than is warranted by the evidence that Raine cites.

3. I say "especially" and "in particular" because of their lack of control over their childhood environments. Arguably they bear, in at least some cases, some responsibility for their current environments; it would be hard to make such an argument with respect to their childhood environments.

4. I am indebted here to P. S. Greenspan (Chapter 10, in this volume), although I am not convinced that the way one is inclined matters as much as she suggests it does.

5. It is important to bear in mind that "cannot control" does not entail the lack of control involved in an involuntary act, as when someone commits what would be an assault but for the fact that she was having an epileptic seizure. The acts of someone who lacks self-control are normally voluntary, meeting the *actus reus* requirement for a criminal conviction. The question is not whether the action is voluntary, but whether the person is perhaps not culpable because excusing conditions apply.

6. But I am not certain. It might be that we think it is impossible for C to refrain from sexually molesting children, but is not impossible for A and B, and on that basis perhaps we would hold that even if they are all trying equally, still C is even more deserving of our sympathy, even less deserving of blame, than the others. Perhaps; we think C couldn't have done otherwise, whereas A and B maybe could have – at least we are not as sure that they couldn't have. Even so, this would bear on the question of whether a genetic predisposition to crime exculpates more fully than other predispositions to crime only if the predisposition really does more fully preclude compliance with the law than do all other factors, and this seems unlikely. It is worth noting, though, that someone who had multiple high-risk factors – genetic predisposition along with an array of terrible environmental factors from childhood (ongoing child abuse, a severely dysfunctional family, very poor schools, a lack of role models, a violent neighborhood, etc.) – might have so many strikes against him as to warrant exculpation, or at least hefty mitigation. (I am grateful to David Wasserman for drawing my attention to this possibility.) Exculpation would more probably be warranted on retributivist than on deterrence grounds, for although arguably this defendant might not be deterrable, it might well serve the purposes of general deterrence to punish him.

7. Although a judgment that person X was less free to refrain from committing a crime than person Y generally inclines us to hold X less responsible than Y (assuming that X and Y committed the same sort of crime), there is a conflicting tendency that should be noted as well. Jurors often treat the criminal who grew up in urban poverty and was under intense pressure to join a gang more harshly than the criminal from the comfortable suburb, whose peers spent a lot of time in the country club to which they and he belonged. Why? The former is seen as more likely than the latter to be incorrigible. This view seems to accept the position that the criminal who grew up in urban poverty is less free to refrain from committing crimes than the criminal who grew up in the suburbs, while at the same time it resists the inference that

"less free" means "less responsible," and thus means "less deserving of punishment." Rather, the view focuses on such pragmatic consideration as, Is the person likely to be a recidivist? If so, adherents of the view say, let's keep him in a very long time (or execute him).

8. Because I limit my discussion to defenses for which a proven genetic predisposition might be directly relevant, I do not consider the possibility that environmental predisposition or "rotten social background" perhaps ought to be a defense. For discussion, see Bazelon 1976; Morse 1976; *United States v. Alexander* 1973; and Delgado 1985.

9. Although it was named for Daniel M'Naghten, who in 1843 attempted to assassinate Robert Peel, then prime minister of England, the test was not applied in his case. (Peel in fact was riding in a different carriage than usual; M'Naghten killed Peel's secretary, who was occupying the carriage he thought would be occupied by Peel.) M'Naghten, apparently a paranoid schizophrenic who believed Peel to be persecuting him, was found not guilty by reason of insanity in a ruling that rejected the view that one should be considered insane, for the purposes of the law, only if one did not know the difference between right and wrong, or did not know what he was doing. In other words, the ruling rejected what we now call the M'Naghten test. Disturbed by the verdict, Queen Victoria asked the House of Lords to survey the views of judges. Under considerable pressure, the judges reaffirmed the old "cognitive" test of insanity, and thus we have the M'Naghten test.

10. The MPC differs from the M'Naghten rule in other respects, as well. The cognitive component from the M'Naghten rule is altered in two ways. First, the crucial verb is not "know'" in the MPC but "appreciate," the notion being that someone might know in a hollow, mechanical way that what he did was wrong but not really appreciate it. The replacement of "know" by "appreciate" thus slightly loosens the cognitive requirement. Second, whereas the M'Naghten test requires that the defendant *did not know* the nature and quality of the act he was committing, or that what he was doing was wrong, the MPC alters this requirement, saying that the defendant must "*lack substantial capacity* either to appreciate the criminality . . . of his conduct or to conform [it] to the requirements of the law" (emphasis added). Unlike the M'Naghten test, the MPC does not preclude that someone with *some* knowledge – even appreciation – both of what he was doing and that what he did was wrong might have a valid insanity defense.

11. It is worth noting, however, that his defense relied on both the cognitive and the volitional elements of the definition of insanity; so it is not clear that his verdict would have been different if the old M'Naghten test were the standard used in his case.

12. See, for example, *United States v. Lyons.*

13. This paragraph draws from Salteburg et al. 2000, 853.

14. Similar questions are discussed in connection with the cognitive component. See n. 10.

15. Not that there was no link at all between the XYY genotype and criminality; it turned out that XYYs did show a significantly higher rate of criminal con-

victions than their XY counterparts, on a percentage basis, but the crimes the XYYs committed were generally petty offenses against property. The incidence of overtly violent or assaultive crimes was actually lower than for the general male population. It also emerged that the XYY males had substantially lower median intelligence levels than XY males, and that there was an inverse correlation between intelligence and criminality in both XY and XYY sampled, suggesting that intelligence was a significant mediating variable between the XYY syndrome and criminality. See Saulitis 1979, esp. pp. 279–282.

16. An exception would be jurisdictions where felony murder (for certain felonies) counts as first-degree murder.

17. For a discussion of the diminished-capacity defense from an era when the defense was taken more seriously than it now is, see Arenella 1977.

18. Unlike the diminished-capacity defense, this would not be an affirmative defense, but would simply be an attempt to convince the jury that the prosecution had not proved the elements of the crime beyond a reasonable doubt. For a helpful discussion of the relationship between diminished capacity and absence of *mens rea*, see *United States v. Schneider*.

19. See *Maher v. People*. The Model Penal Code has a different version of the defense, and several states have adopted the MPC version into their statutes. Murder can be mitigated to voluntary manslaughter if, in the language of the MPC, the "homicide, which would otherwise be murder is committed under the influence of extreme mental or emotional disturbance for which there is reasonable explanation or excuse" (sec. 210.3).

20. This is hard to state; I don't think it is altogether clear what the key idea is. What is it to lose self-control, anyway? And must the provocation be such that the reasonable person would very probably have lost self-control in the situation? Or is it enough that it would have been extremely upsetting? (And, if so, must it be upsetting in the sort of way that would tempt one to go after the offending person with a weapon, rather than to weep or to write an angry letter?) I hope to address these issues (and the implicit assumptions about both gender and moral psychology that lie just beneath the surface) in another paper, tentatively entitled "Killing in the Heat of Passion."

21. As noted earlier, that this was genetic would not appear to strengthen the case at all; what is needed is evidence that the person has the problem. Its origin would not matter (unless, perhaps, it was brought on by the agent him- or herself [and as an adult]).

22. If so, this would suggest that it might also be relevant to a different defense: the self-defense justification. If someone suffering from impulsivity due to (an uncorrectable) problem such as a serotonin deficit killed someone in self-defense, but in fact was (by the standards of the ordinary reasonable person) rash in thinking that the supposed aggressor really did pose a serious threat to his well-being, perhaps his impulsivity would be relevant to judging whether a reasonable person would have seen the situation as he did. The reasonable person would be the reasonable impulsive person. As I explain later, the courts have been, understandably enough, quite reluctant to substitute for "reasonable person" "reasonable impulsive person."

23. Courts rarely discuss this issue, and therefore there are not enough cases to provide a basis for stating a general rule. Where courts have discussed it, however, this has been a common response.

24. Cf. *Masciantonio v. R*, an Australian high court case in which the defendant argued, unsuccessfully, that the jury should have been instructed to consider how a reasonable Italian man would have reacted to the provocation. The case is discussed by Herring 1996.

25. The quotation is from a "Viewpoint," written by T. J. Mooney (1992), a former Maryland state delegate.

26. Although Congress has moved slowly in this area, at least twenty-six states have adopted laws regulating the use of genetic test results to prevent discrimination by insurers and employers. See *New York Times*, October 18, 1997, A1. President Clinton's first executive order of 2000 prohibits federal employers from requiring genetic tests as a condition of being hired or receiving benefits. See *Professional Ethics Report* 13 (Winter 2000): 3.

27. For an excellent discussion of these issues, see Kitcher 1996.

28. See Saulitis 1979 for discussion of similar problems in conducting research on XYY Syndrome.

29. For examples of such changes, see Kotlowitz 1991.

REFERENCES

Arenella, P. (1977). The diminished capacity and diminished responsibility defenses: Two children of a doomed marriage. *Columbia Law Review* 77: 827–865.

Bazelon, D. L. (1976). The morality of the criminal law. *Southern California Law Review* 49: 385–405.

Delgado, R. (1985). "Rotten social background": Should the criminal law recognize a defense of severe environmental deprivation? *Law and Inequality: A Journal of Theory and Practice* 3: 9–90.

Hernstein, R. J., and Murray, C. (1994). *The bell curve: Intelligence and class structure in American life*. New York: Free Press.

Herring, J. (1996). Provocation and ethnicity. *Criminal Law Review*: 490–493.

Kitcher, P. (1996). *The lives to come: The genetic revolution and human possibilities*. New York: Simon and Schuster.

Kotlowitz, A. (1991). *There are no children here*. New York: Doubleday.

Maher v. People, 10 Mich. 212 (1862).

Masciantonio v. R, A.J.L.R. (1995).

Millard v State, 8 Md. App. 419, 261 A. 2d 277 (1970).

Model Penal Code and Commentaries. (1985). Part I, pp. 171–172. Philadelphia: American Law Institute.

Mooney, T. J. (1992). Viewpoint. *Prince George's Journal*, November 3, A4.

Morse, S. J. (1976). The twilight of welfare criminology: A reply to Judge Bazelon. *Southern California Law Review* 49: 1247–1268.

People v. Tanner, 13 Cal. App. 3d 596, 91 Cal Reptr. 656 (1970).

Raine, A. (1993). *The psychopathology of crime*. San Diego: Academic Press.

Regina v. Newell, 71 Crim. App. 331 (1980).

Saltzburg, S. A., Diamonds, J. L., Kinports, K., and Morawetz, T. H. (2000). *Criminal law: Cases and materials.* 2nd ed. New York: Lexis Publishing.

Saulitis, A. (1979). Chromosomes and criminality: The legal implications of the XYY syndrome. *Journal of Legal Medicine* 1: 269–291.

State v. Roberts, 14 Wash. App. 727, 544 P. 2d 754 (1976).

United States v. Alexander, 471 F.2d 923 (D.C. Cir.), *cert. denied,* 409 U.S. 1044 (1973).

United States v. Freeman, 357 F. 2d 606 (D.C. Cir. 1966).

United States v. Lyons, 731 F.2d 606 (D.C. Cir. 1966).

United States v. Schneider, 111 F. 3d 197 (1997).

Chapter 9

Genes, Statistics, and Desert

PETER VAN INWAGEN

Suppose there is a population in which a certain type of criminal behavior is much more common than it is in most other populations that have been studied. To what extent can the relatively high frequency of that type of behavior in that population be ascribed to genetic, as opposed to environmental, factors? In the real world, this is always a very difficult question.

Let us suppose that – in some case, in respect of some type of behavior – this difficult question has been answered. Let us suppose that the high frequency of a certain type of criminal behavior in "population A" has been shown (to the satisfaction of all of the statisticians, criminologists, sociologists, and so on – of all political persuasions and ideologies – who have studied the matter) to be, to a significant degree, a product of genetic factors. Before investigating the consequences of this supposition for certain questions about punishment and desert, however, let us consider how such a conclusion could be established.

Suppose that, starting at a certain date, the babies born to parents belonging to population A and the babies born to parents belonging to a second population, population B – in which the incidence of the type of criminal behavior under investigation is significantly lower than it is in A – were exchanged in their cradles (the exchange being stealthy enough that the parents do not notice), the statistical profile of population A would, if no important changes occurred in the environmental conditions under which its members live, become significantly more like the present profile of population B after an appropriate amount of

I wish to thank the following people for valuable comments and criticism: Marcia Baron, Jorge Garcia, Patricia Greenspan, Michael Slote, Alan Strudler, Laurence Thomas, and David Wasserman.

time had passed (and, of course, vice versa). And let us suppose that a similar result has been obtained with respect to A and "population C," with respect to A and "population D," and so on, for a large and varied family of populations, populations in which the environments in which children are raised vary widely with respect to all of the environmental factors that it is reasonable to suppose have consequences for the incidence of criminal behavior in a population. (And we suppose that we have found no population for which this result has not been obtained.)

Even if there were investigators who had established the results I have imagined, they would have found no absolute proof that if they had gone on to examine one more population, they would have obtained a similar result. For all they could show without collecting further data, if they had gone on to compare population A with "population Q" (another population in which the type of criminal behavior we are interested in is low), was that exchanging babies from A with babies from Q would leave A with the same high incidence of this behavior and Q with the same low incidence of it. Thus, even our fantastic imaginary evidence would not rule out the following possibility: the high incidence in A of the criminal behavior under investigation is due, to a significant extent, to environmental factors that are common to A and all of the populations with which A has so far been compared – but which might well not be features of the environment of some populations. It does seem, however, that as the number and variety of the populations with which A has been compared in the way imagined and with the result imagined increase, the probability increases that a population with the genetic makeup of A would, in any possible human environment, exhibit a relatively high incidence of the criminal behavior being studied. And this, I suppose, is what it would mean for the relatively high incidence of some sort of criminal behavior in population A to have a genetic cause.

To suppose that we have collected evidence of the kind I have imagined is to make an extravagant supposition, but not an impossible one. It would be possible, in principle, to collect such evidence, and therefore, possible in principle to demonstrate that it is highly probable that the high incidence of a certain sort of criminal behavior in a population is due to genetic features of that population. And it would not be necessary to exchange babies in their cradles to carry out the demonstration. When I say that it is in principle possible to collect such evidence, I am putting forward a more interesting thesis than that it is "in principle possible" actually to exchange the babies belonging to two whole populations. I am saying that it is in principle possible to determine what the results of such an exchange would be without making it. Justifying conclusions of that sort is just what statistical inference is for.

I must emphasize that I am saying only that it is *in principle* possible to determine whether, for two populations that exhibit significantly different statistical profiles as regards criminal behavior, the result of a "baby exchange" would be eventually to "reverse" (to some degree) those profiles. What is possible in principle might be forever impossible in practice. It might be that, although it is possible to say *a priori* what sort of evidence would establish or refute "reversal hypotheses" – and the even more ambitious hypotheses that ascribe the incidence of types of behavior in a population to the genetic peculiarities of that population – it is in practice impossible to collect evidence that satisfies these *a priori* requirements. However this may be, what is only in principle possible is often of considerable philosophical interest. I want to imagine a population in which a certain type of objectionable behavior is significantly more frequent than it is in most populations, and to imagine that it has been uncontroversially established that the high frequency of this behavior in this population is due to genetic features of the population. I want to imagine this because it constitutes a kind of "worst-case scenario" for those who worry about the relations between the genetic makeup of human beings and questions of punishment and desert. I want to investigate the consequences of the worst-case scenario. It might, after all, turn out that some version of the worst-case scenario is true, and there seems to be no reason to wait till some possibility materializes to start worrying about what to do about it. And even worst-case scenarios that, so to speak, never make it to the screen can be useful to theorize about, since they provide material for *a fortiori* arguments. ("If we shouldn't use nuclear weapons even if the other side attacked us with them without warning, we certainly shouldn't use them in any other case.")

Let us fill out our worst-case scenario a bit – at least to the extent of supplying a crime and a few figures. Let us say that within population A, rape is very common. One man in twenty, let us say, has at least attempted to rape someone. In population B, however, only one man in a thousand has attempted rape. We are, of course, supposing that the two populations are ones that people belong to from birth – or, better, from conception – or not all. Population A, for example, is not supposed to be something like "inmates of federal penitentiaries." We are supposing, moreover, that if the male babies born to the parents of the two populations were (covertly) exchanged, then, after a suitable interval, the proportion of population A that were rapists would begin to decline and would eventually level out at a figure substantially lower than one in twenty; and, of course, we are supposing that the proportion of population B that were rapists would begin to rise and would eventually level out at a figure substantially higher than one in a thousand. And let us suppose that we have established similar results for A and C, A and D, and A and a great many other populations in which rape is significantly less common than it is in A – all we have been able to compare A with. If we had such evidence, we should have very good reason to believe that there was a genetic explanation for the abnormally high proportion of rapists in population A. We should not, of course, know *what* the genetic explanation was; that would be a matter for further investigation, investigation that would probably have to be carried out partly by examining human genetic material and not simply statistics about human behavior.

So much for the question how one might establish the conclusion that the high frequency of a certain type of criminal behavior in a certain population might have a genetic basis. I will now, as I have promised, turn to "certain questions about punishment and desert."

The questions that interest me are these. To what extent would the facts I have imagined, if they were real facts, provide the rapists who belong to population A with an excuse for their crimes? Should we (in that case), in writing our criminal code, be "population-blind"? Would it be *fair* to write laws that prescribed the same criminal penalties for anyone convicted of (a certain type of) rape, when we know that the proportion of the members of population A who commit rape is, because of the genetic makeup of that population, significantly higher than the proportion of the members of most other populations who commit rape? Do the members of A *deserve* to be treated the same way under the law as the members of (for example) B?

There seems to me to be one possible circumstance in which it would be absolutely clear that our laws regarding rape should in no way take into account the genetic peculiarities of A. Suppose we did indeed identify the specific genetic factor that was responsible for the relatively high proportion of rapists in A. Suppose it was discovered that the possession of "Gene Combination Alpha" was much more strongly correlated with rape than was membership in population A, and that it had been proved that almost all of the "A" rapists possessed this gene combination. And suppose it had been proved that this combination of genes was so rare in population B and most other populations as to be almost non-existent. Suppose it was shown that the distribution of this genetic factor in the various populations studied (together with the very strong correlation of this factor with rape) accounted very well for the differing proportions of rapists in those populations. Suppose that there was an easy-to-perform and reliable test that could be used to determine whether a given man possessed Gene Combination Alpha. Then it would seem to be undeniable that our laws should not take the either the high proportion of rapists in A or the fact that this proportion is known to have a genetic basis into account. It is individuals and not populations that are brought to trial, at least under civilized legal systems, and any given rapist either possesses Gene Combination Alpha or he doesn't. Whether he does or does not might be a relevant matter to bring up at his trial. Whether he belongs to a population in which that combination is frequent or infrequent is certainly irrelevant.[1]

If, however, we added this kind of knowledge to our imaginary case, this would in a sense change nothing about it that was of philosophical interest. What the addition of such knowledge would change would be only the question that was the focus of philosophical interest. It would be the question, To what extent would having Gene Combination Alpha provide an excuse for rape? rather than the question, To what extent would belonging to population A provide an excuse for rape? that was the focus of philosophical interest. Let us therefore simply assume that we do not know what the genetic factors are that explain the high proportion of rapists in population A.

The interest of the question "To what extent does belonging to population A provide an excuse for rape?" lies in the fact that members of population A do not invariably commit rape. In fact, most men who are members of A get through their lives without trying to rape anyone, even when – let us suppose – they are in circumstances in which it would be reasonable for them to believe they could get away with it. Consider, by way of contrast, "population X," in which *all* men attempt rape whenever they think they have a reasonable chance of doing so with impunity. (We suppose, again, that there is good evidence that these men would behave that way no matter what environment they were raised in.) The question, To what extent does belonging to population X provide an excuse for rape? is much less interesting than the question I have raised. It seems fairly clear that belonging to population X provides the rapist with a really excellent excuse, for it seems fairly clear that in that case the rapist's behavior is genetically determined.[2] And it seems fairly clear that we do not want to blame people for engaging in behavior that is genetically *determined* (as opposed to genetically influenced). The men of population X will, it is true, have to be regarded as dangerous, but the proper attitude toward them, it would seem, ought to be like our attitude toward a typhoid carrier. We should restrict their freedom of movement, as we do with typhoid carriers, but we should feel sorry for them; we should feel sorry for them because we should feel that, however necessary it might be for us to restrict their freedom of movement, they do not *deserve* it. Just as being a typhoid carrier is a misfortune, so – I believe this would be our reaction – being genetically determined to commit rape would be a misfortune. (Of course, we might blame some member of population X for being indifferent to the consequences for others of his condition, just as we blame Typhoid Mary for being indifferent to the consequences for others of *her* condition; but, presumably, we should do this only if we did not regard indifference to the consequences of one's genetically determined condition as being itself genetically determined.)

What can be said in defense of the thesis that membership in a population like population A – a population in which, for genetic reasons, the proportion of rapists is high, but in which, at least as far as anyone knows, no one is genetically determined to commit rape[3] – is at least some sort of excuse for rape? How might a convicted rapist belonging to such a population try to use this fact in court to his advantage?

Well, suppose that a man who belongs to population A has been convicted of rape, and I, who belong to population B, am presiding at his trial. The jury has just delivered its verdict, and I am about to pass sentence on him. I give him the most severe sentence the law allows, and accompany the sentencing with some remarks about the horror of rape and how mercy is entirely out of place when one is dealing with rapists. Suppose he replies, "It's not your place to judge me. I am a member of population A, and you are not. I am thus laboring under a genetic burden that fortune has placed on my back and not on yours. Since you don't share my genetic burden, you are not in a position to pass moral judgment on me. What is more, it's not at all fair that I should be given the most severe sentence the law allows. If you give me that sentence, what sentence will you reserve for a member of *your* fortunate population – there are a few – who commits the same crime?" Note that in this speech the rapist presents two arguments for two different conclusions. One argument is *ad hominem*, and its conclusion is that the moral condemnation is out of place. The conclusion of his second argument, which could be addressed to any judge, is that he should not receive the most severe sentence possible.

I want to approach these two arguments by looking at some analogies. Let us look at some quite different cases of statistical correlation between criminal behavior and various genetic and environmental factors. (But *can* environmental factors provide a suitable analogy? I don't see why not. If we are interested in matters of excuse and desert, the only relevant questions to ask about a factor that has somehow influenced an agent's behavior are, Had the agent a choice about whether that factor was present? and, Had the agent a choice about whether that factor, if present, influenced his behavior? That someone who did something objectionable was drunk at the time is not much of an excuse if the agent had a choice about whether to be drunk, or if it was within his power to place himself in circumstances in which his being drunk would not have led to that sort of behavior. And features of one's environment can as easily be things that one has no choice about as the sequence of base pairs in one's DNA. No one has any choice about whether he was sexually abused at age four or was raised in grinding poverty or was born a member of a despised and visible minority.) In devising examples turning on environmental factors, I shall not, of course, assume that the populations that figure in the examples are ones a person has to be born into to belong to. In presenting the analogical cases, I shall assume that we know what explains the statistical differences between the populations that are contrasted. I shall feel free to do this because I am interested in the question, What *follows* about the responsibility of individual members of various populations between which there are statistical differences that are due to factors outside the control of their members?

Now the analogies.

There are two islands. Bank robbery is much more common on one of the islands than the other: an inhabitant of Island A is in fact about twenty times more likely to rob a bank than is an inhabitant of Island B. It turns out that the explanation is not far to seek. On Island A, there are hundreds of small banks that are (as banks go) pretty easy to rob. On Island B, there are only a few large banks, and they are equipped with all sorts of state-of-the-art antirobbery devices.

Suppose that someone who lives on Island A has been convicted of bank robbery, and I, who live on Island B, am the judge at his trial and I am passing sentence on him. I give him the strictest sentence the law allows, and accompany the sentencing with some remarks on the horror of bank robbery (I was educated in Switzerland) and how mercy is entirely out of place when one is dealing with bank robbers. Suppose the convicted bank robber replies, "It's not your place to judge me. I am a native of Island A, and you are not. I am thus laboring under an environmental burden that fortune has placed on my back and not on yours. Since you don't share my environmental burden, you are not in a position to pass moral judgment on me. What is more, it's not at all fair that I should be given the strictest sentence the law allows. If you give me that sentence, what sentence will you reserve for a member of your fortunate population – there are a few – who commits the same crime?"

I do not think that most of us would regard these arguments as very convincing. It is interesting to ask why we don't, however.

Let us look at a second case.

There are two islands. Bank robbery is much more common on one of the islands than the other: an inhabitant of Island A is in fact about twenty times more likely to rob a bank than is an inhabitant of Island B. It turns out that the explanation is not far to seek. There are genetic differences between the inhabitants of the two islands, differences that have the consequence that people with the mental and physical capacities that make skilled bank robbers are much more common on Island A than on Island B. These are, let us say, manual dexterity, nerves of steel, mechanical ability, a good memory for detail, excellent spatial intuition, exceptional athletic ability . . . whatever. For genetic reasons, the inhabitants of Island B tend to be nervous, clumsy, scatterbrained couch potatoes.

If we imagine a convicted bank robber from Island A arguing that he ought to receive some sort of special consideration from the court because he was born into a population that is deficient in nervous, clumsy, scatterbrained couch potatoes (or even because he himself demonstrably lacks these genetic advantages for growing up to be a non-bank robber), we shall find it difficult to imagine anyone's being convinced by his argument.

Now a third example, again turning on environmental factors. On Island A, there are secret criminal societies (like that presided over by Fagin, but with loftier criminal ambitions) that kidnap children and raise them to be bank robbers. On Island B, there are no such societies; as a consequence, bank robbery is much more common on A than on B.

A fourth example, this time involving a genetic factor. The inhabitants of A are, for genetic reasons, much harder to "socialize" than is the human norm. As children they, or at least a significant proportion of them, have a much greater tendency toward bullying, petty theft, and vandalism than the children of most populations. (The inhabitants of the island are aware of this unfortunate feature of the island's gene pool and, if possible, adopt children from off-island rather than conceive their own; well-conducted empirical studies of the careers of these adopted children confirm the intuitions of their foster parents.) The inhabitants of A who exhibit these tendencies as children tend to rob banks when they grow up, for the simple reason that, as Willie Sutton put it, "That's where the money is." As a consequence, bank robbery is much more common on A than on B.

In these two cases, we should probably find a convicted bank robber's plea for some sort of special leniency to have some plausibility, whether or not we in the end allowed it to influence our decisions about how he or she ought to be treated by the court.

What is the lesson of these cases? It seem to me to be something like this. A factor, whether genetic or environmental, that explains why it is that a certain type of criminal behavior is more common in a certain population is not perceived as providing any sort of excuse for those who engage in that type of behavior if its effect is due to its increasing the prevalence of the skills required for that kind of behavior or the opportunities available to members of that population to engage in that sort of behavior with impunity. (The same point would apply to the easy availability of means. A convicted bomber could not plausibly ask for mercy on the ground that high explosives were easily available in his society, although he might, with some plausibility if it were true, plead that he was raised in a society in which bombers were lionized by the news media. Nor could he offer as an excuse the fact that he had inherited the – rare, let us suppose – mechanical skills necessary for constructing bombs, though he might plead that he had inherited a disposition to violence or a sociopathic disregard for human life.)

But if a factor works by creating or strengthening a desire such that to act on that desire would be to engage in a certain sort of criminal behavior, we tend to regard the plea that that factor is prevalent in some population to which one belongs to be at least some sort of excuse for having engaged in that behavior. If, for example, there were a genetic factor that could be shown to produce in males an inordinately strong desire for immediate sexual release, or a desire to degrade women, the fact that this genetic factor was present in a given man who had been convicted of rape would probably be regarded as at least relevant to the question what sort of moral attitude we should take toward him and what sort of action a court should take in passing sentence. If, moreover, a factor tends to have adverse effects (adverse in our judgment) on an individual's abstract or second-order desires – the desire not to cause pain, say, or the desire not to desire to degrade women – we should probably regard the presence of that factor as relevant to questions of excuse and desert. If, for example, it could be shown that a convicted rapist had been raised by parents who taught him always to seek immediate gratification of the desires of the moment and never to consider the consequences for himself or others of his actions, we might well regard this fact about his nurture as a mitigating circumstance. If, finally, a factor tends to interfere with an individual's ability to implement his or her abstract or second-order desires – if it produces a lack of self-control, low intelligence, ignorance of generally available ways of dealing with situations in which one's momentary desires are in conflict with one's abstract or second-order desires – we should probably regard the presence of this factor as relevant to questions of excuse and desert.

Perhaps we could sum up these tendencies in the following formula. Suppose a certain kind of criminal act is significantly more prevalent in a certain population than in most other populations. If whatever factor produces this effect is "external," if it produces its statistical effect only by placing some of the members of that population in certain *circumstances*, if it leads them into temptation, we do not regard it as providing any sort of excuse for those members of the population that engage in that behavior – and this despite the fact that the members of the population have no choice about whether they are members of a population in which that effect is present. But if the factor is "internal," if it produces its statistical effect wholly or partly by acting on the desires and values of the members of the population (or on their ability to alter, or to act or refrain from acting in accordance with, certain desires and values), then we tend to regard this factor as something that should at least be considered when we are judging the members of the population that engage in that sort of behavior. It should be remarked that this "formula" is only a formula – an easy-to-remember device for summing up certain tendencies we have. I do not want to place too much weight on the particular terms of this formula. I particularly warn against placing too much weight on the words "external" and "circumstance": in the sense I am giving to these words, an agent's size and bodily strength or his possession of certain items of purely factual knowledge could count as external factors, as a component of the circumstances in which fortune has placed him.

Let us look at the case of rape. Does this tendency that I have alleged to exist manifest itself in the case of that crime? I think so. Imagine a society in which – owing to some economic necessity – women are more frequently alone and far from help than is common in most societies. For good measure, imagine that in this society, it is customary for men to cover their faces when they go out in public, like women in traditional Islamic societies. It would not be surprising if rape were markedly more common in that society than in most. But we should hardly regard the plea, "I live in a society in which it's relatively easy to find opportunities for rape, and in which it's hard for the victim of rape to note any features of her assailant that might later serve to identify him – and I have no choice about whether I live in such a society" as a very effective one. Or imagine a society in which men were much larger and stronger than women – significantly more so than is in fact the norm in human populations. In this society, too, it would not be surprising if rape were more frequent than is the norm. But, "I have to live in a society in which I am surrounded by women whom it is physically easy for me to force myself on," is not an excuse that we should be likely to find convincing.

It might be argued that the tendency that I have alleged is less clear than I have made it out to be. One could think of cases that might tell against it. Drug addiction is more common among doctors and nurses than it is among the members of other high-stress professions (such as airline traffic controllers). The usual explanation is simply that it is much easier for doctors and nurses to get drugs than it is for most people. Assuming this to be the case, cannot doctors and nurses who are drug addicts offer the general easy availability of drugs in the medical professions as an excuse for their addiction? What about bank clerks who have embezzled money? Can't a reformed alcoholic who has relapsed plead (supposing this to be true) that people were always offering him a drink? I am inclined to account for our sympathetic reaction to these proposed excuses by pointing out that widespread opportunity can mean frequent temptation, which can, in time, increase the strength of one's desires, or weaken one's will with respect to resisting them. The most convincing of the three cases is that of the reformed alcoholic who relapses; we should note that in this case an "internal" debility was present from the start and that the frequent episodes of temptation could plausibly be supposed to have been gradually strengthening it. In short, these are not cases in which the greater-than-normal frequency of objectionable behavior in a population is due *only* to "external" factors.[4]

I continue to believe, therefore, that we do have this tendency. Is it justified? I will argue that it is not. I will begin by presenting two pairs of cases (one "environmental" pair and one "genetic" pair). Each pair will be constructed to bring out pretty strongly our tendency to regard "external" and "internal" factors as being importantly different. I will argue that there is nothing about internal and external factors that justifies us in treating them differently. Here is the "environmental" pair.

We have two societies in which rape is significantly more common than in most societies. This can be explained (we have somehow shown) entirely by features of the environments in which the members of the two societies live. The two operative environmental factors are these:

1. In Society One, there is, and has been for more than a generation, legal, ubiquitous, and very well produced pornography that is essentially a glorification of rape. Even parents with the best wills in the world find it extremely difficult to prevent adolescent boys from being continually exposed to this pornography.
2. In Society Two, there is an illegal but cheap and easily available drug that facilitates rape: it is tasteless, odorless, fast-acting, and easy to administer surreptitiously. It renders the victim semiconscious and pliable. The human metabolism breaks it down into untraceable residues very fast: a few hours after it has been administered, it is undetectable by any medical test. Those who have been given this drug have afterward only the vaguest and most confused memories of what happened while they were under its influence.

In the "genetic" pair, we have two societies in which rape is significantly more common than in most societies. This can be explained (we have somehow shown) entirely by differences in the genetic make-ups of the members of the two societies. The two operative genetic factors (they have figured in cases we have already considered) are these:

3. Among the male members of Society Three, a certain gene-sequence is very common; it has the following phenotypic effect on those men whose genotype contains it: they experience an inordinately strong urge for immediate sexual release.[5]
4. In Society Four, the men are (for genetic, and not dietary or other environmental reasons) much larger and stronger than the women, significantly *more* so than in is the human norm.

I take it that most of us would regard being a member of Society Two or Four (the two "external factor" societies) as providing no sort of excuse for rape, and that we should experience at least some tendency to regard membership in Society One or Three (the two "internal factor" societies) as providing at least some sort of excuse; membership in either of the latter two societies, we are inclined to think, is a mitigating circumstance that should be taken into account when we determine the rapist's penalty or pass moral judgment on him. But what justification could be given for this difference in attitude?

In each of the four societies, the rapist has certain desires, and – whether or not he struggles against them – he acts on them and commits rape. In most cases, the more typical cases, the rapist will also have had certain desires and tendencies that pulled him in the opposite direction. If he is a hardened, habitual rapist, he may not have had any opposing desires, for repeatedly acting on certain desires tends to extinguish any desires or tendencies that oppose the desires that are repeatedly acted on. If the hardened, habitual rapist's desire to force himself on a particular woman on a particular occasion is really *unopposed* – by any values or feelings of human sympathy or preference for a sexual partner who is actually sexually aroused or even by fear of punishment – then perhaps he can't do otherwise than act on that desire. (This would be a consequence of the conclusions of my 1989 paper, "When Is the Will Free?") But let us consider those much more common and typical cases in which something – human sympathy, childhood moral training, fear of punishment – opposes the rapist's momentary desire to commit the rape he is contemplating, and let us suppose that in these cases he is *able* not to act on his desire to commit the rape he is contemplating.[6]

If a man contemplating rape is indeed able to refrain from acting on his present desire, if he is indeed able to refrain from committing the rape he is considering, then I do not see why the fact that he had had that desire should, afterwards, provide him with any sort of excuse for what he has done. (And this even if his having that desire is not something he has any choice about.) The presence of this desire in his psychological economy is not a mitigating circumstance. If, moreover, the man contemplating rape is able to refrain from acting on his present desire, then I do not see how any facts about the *source* of that desire can provide him with any excuse if he decides to act on it. In Society One, the desires on which many rapists act are due to their repeated exposure during their formative years to a certain kind of particularly vicious pornography. In Society Two, the society in which the drug that facilitates rape is easily available, it is *opportunities* to commit rape, rather than momentary desires to commit rape, that are due to a corrupt environment. In Society Two there may be no one cause that produces all or most of the momentary desires to commit rape that are experienced by the men of that society; still, each particular momentary desire will have *some* cause – one that will quite possibly be outside the individual's control. In Society One, a large number of the momentary desires that issue in rape have a common cause; but why should that fact be relevant to the question how we should judge the men who act on them?

The following speech, surely, is not an excuse a rapist could plausibly offer, even if everyone were wholly convinced of its truth: "I admit that I raped the woman who has accused me. But before I attacked her, I experienced a strong desire to rape her. And I was born a member of the Ruritanian lower-middle class, in which a higher-than-normal proportion of men experience such desires, and it has been proved that there is a common cause – some factor widespread in the Ruritanian lower-middle class – for many of these episodes of desire." If the momentary desire itself does not provide the rapist with an excuse, why should he be provided with any excuse by the existence of a factor that caused the desire, is widespread in a population to which he belongs, and produces similar desires in other members of that population?

It seems to me that internal factors like desire do not have importantly different implications for questions of excuse and desert from external factors like opportunity. Every rapist has of course had opportunities to commit rape, and an opportunity, we all agree, is no sort of excuse. If many of the opportunities to commit rape that are available to the rapists in some population have a common source (a common environmental source, as in Society Two, or a common genetic source, as in Society Four), we do not regard the existence of this common source as relevant to questions of excuse or desert. And desire would seem to be no different from opportunity in this respect: the existence of neither a (resistible) desire to commit rape nor of an opportunity to commit rape is any sort of excuse for the act; the discovery of a source of desires or a source of opportunities (whether a genetic or an environmental source) that operates across a population to which a rapist belongs would add nothing of relevance to the deliberations of those deciding how to punish or judge him. Just as a rapist cannot put forward a common, population-wide source of particular opportunities to commit rape as a mitigating circumstance, so a rapist cannot put forward a common, population-wide source of momentary desires to commit rape as a mitigating circumstance.

There are, of course, other internal factors than desire that are relevant to questions of excuse and desert. There are, for example, the agent's internal resources for dealing with desires that are in conflict with his or her values or higher-order desires. But the point I am making is quite general: if the presence of some particular factor in an agent's internal economy is not an excuse for some act of the agent's, why should the fact that the agent belongs to a population in which that factor is more common than in most other populations be an excuse for that act?

Perhaps there are some who will not find this argument convincing. We might try to articulate their reservations by imagining someone, who because of his or her special relationship to a convicted rapist is inclined to regard any circumstance that could conceivably be regarded as mitigating as really being so. We might imagine a mother who appeals on behalf of her son to the court (either a court of law or the court of public opinion) along the following lines. "You should regard my son's having grown up in a society that permits vicious pro-rape pornography as a mitigating circumstance. You should be merciful in passing sentence on him [in making moral judgments about what he did]." The plea is – to my ears, anyway – a poignant one, but I don't think we should allow it. (Unless, of course, the effect on the young man of his having grown up in an environment in which such pornography was prevalent was to render him literally *unable* to refrain from acting on the desires the pornography generated. But if that were the case – and we are supposing that it isn't – we should not have an example of a mere mitigating circumstance: we should have a case in which the rapist should be absolved of all blame, a case in which he should simply be regarded as a "rape carrier.") As long as we are convinced that the rapist had a choice about what he did, we should not reduce his sentence or soften the moral judgments we make about his act. What we can do, and what I believe we *should* do, is feel sorry for him.[7] And it would certainly do us no harm – the men among us, that is – to reflect that we might well ourselves have done what he did if we had been raised in the same corrupt environment. (Come to that, it would probably do us men no harm to ask ourselves seriously how we should have behaved if we had been raised in a society like Two or Four in which it was absurdly easy to commit rape with impunity.)

To have to deal with a recurrent desire (or, for that matter, with recurrent opportunities)[8] to commit some wrong act is a misfortune, a burden. We can, and should, feel sorry for those who have to bear burdens that we don't, and we may profit from asking ourselves how we should have borne up under them. We should not, however, regard them as mitigating circumstances. If a mother steals because she and her children are starving, that is a mitigating circumstance. If I betray my country or the Revolution (or whatever) under torture or because my family is being held hostage, that is a mitigating circumstance. If someone commits rape as the alternative to the murder of his family (one can easily imagine this alternative being forced on someone in one of the nasty little ethnic wars of the present decade), the fact that he faced this alternative is a mitigating circumstance.[9] Having a (resistible) desire to do ill that most other people do not have is, however, no more a mitigating circumstance than is having an opportunity to do ill that most other people do not have.

This general judgment applies if the (resistible) desire to do ill has a genetic cause; the source of a desire is irrelevant to the question whether its presence in an individual should be regarded as a mitigating circumstance. And it applies if the desire is significantly more common in some population to which the agent belongs than it is in most other populations; the presence of a desire in other individuals is irrelevant to the question whether its presence in a given individual should be regarded as a mitigating circumstance. It therefore applies if the desire is significantly more common in some population to which the agent belongs than it is in most other populations owing to genetic differences between that population and the populations in which it is less common. I conclude that even if it could be proved beyond the shadow of a doubt that the high incidence of some type of criminal behavior in a certain population was due to genetic causes, causes that operated by affecting "internal" factors – by producing resistible desires; by warping values that the agent could see to be warped by reflecting on other values that he or she has; by diminishing (but not eliminating) the agent's capacity to deal with desires he or she wishes not to act on – this discovery would be morally and legally irrelevant. The laws governing that sort of criminal behavior ought to be the same for, and applied with the same degree of rigor to, the members of that population as everyone else.[10] And we should make the same moral judgments about those who are members of that population and engage in that behavior that we make about those who are not members of that population and engage in that behavior.

NOTES

1. This point applies not only to gene combinations, but to any factor that might be a cause or partial cause of bad behavior in an individual. We shall later apply it to desires and other psychological factors.

2. Marcia Baron has asked me how it can be that the behavior of the members of X is genetically determined if they are able to take into account the possibility of being punished, and (by implication) sometimes refrain from an act of rape they would otherwise have committed if they believe the risk of punishment is too high. I think we must distinguish between being determined and being irrational (i.e., not being rational in the "value-free" or "Humean" sense). The men who belong to X, as I am describing them, have the following dispositional property: whenever they see an opportunity to commit rape and believe that they could get away with acting on it, they do act on it. I am supposing, moreover, that there is good evidence that it is genetically determined that they have this dispositional property.

3. It is consistent with the evidence we have imagined that each rapist in population A was genetically determined to commit rape on the particular occasions on which he did. But the evidence would provide no reason to suppose that this was in fact the case.

4. The fact that "external" factors (like temptation) can reinforce or otherwise affect "internal" factors (like desire) suggests that the distinction between the prevalence of a kind of behavior in a population being due to external factors and its being due to internal factors is considerably more complicated than what I have said in the text allows – perhaps even that it is a dubious distinction. But if this is so, it can only strengthen the case for the conclusion that we ought to resist our tendency to regard this distinction as morally significant.

5. I am not supposing that "an inordinately strong desire for immediate sexual release" is normally or ever the "cause" of rape. I am supposing that if an inordinately strong desire for immediate sexual release was much more common in some population than in most, this could explain why a higher-than-normal proportion of the men in that population were rapists. This is like supposing that the fact that the summer of 1982 was very dry could explain why there was a higher-than-normal number of forest fires that summer. It certainly does seem plausible to suppose that if terrorists were to add to the New York City water supply a drug that causes men to experience an inordinately strong desire for immediate sexual release, the number of rapes committed in New York could be expected to increase: no doubt many men whom various factors predisposed to rape, but who would, nevertheless, not have committed rape in the normal course of events, would be "pushed over the edge" by ingesting the chemical.

6. Even the man who is now a hardened, habitual rapist will almost certainly have been in this state "at first" – when he committed his first rape or his first few rapes. We may therefore hold him responsible for the rapes he commits in his present state, for we may hold him responsible for the fact that

he now lacks the ability to resist those desires. Or at least this seems reasonable to me. I have defended a position of which this thesis is a special case in "When Is the Will Free?" (1989).

7. I am here discussing only questions concerning the sorts of judgments we should make about, and the attitudes we should take toward, a particular individual and a particular act he has performed. I do not mean to imply that we have no obligation to try to find a way to lighten or remove the psychological burden that an individual bears. And I certainly do not mean to imply that we have no obligation to try to find a way to reform the corrupt environment that has placed that burden on him.

8. Recurrent opportunities to commit some wrong act will be a misfortune only for those who have some "standing" desire to commit that act. But then being subject to recurrent desires to commit some wrong act will not be much of a misfortune for those who have no opportunity to act on them.

9. These examples are cases in which circumstances mitigate the wrongness of an act because they are cases in which circumstances dictate that the alternative to performing the wrong act is to cause or allow something very bad to happen. (Indeed, if the alternative is bad enough, most of us will want to say that the act was not, in the circumstances, wrong; most of us would probably judge that it is not wrong to steal food if one's children are starving. But the examples can easily be modified so that they are clear cases of wrong acts whose wrongness is mitigated by the circumstances under which they are performed. Suppose, for example, that the children in the "starving children" case are in fact not starving but are nevertheless painfully thin and ill-nourished, and that the mother steals food from a family even more needy than hers.) But there is no "bad alternative" to rape – except in extremely rare cases like the one imagined in the text. It may be that the rapist would regard the existence of an unfulfilled desire to commit rape as a "bad alternative," but most of us will not, and we shall therefore say that the presence in him of a desire to commit rape was not a mitigating circumstance.

10. At least if the only thing the legislatures and the courts are considering is the *fairness* of the laws and the sentences. If deterrence is a factor in their considerations, it might be advisable for them to adopt a different legal strategy with respect to members of that population.

REFERENCE

van Inwagen, Peter. (1989). When is the will free? In *Philosophical perspectives* vol. 3, ed. J. Tomberlin, 399–402. Atascadero, Calif.: Ridgeview.

Chapter 10

Genes, Electrotransmitters, and Free Will

P. S. GREENSPAN

There seems to be evidence of a genetic component in criminal behavior. It is widely agreed not to be "deterministic," by which discussions outside philosophy seem to mean that by itself it is not sufficient to determine behavior. Environmental factors make a decisive difference – for that matter, there are nongenetic biological factors – in whether and how genetic endowment manifests itself phenotypically.[1] Moreover, even if its manifestation were inevitable, its bearing on criminal behavior apparently turns on general personality traits on the order of "impulsivity" that under different conditions of life could take very different forms. My concern in this chapter is not with "genetic determinism"; it should be obvious that the notion makes sense only if determinism admits of more limited (and less worrisome) forms than the one involving universal causal necessitation that philosophers have in mind by the term. Instead I want to ask whether current accounts of the link between genes and criminal behavior would manage to undermine free will anyway, even apart from worries about determinism (Greenspan 1993, 31–43).

I eventually suggest that some of the other terms in this area should also be pried apart. In particular, the genetic accounts in question may

I owe thanks to David Wasserman and Jorge Garcia for comments on the first draft of this chapter in April 1995. I received further comments on later drafts of all or part of the paper from Daniel Dennett and others attending philosophy department colloquia in New Zealand and Australia from June through October, 1995, at the following universities: Auckland, Otago, Canterbury, Australian National University, and the University of Queensland. Let me also thank the Psychology Department at Sydney University for their reactions. I made some last-minute corrections after reading the penultimate version at the June 1997 meetings of the Canadian Philosophical Association with John Baker as commentator, but have left some further thoughts for another paper.

well be thought to undermine freedom and yet to allow for responsibility in the sense of blameworthiness. Omitting reference to issues raised by the philosophic literature (but inserting some parenthetical cautions against misinterpretation), my discussion can be summed up in capsule form as follows:

The implications of genetic research for issues of responsibility depend on what sorts of causes of behavior it yields evidence for (assuming it yields evidence for any). One distinction commonly made is that between causes of normal agency and "interfering" causes. A simple sort of psychological model in terms of interfering causes is provided by the usual construal of kleptomania and similar (putative) cases as involving irresistible impulse. I make these out as cases of "psychological compulsion" insofar as they involve a kind of internal interference: a threat of continuing mental disturbance that is sufficient to interfere with the ability to do otherwise.

However, there is an alternative picture guiding current research on criminal behavior (among other things) – of "inadequate resources" for behavioral control. This substitutes for interfering causes an absence of "enabling" causes of normal control, such as adequate supply of serotonin and other electrotransmitters. The effects of serotonin shortage on criminal behavior according to the current accounts would seem to involve a kind of localized learning disability. But even if this undermines freedom, the relation of the incapacity in question to norms of motivation may still allow for an element of moral responsibility. Blame as a reactive attitude may still be warranted (even if it also ought to be offset by a contrary emotion) toward an agent whose unresponsiveness to social learning manifests itself in patterns of voluntary harm to others.

My argument divides into three main parts: first, I discuss standard philosophic approaches to free-will issues, including some work of my own; second, I attempt to make a different kind of sense of current genetic (and other biochemical) accounts of "aggressive impulsivity" as problematic for freedom; and, third, I indicate how we might defend an element of responsibility in the cases in question even without free will.

INTERNAL CAUSES

According to most philosophers causal determination of behavior or behavioral traits does not pose any particular problem for freedom as long as it works within normal motivational channels of the sort involved in moral agency. This view of free will, dominant in modern philosophy, is known as "compatibilism" (because it understands free will as compatible with determinism), or sometimes "soft determinism" (because it effectively "cushions" our ordinary notions of moral responsibility against the effects of determinism). As a general approach to free-will issues the view rests on a distinction between two possible ways of influencing an agent's behavior: first, one that allows the agent the usual causal role in generating behavior, via deliberation, choice, and other exercises of what we call "the will," even if these are ultimately traceable to external factors; and, second, another mode of influence that essentially shortcuts the will and bypasses normal agency, by linking behavior more directly to external causes.

An easy illustration of this distinction might be provided by contrasting, on the one hand, a typical sort of physicalist account of action in terms of desires, taken as equivalent to brain states and traced back to physiological and environmental causes, and, on the other hand, action produced by experimental intervention in the normal course of desire formation, say by electrical stimulation of the brain. On a compatibilist account, only the latter sort of influence would undermine the agent's responsibility, by attributing responsibility to something else instead of the agent, a cause that interferes with normal agency, and not just by explaining normal agency in terms of something further that allows it the same pivotal role in the chain of events.

Whether or not the compatibilist account yields a plausible approach to free-will issues generally, the intuitive distinction on which it rests does seem to ease worries about genetic explanation in particular. Current advances in genetic research – the prospect of coming up with more and more genetic explanations of human malfunction, possibly including behavioral malfunction – are treated in the popular and scientific press as if something special were at issue beyond the usual worries about *any* form of scientific explanation. But it is unclear why genetic causes would be more problematic in principle than any other causes we might discover – social and environmental causes for criminal behavior, say – unless they in some way shortcut normal agency in the manner of the brain experimenter. If they leave our usual picture of moral decision making intact and simply enlarge it with an account of the sources of some particular patterns of decision making – behavioral traits like those we sum up as individual moral character – then there are two possibilities: either compatibilism is true, and our normal responsibility ascriptions still apply; or compatibilism is false, and causal explanation would undermine our normal ascriptions. But the same would also have to be said for nongenetic causal explanation, in terms of social patterns, events in the environment, and similar factors. So news of advances in genetic explanation is at any rate no *worse* news for free will than potential advances in the other forms of scientific explanation that it replaces or supplements.

Perhaps it is worse news in something like a "public relations" sense: it looks worse, just to the extent that genetic explanation combines so simply and readily with nonscientific views of personality and personal causal influence of the sort summed up by the notion of moral character. A common philosophical alternative to determinism, a form of incompatibilism known as "agent-causation," understands "the agent" as a distinct type of cause, to be distinguished from putative "event-causes" of action (Donagan 1977). But genetic explanation of personality traits would seem to insert the agent – or his nature as an agent, his moral personality or character – into a deterministic chain of events, by way of a causal account of personality formation. The appeal to genetic causes of character may look worse for free will, then, to the extent that it is capable of co-opting the explanatory model assumed by a standard incompatibilist defense of free will. However, there also are more selective ways of understanding free will, without treating our ordinary talk of agency and decision making as unanalyzable into simpler causal terms, so that at most only formative causes can get a grip.

A more selective account could allow for our intuitive distinction between acts that are within an agent's control and others that may not be, on the model of the distinction I outlined earlier between a normal act and an act resulting from electrical brain stimulation by an experimenter. The distinction depends on the particular causal mechanism appealed to in explaining action rather than on causal explanation per se. The case of experimental brain stimulation is a relatively easy one, because it picks out another agent, the experimenter, as a cause acting earlier that can serve as an alternative locus of responsibility. But there are more problematic cases, where the causal mechanism explaining agency seems to shortcut normal self-control in favor of something impersonal, on the order of a gene. On a more selective account, however, free will depends on *how* a gene (or other putative cause beyond the agent) gives rise to its behavioral effect. If it works through normal channels, as it were, allowing for normal exercise of choice or will in the agent, then genetic causation apparently falls on the unproblematic side of our intuitive distinction and leaves free will intact. In this area, genetic explanation indeed may pose a threat to our ordinary view of freedom, particularly as illustrated by current research on criminal behavior. But the threat is not genetic causation *simpliciter* but rather the combination of that (or some other) appeal to impersonal causes of character or temperament with a particular understanding of the causal mechanism as one that essentially takes self-control out of an agent's hands.

Compatibilist and incompatibilist defenses of free will agree in locating the causes of free action within the agent in some sense, in contrast to causes operating on the agent from without. The distinction follows Aristotle's treatment of compulsion as a species of involuntary action, although Aristotle was not concerned particularly with the possibility of causal chains leading back beyond the present situation or with psychological variants of compulsion (*Nicomachean Ethics*, bk. III, sec. I). The more problematic cases arise when we allow for internal (in the sense of "mental") causes that still might seem to be external to what philosophers think of as "the agent," or the bearer of moral responsibility: the agent's core self, or "will" as an active principle, distinct from at least some of the various mental states an agent might exhibit. There are impulses, say, on the order of the kleptomaniac's urge to steal (to take a familiar problematic case), that might be seen as "coming over" an agent, characterizing him only in a passive sense, even if they do so regularly enough to be said to be "in character."

An account along Aristotelian lines might treat some such cases as voluntary, to the extent that the will may be seen as actively yielding to impulse, on the model of a robbery victim coerced to surrender his money. He hands over the money because he wants to (and reasonably so) under the circumstances – circumstances of external threat – though he does not want to be in those circumstances in the first place or to have to make the choice he does. I at one point proposed a different application of the model to psychological compulsion, understood as involving internal constraint, presumably by a threat of emotional discomfort (Greenspan 1978). This would make the dividing line between free and unfree action a matter of degree: whether a threat is disturbing enough to interfere with the ability to do otherwise.

The model may initially seem strained in application to kleptomania, if there really is such a thing, so that we do want to say that the agent's action in such cases is unfree. The only threat a kleptomaniac faces if he resists the urge to steal would seem to be the continued agitation of that unsatisfied impulse. Of course, frustration is unsettling; but is it disturbing enough in these cases to be said to make the agent unable to refrain? Does the kleptomaniac really act on an "irresistible" impulse, an impulse that, if not strictly impossible to resist, is at any rate hard enough, in light of the threat, that resistance is unreasonable for us to expect of the agent under those circumstances – on the model of a spy who gives up information under torture? He may have overriding reason to resist, that is, but on this account he is overcome by a threat of continuing distress that undermines his ability to act on the balance of reasons.

The account does not really turn, then, on how bad a threat is (or even how bad it seems in prospect) but rather on how *upsetting* it is, how much it tends to disrupt deliberation, or more specifically, self-control, the ability to resist acting on impulse. So the model of psychological compulsion or internal interference might be extended to fit kleptomania and like cases, along with more usual instances of cravings and addictions that similarly might not involve intense discomfort. But there is another possible approach, distinct from psychological compulsion, that is exemplified by the current research on criminal behavior. Instead of looking at the element of internal threat – the urges that arguably overcome the will to resist and hence make action less than fully voluntary – we might shift our explanatory focus to the will itself, the psychological resources that enable an agent to resist: what is commonly called "strength of will." It is also commonly thought of as something up to us, but that is what a genetic basis seems to call into question.

INHERITED IMPULSIVITY

The current research attempts to locate causes of criminal behavior in a genetic abnormality in the supply of electrotransmitters and other biochemical factors that regulate self-control. The usual appeal is to serotonin shortage; "impulsivity" is the associated trait, manifesting itself in aggressive behavior where other factors are in play, but also at issue in other forms of behavioral or mood malfunction such as depressive tendencies (Raine 1993). In the case of criminal behavior, then, the explanation apparently offers impersonal causes (genes and their biochemical effects) for processes that limit the ability to exert behavioral control.

This explanation is an alternative to representing self-control as overcome by an internal threat. The agent lacks the resources to withstand certain normal kinds of mental upset – the frustration of an impulse or other desire – rather than facing an abnormal kind or degree of upset, as suggested by the phrase "irresistible impulse" in the kleptomania example. So an impulsive agent's will is undermined from inside, as it were, by the absence of an enabling factor presupposed by normal agency, rather than being defeated or counteracted by a powerful interfering factor (even one that is internal to the agent in some sense) on the order of an impulse. His will or his character is defective, in the sense of just not being up to the task imposed on all of us, the task of normal behavioral self-control. But if the defect is inherited, one might ask, who is to blame?

The point is not that genetic explanation of criminal behavior is itself problematic for free will but rather that it may pose problems in combination with a shift away from our intuitive view of the causal mechanism involved in failures of self-control. One might say that the current research model takes the "mania" out of kleptomania – the suggestion of "irresistible impulse" as a stirred up counterforce to self-control, like a demon fighting oneself – and substitutes personal inadequacy, a shortage of the means of self-control, with an impersonal cause. The cause on this account, rather than shortcutting the will, *shortchanges* it: it denies the agent the very stuff of self-directed agency, on the model of a spy under torture who just becomes too exhausted to resist after a long period of sleep-deprivation. The cause of uncontrol here is something more like depletion than disturbance or discomfort.

I do not mean to suggest that such a simple account of the postulated causal mechanism is adequate to capture it. Nor am I in a position to assess it fully. It may or may not replace our intuitive view of these cases. But for purposes of argument I want to keep it in simple terms and just suppose that it is essentially on target. It does seem to have had some successes, as indicated by practical criteria, most notably the effects of medication such as Prozac in controlling depression and related forms of impulsivity (Kramer 1993). I think we can see that, if this should indeed turn out to be the appropriate model for impulsive (or even just some impulsive) criminal behavior, it would essentially set up the relevant trait as an inherited inability. By virtue of the genetic bases of personality, according to the model, certain agents simply lack the biochemical equipment for normal self-control.

The "impulsive" criminal has the same impulses as the rest of us, at least initially, on this account; he simply lacks our capacity to suppress them, or to let them pass unsatisfied. He is no more disturbed at suppressing them but just cannot manage it, or cannot manage to learn to do it, because he lacks the requisite supply of serotonin. What serotonin does is to facilitate transmission of impulses (in the electrical sense: nerve impulses) across synapses. This might be seen as a graphic depiction in neural terms of failure to "let go." It amounts to a defect of thought processing. In application to criminal behavior, according to the current accounts, it apparently involves an inability to learn from punishment (Raine 1993, 93). But such inability is no more within control, if it is genetic, than other forms of learning disability.

To take a case we can understand from the inside, consider the effects of memory loss due to age. Once I have lost a certain number of brain cells, let us suppose, I no longer have the resources (on a comparably primitive picture of mental storage) to memorize this essay for oral presentation – or rather, perhaps, to do so without undue distraction, in the way required by a reasonable performance. So there is a clear sense in which I cannot manage the task: I cannot manage it easily enough to make it reasonable to expect of me, even on the assumption that it is possible, strictly speaking. We might imagine a situation where I am asked to give a paper extemporaneously. If my brain cells really are just not up to it at this point, then the demand is unreasonable.

This explanation is compatible with the acknowledgment that there may be various indirect or extraordinary measures, beyond the normal means of deliberative self-control, that I might take, or might have taken, to build up my mental storage capacity. Similarly, one might say, for someone with an inherited tendency to impulsivity who commits a crime because he does not have the thought-processing capacity needed for normal resistance to impulse: his will is impaired, or the cognitive underpinnings of will or decision making are impaired, just as we are supposing my memory is. We both fall short of a general human norm, though my deficiency counts as a normal effect of age, a normal process whose unwelcome effects we might like to change but have more or less learned to live with, whereas his is traced back to a genetic abnormality, with behavioral results we cordon off as unacceptable because of their moral significance.

Besides this reference to norms, there are other likely differences between the two cases. A crucial difference is offered in my next section. But, first, one might want to say that the cases differ just insofar as memory defects tend to frustrate an agent's ability to act in his own interests, so that they mainly are unwelcome to us. However, the same may be true of the aggressively impulsive agent. He should not be assumed to be a psychopath, in the sense of someone without guilt or other moral feelings based on empathy with his victims. At any rate, he may have such feelings after the fact, when he manages to reflect on his behavior. If aggression is something he cannot help, though, in the sense I have tried to capture, I think we would take his behavior to be *unfree* even if it did not involve regret or other motivational conflict.[2]

Another thing to consider is the role of differences in social placement of the agents in the two cases, as exemplified by the likelihood of my knowing about various self-training practices and memory aids. This means that short-range unfreedom in a case like mine – the genetically based inability to do otherwise, at least in immediate terms, as things stand – does not imply that my situation cannot be changed by my own efforts. By contrast, the impulsively aggressive agent is likely to be someone who, by virtue of upbringing and current situation, just would not know how to change his habits of thought processing, or even that he has reason to do just that. Nor would he find out soon enough and with enough external resources to manage the requisite behavioral change.

If society does know these things and does have the relevant resources and is partly responsible for the situation where the agent does not, it might be said to bear responsibility for such cases, contrary to the moral often drawn from popularized genetic accounts, which tend to be politically conservative. However, to say that society is responsible need not be to deny all moral responsibility to the agent, even supposing that an agent is not *free* if his action results from an inherited deficit of the electrotransmitters needed for normal self-control. The notions of freedom and responsibility, along with other elements of the debate on these issues, need to be pried apart to yield an expanded conception of the alternatives (and with it, a less polarized picture of the political options) than the debate usually assumes.

REACTIVE BLAME

Even if we make all the assumptions I have allowed so far about the success and free-will implications of current causal models of criminal behavior, it is possible to detach the question of freedom from the backward-looking or "reactive" component of responsibility (Greenspan 1987, 63–80; esp. 78–79; cf. 70).[3] The will may be subject to a disability – internal limitation as opposed to interference – but some form of blame may still be warranted to the extent that the act in question does depend on the agent's will and hence counts as voluntary (Gert and Duggan 1986).

In Aristotelian terms, the immediate cause of action in these cases is within the agent, though it undermines his ability to choose, in the sense that implies deliberation (Aristotle, *Nicomachean Ethics*, bk. III, sec. 1). I would call these cases of weak character, with impulsivity seen as a relatively stable trait of an agent, a trait of temperament liable to moral blame; in philosophers' terms it yields weakness of will, whether or not it is ultimately caused by some shortage in the prerequisites of normal agency (*Nicomachean Ethics*, bk. V, sec. 8, 1135b18 ff).[4] Unlike some contemporary proponents of virtue ethics, Aristotle of course has no intention of banishing blame from ethical discourse; he distinguishes virtue and vice, as states of character, from passions or feelings by the fact that we do consider them fit subjects of praise and blame (*Nicomachean Ethics*, bk. II, sec. 6, 1105b28–106a2; cf. esp. Williams 1985). What he has in mind by blame is itself something other than a feeling; an alternative translation is "censure." But it is worth noting that Aristotle's definition of anger in the *Rhetoric,* in terms of pain at an unjustified slight to ourselves or our friends, would allow for a corresponding moral sentiment in cases like those we are concerned with, supposing that the aggressively impulsive agent, with whatever incapacities we have allowed him, is at any rate capable of *insult* (Aristotle, *Rhetorica:* bk. II, ch. 2). I think there is something right about this, as a point about emotional blame – and ultimately about warrant for a corresponding judgment of responsibility.

With aggressively impulsive behavior taken as unfree but voluntary, we can pinpoint the main way such cases differ from genetic impairment of memory – whether due to age, as in my earlier analogy, or to some sort of lifelong mental handicap. Volitional defects seem to be internal to an agent in a sense in which purely cognitive deficiencies of the usual sort are not. The distinction between them makes sense if we think, along more Humean lines, of the notion of an agent's "character" as set up to provide a proper target of emotion in its role in enforcing moral norms.[5] Retributive emotions like anger and resentment are justified in general terms, counting as appropriate in the sorts of cases where they normally serve as ways of modifying behavior, such as cases of volitional failure, even if the behavior turns out not to be modifiable in the particular case at hand.[6]

The question of responsibility often takes the form in contemporary discussion of a question about the justification of retributive emotions because of P. F. Strawson's influential treatment of free will as a question about our "reactive attitudes": whether they (and the social practices built on them and supporting moral judgment) could survive a serious belief in determinism (Strawson 1962, 1–25). Although my concern here is not with determinism (in the philosophers' sense of universal causal necessitation), Strawson's appeal to emotional and other reactive attitudes helps in addressing a separable worry: that at most we would be "going through the motions" of responsibility attribution if we imposed punishment just to influence behavior.

Although society has to protect itself from aggressive crime, typically by imposing legal penalties, whether the penalties amount to anything like justified punishment would seem to depend on what sort of attitude toward the lawbreaker is warranted (Feinberg 1970, 95–118). The thought that inflicting punishment on someone would avert future harm is notoriously inadequate. Punishment is supposed to express blame. This is not meant to say that we necessarily ought to feel blame when we punish or that expressing blame is the prime purpose or justification of punishment (or even a consideration that ought to affect the penalties we inflict) but just that the penalties are supposed to be justified *as* expressions of blame. That would seem to be the test of whether it is punishment we are justifying, or something else – some use of the agent for others' good or her own that is not in the same way directed toward her as a person.

Strawson (1962) thinks of blame as involving a "participant" attitude toward its object – treating the object of blame as a fellow participant in the moral institution. The alternative is the "objective" attitude, which is essentially impersonal. Although his account allows for a continuum between these two poles of reaction, thinking of some agent as abnormal is supposed at least to incline us toward emotional distance. (Strawson 1962, 11; cf. pp. 9, 19).[7] However, I think that a fuller picture of our emotional alternatives in such cases would still permit a form of blame for aggressively impulsive agents, as cases of volitional impairment that involve defective character.

Consider again the contrasting case of impaired memory. On plausible versions the problem is not localized in the way that aggressive impulsivity is. We would indeed be inclined to blame someone with a comparably selective form of memory failure – for instance, someone who could not seem to keep track of any commitments to other people while absorbed in his own pursuits. What would be subject to moral criticism in such cases is something like the agent's personal priorities, a basic fact about his character, even if not itself (or currently) open to control. Similarly, what is blameworthy about the aggressively impulsive agent is the way his impulsive behavior targets others, rather than being channeled into harmless or, at any rate, self-destructive avenues like so many of the rest of us. If he were impulsive in all ways at once, he would not represent a coherent object of blame; but the fact that his character developed in the particular direction it did requires a further explanation that may be extenuating but not exonerating, depending on details.

There are different possible ways of filling out the cases in question, that is. Some may rest on a degree of delusion – not immediately a problem of control, but rather of knowledge. In Aristotle's terms the excusing condition might not be external causation (compulsion) but instead a form of ignorance. One possibility mentioned in the literature as linked to serotonin deficit is a pattern of attributing hostile intent to others. An agent whose aggressive behavior stems from this limited sort of delusion might not be fully responsible in the usual terms to the extent that he misconstrues the situation in which he acts – if his delusion is not willfully self-induced, on the model of cases of harm done while intoxicated.

Some other possibilities that come up often in discussion seem to involve an appeal to childhood suffering on the part of the agent. If the cause of serotonin deficit were some sort of childhood trauma – as it could be in other cases, such as cases of abusive punishment during childhood – we might want to withhold a punitive reaction on the grounds that the agent has already suffered enough. But a genetically induced imperviousness to punishment may not always lead parents and other caretakers to pile on more. There may be other cases where adults just give up on punishing a child who is unresponsive to it, so that he actually manages to escape some of the normal forms of childhood emotional suffering.

More to the point would be a simpler appeal to the age of character formation. The agent's priorities presumably were set at a stage of development too early for intelligent choice. They might be treated as basic beliefs about what is important, what matters and at what cost – when other agents' alleged hostility deserves a violent response, say. But note that these are evaluative beliefs, encoding moral error rather than ignorance of fact. In Aristotle's framework this does allow for responsibility – at least of the sort that goes with weakness of will, where the belief encoding an agent's priorities is in conflict with his "better judgment."

Sympathy may still be warranted toward such an agent, in the way that we sometimes have in mind when we attribute someone's immature adult behavior to the fact that he was "spoiled" as a child, even if blame is warranted too. We feel sorry for him in part just because he is unfree: "He can't help himself." But blame can also be seen as a legitimate response to such an agent by virtue of his character. This need not mean that we hold him responsible *for* his character, certainly not in a sense that entails control at some earlier stage over its formation. Rather, we hold him responsible for action – not for his character per se but for a consequence or expression of it – insofar as we attribute what he does to something basic about who he is: a self-defining pattern of response that at this point sets his priorities.

Familiar cases, in philosophy and in everyday life, indicate that we do blame people for inadequate emotional resources of the sort that affect moral agency. One of the advantages often cited of an Aristotelian or Humean over a Kantian approach to ethics is the way the former allows for our intuitive view of the moral status of character traits involving emotional response. The image of an unsympathetic person, a moral Scrooge, awakens feelings of censure, which may be softened but are not stilled by the acknowledgment that Scrooge was shaped by his background and in current terms (under more ordinary circumstances) is not open to change. The same applies to the sort of lawbreaker who is commonly called "incorrigible," a word that in common parlance can itself be used to express blame.

As to our own responses to such a person, note that anger is not the only available emotion that might be proposed as a form of blame. If the impulsively aggressive agent has a defective character, more passive emotions on the order of contempt – scorn, repugnance, horror, and so on; the kind of disgust that goes with "He's incorrigible!" – would provide a version of blame that is sufficiently punitive for most purposes. A feeling of contempt on the part of others (or, for that matter, oneself) is aversive for most agents and hence can be used to control behavior, if we assume a modicum of control over emotions themselves via acts of attention. Whether or not retributive, then, contempt is in that sense a punitive reaction. However, it would not be ruled out – if anything, it might be reinforced – by the admission that its object is unfree. He is more to be pitied on that account, perhaps, but in the sense of pity that is not at all incompatible with looking down on someone. In any case, anger itself (including as a variant Strawson's resentment) is not incompatible with pity – or, for that matter, with various more charitable emotions like compassion – even if the content of these opposing reactions might be said to pull in opposite directions, one pressing for retribution, the other for some form of aid or personal support. Ambivalence would seem to provide an answer to the worry that our usual range of moral reactive attitudes may be undermined by unfreedom (Greenspan

1980, Watson 1987). Instead, the range may be expanded to include various opposing emotions. This broader perspective is not the same as an "objective" stance, even on the assumption that the conflict ultimately should settle into a more balanced attitude, of "mitigated" blame, as the proper all-things-considered backdrop to legal punishment, which of course has to involve a blend rather than simple juxtaposition of opposing considerations.

What emotional or "reactive" blame registers, on this account, is the salience of the agent as a personal cause of wrong, something that is not erased by the acknowledgment that he was unfree to do otherwise. In terms of Aristotle's definition of anger, the agent who cannot control his aggression deserves anger to the extent that he "slights" others. If we cannot expect change of him, then instead of forgiveness in the sense that entails social acceptance or reconciliation, blame fades into a tempered attitude that resembles objectivity in its lack of personal involvement but still involves a negative moral evaluation. My conclusion is that we do have a foundation here for a form of moral blame.

NOTES

1. For a sample of the recent discussion of these issues, with particular stress on the positive implications of genetic findings for developmental research, see Gottesman and Goldsmith 1994.
2. The interpretation of unfreedom in terms of motivational conflict is usually associated with the two-level view of volition in Frankfurt. On my current reading, however, Frankfurt would make a similar point to that expressed here in terms of reactive attitudes, by appeal to his distinction between free will and free action. See Greenspan 1999.
3. My initial argument for this point appears in Greenspan 1987.
4. Here Aristotle discusses action from anger and his distinction between unjust acts and unjust character. Because the latter implies choice in Aristotle's sense of deliberative decision, my use of "character" departs from his framework.

   I also ignore some other fine points of Aristotle's discussion, for example, his distinction in bk. VII, sec. 7, between weakness and impetuosity, as a kind of incontinence involving hasty action (prior to deliberation rather than ignoring its results). The term "impulsivity" suggests this possibility, although its application to depressive tendencies and the like argues for a broader interpretation.
5. For the treatment of blame as a "feeling or sentiment," cf. Hume 1978, 469. I do not take emotional blame to amount to a particular emotion, as will become evident. Nor do I take all blame to be emotional. The most basic form would seem to be linguistic, if we judge by etymology: the term comes from a Late Latin contraction for "blaspheme." I stress variants of anger in this dis-

cussion, as opposed to Hume's basis in personal hatred. For further discussion of Hume's view on these issues, see my "Humean Sentimentalism and Free Will" (in progress).

6. For my general account of these issues see Greenspan 1988. Its fuller application to ethics is given in Greenspan 1995.

7. Strawson (1962, 17) briefly suggests a different line (though not quite the one defended in this chapter) on specifically moral reactive attitudes. However, note that what he has in mind are essentially forward-looking attitudes – holding an agent to moral demands – as opposed to responsibility in the backward-looking sense I attempt to isolate (though cf. his remarks on p. 19). At any rate, Strawson's concern with universal determinism makes his discussion at points inapplicable (as he frequently acknowledges) to the more complex sorts of cases at issue here.

REFERENCES

Aristotle. *Nicomachean Ethics*. Trans. D. Ross. Oxford: Oxford University Press.
  Rhetorica. Trans. W. R. Roberts. In *The Basic Works of Aristotle*, ed. R. McKeon, 1318–1451. New York: Random House.
Donagan, A. (1977). *The theory of morality*. Chicago: University of Chicago Press.
Feinberg, J. (1970). *Doing and deserving*. Princeton, N.J.: Princeton University Press.
Frankfurt, H. G. (1971). Freedom of the will and the concept of a person. *Journal of Philosophy* 68: 5–20.
Gert, B., and Duggan, T. J. (1986). Free will as the ability to will. In *Moral responsibility*, ed. J. M. Fischer, 205–224. Ithaca, N.Y.: Cornell University Press.
Gottesman, I. I., and Goldsmith, H. H. (1994). Developmental psychopathology of antisocial behavior: Inserting genes into its ontogenesis and epigenesis. In *Threats to optimal development: Integrating biological, psychological, and social risk factors*, ed. C. A. Nelson, 69–104. Hillsdale, N.J.: Erlbaum.
Greenspan, P. S. (1978). Behavior control and freedom of action. *Philosophical Review* 87: 22–40. Reprinted in *Moral responsibility*, ed. J. M. Fischer, 191–204. Ithaca, N.Y.: Cornell University Press, 1986.
  (1980). A case of mixed feelings: Ambivalence and the logic of emotion. In *Explaining emotions*, ed. A. O. Rorty, 191–204. Berkeley: University of California Press.
  (1987). Unfreedom and responsibility. In *Responsibility, character, and the emotions: New essays in moral psychology*, ed. F. Schoeman, 63–80. Cambridge: Cambridge University Press.
  (1988). *Emotions and reasons*. New York: Routledge, Chapman and Hall.
  (1993). Free will and the genome project. *Philosophy and Public Affairs* 22: 31–43.
  (1995). *Practical guilt*. New York: Oxford University Press.
  (1999). Impulsivity and self-reflection: Frankfurtian responsibility versus free will. *Journal of Ethics* 3: 325–340.
Humean sentimentalism and free will. In progress.

Hume, D. (1978). *Treatise of Human Nature*. Ed. L. A. Selby-Bigge. 2nd ed. Oxford: Clarendon.

Kramer, P. D. (1993). *Listening to Prozac*. New York: Viking.

Raine, A. (1993). *The psychopathology of crime: Criminal behavior as a clinical disorder*. San Diego: Academic Press.

Strawson, P. F. (1962). Freedom and resentment. *Proceedings of the British Academy* 48: 1–25. Reprinted in J. M. Fischer and M. Ravizza, *Perspectives on Moral Responsibility* (Ithaca, N. Y.: Cornell University Press, 1993), 45–66.

Watson, G. (1987). Responsibility and the limits of evil. In *Responsibility, character, and the emotions: New essays in moral psychology*, ed. F. Schoeman, 256–286. Cambridge: Cambridge University Press.

Williams, B. (1985). *Ethics and the limits of philosophy*. Cambridge, Mass.: Harvard University Press.

Chapter 11

Moral Responsibility without Free Will

MICHAEL SLOTE

The title of this chapter sounds like a contradiction in terms, but let me try to persuade you otherwise. The idea that moral or ethical value judgments can be made about people who are metaphysically unfree is, to begin with, familiar from Spinoza's *Ethics*. If we properly understand the notion of justice, we can make room for judgments of moral responsibility and accountability as well, without having to reassure ourselves that human beings have free will. In a virtue-ethical framework, I argue that if we understand moral obligation and social justice as calling for certain sorts of practical attitudes, or motivations, on the part of individuals and citizens, we can obviate the question of free will and still deal credibly and, if I may say, responsibly, with issues of genetic predisposition that have put such stress on recent thinking about individual and social morality. But let me first discuss the question of free will in a more general way.

The idea of moral responsibility without free will is not a contradiction, if the idea of a good society can *substitute* for the idea of free will in providing a basis for the punishments, penalties, incentives, and rewards that are the hallmark of moral and social accountability. Human beings, whether metaphysically free or not, can be placed in morally better or worse social conditions; and the laws instituted by a good or just society have or can have a legitimate authority over its members whatever the metaphysical status of their wills. Let us continue, however, to develop a fuller argument and, to that end, consider first what it would be like for moral value judgments to have validity in a world where people lacked freedom of will.

I am indebted to Jorge Garcia, Pat Greenspan, Judy Lichtenberg, Alan Strudler, Peter van Inwagen, and, most especially, David Wasserman, for helpful suggestions.

Spinoza considers such a universe and is happy to describe it for us. He holds that people can be more or less rational or virtuous even under conditions of total causal or metaphysical determinism, and because Spinoza is an incompatibilist about free will and determinism, this means people can be morally better or worse without any of them being metaphysically free.

How is this possible? Well, we can think better or worse of people's motives or character quite apart from whether either ever was or is under their control. A serial killer may be judged morally vicious even if we believed he or she was *made* vicious by the cruel treatment of his or her parents. What the person is now like is bad, ethically or morally bad, even if we judge him or her to lack all responsibility for being that way; and Spinoza himself spoke of virtuous or vicious traits in people while denying that anyone was free or morally responsible for his or her moral character.

Moral characterizations, then, seem feasible under conditions where free will is either doubted or denied, and under such conditions, they are not (as has sometimes been said) merely forms of aesthetic evaluation. Aesthetic evaluations, to be sure, are not tied to moral praiseworthiness and blameworthiness, and to that extent they resemble the moral characterizations I have just been talking about. But when we say that the serial killer is a "moral monster," that is no merely aesthetic judgment. It arises out of a concern for human life and well-being and is distinctively moral; and the use of "monster" indicates that although we think morally worse of such a person, blameworthiness or responsibility is probably not appropriate. Monsters (as in stories) are monsters by their very nature, having no choice in the matter, and in characterizing someone as a moral monster we make a moral criticism without (necessarily) imputing blameworthiness or responsibility for having or developing the traits that ground that criticism.

But having in this way assured the possibility of moral judgment in the absence of free choice or control, I may well seem to have backed myself into a corner in connection with the larger goals of this chapter. For my purpose here is to show that *moral responsibility* is possible without (considering) free will, and all I have shown so far is that moral criticisms that *evade all issues of responsibility* are possible in a world without metaphysical freedom. In distinguishing, as I did, between the viciousness of a serial killer and his or her responsibility and blameworthiness for being that way, the assumption that moral responsibility requires free will was never once doubted and was perhaps even essential to the distinction I was attempting to draw. So I seem as far away from making sense of the title of this chapter as I was at the start.

There is this difference, however. It may now be clearer how moral criticism and evaluation can occur and be valid in the absence of freedom and responsibility. Now we can develop interesting conceptions of social justice in particular without recourse to any metaphysical assumptions about freedom and be in a position to derive a proper conception of moral responsibility or moral accountability from considerations about social justice – again, without any recourse to ideas about free will. This, in turn, will enable us to view issues about genetic predispositions toward criminal behavior in a distinctive new light.

Now it has often seemed natural to base the idea of a just society on ideas about desert. We think certain crimes are worse than others and that their perpetrators deserve more severe punishments as a result. And social justice is supposed to enshrine these distinctions, so that, for example, any just society will insure that people are punished as they deserve for certain kinds of criminality. However, desert-based conceptions of justice, like Aristotle's in the *Nicomachean Ethics,* also assume that certain kinds of virtue and virtuous activity have to be *rewarded* in a society that is truly just. And so, more generally, such views hold that just societies and just laws must be responsive to and shaped by facts about individual desert that are antecedent to or independent of any institutions.

But such desert-based theories are no longer prevalent and are even in disrepute, in part because they reflect a strong feeling on the part of political philosophers that we can't really make sense of the idea of criminal or moral desert apart from particular institutional arrangements. It is thought, for example, that we can't make sense of the idea that a particular good person deserves certain social rewards or advantages or that certain criminal activity deserves so much punishment (but not more) in abstraction from particular social or legal institutions.

Interestingly, despite Kant's decided views about preinstitutional moral desert, his account of social justice is not anchored in such considerations. He explicitly holds that a legal system should not be attempting to apportion rewards and punishments in accordance with people's moral deserts. Rather, there is some evidence that Kant regards the fairness of punishments – for example, as a matter of the fairness or justness of the basic institutions and procedures of the society in which the punishments are being meted out (Hill 1992, 176–195). On such a conception, if a just society via just procedures institutes and applies certain penal laws, then the penalties and punishments thus generated are ipso facto just. And recent contractarians like John Rawls and Thomas Scanlon have to a large extent followed Kant's lead here and sought to understand social justice in terms independent of intuitive, preinstitutional notions of desert (or merit) (Rawls 1971, 103, 314–315; Scanlon 1988). But other approaches to social justice also allow us to proceed in this fashion, and in particular I think various forms of virtue ethics allow for it. Because I am a virtue ethicist and think such an approach more promising at this point than any form of contractarianism, let me explain how – or at least one way in which – the notion of just criminal laws and penalties, and the attendant idea of moral, or valid legal, accountability or responsibility, can be developed in virtue-ethical fashion.

Most virtue ethics has focused on individual morality and left larger social issues to other theories, and of course the forms of virtue ethics that flourished in the ancient world have a notorious record of siding with antidemocratic social ideals; Stoicism is a partial, but only a partial, exception to this criticism. In the modern world, however, we ethicists appeal to motives and character traits that did not much occupy or interest ancient thinkers – traits or motives like kindness, sympathy, compassion, and benevolence. And I would like to sketch for you very, very briefly a virtue-ethical ideal of individual morality and social justice that appeals to a certain kind of ideal of benevolence.

Utilitarianism famously appeals to the idea of benevolence as a touchstone of moral evaluation. Acts are right if their consequences are those that a foreseeing and impartial benevolence would have favored. But it is also possible for benevolence to play a less indirect and more foundational role in a moral theory. Virtue ethics in its most radical "agent-based" forms treats motivation or inner character as the basis for act evaluations, and an agent-based virtue ethics that treated one or another kind of benevolence as inherently the best or highest of motives could claim that the rightness or wrongness of acts depends on whether they sufficiently express or reflect such benevolence, rather than on their actual or objectively likely consequences. (Any such view will have some difficulty accommodating ordinary deontology, but that is a problem shared with utilitarianism and consequentialism generally, and it is a problem all these theories may somehow be able to surmount.)

Of course, whether someone is benevolent or not may not be subject to her control or free will, but that is just the point. An agent-based criterion of rightness and wrongness makes no assumptions about free will, and yet it does seem to distinguish between actions in a morally relevant way, because those who are benevolent and act accordingly seem to be morally better people and to act in a morally better way than those who are lacking in benevolence (e.g., selfish or malicious) and whose actions exhibit such deficient motivation. This is merely the sketch of a possible theory or set of theories, but the relevance to ideals of social justice might be put briefly as follows.

An agent-based view of social justice grounded in the idea of benevolence (and avoiding notions of free will) can say that a society is just to the extent its members, or those living in it, are benevolent (toward one another); and it can treat the evaluation of particular institutions or laws in a manner analogous to the agent-based treatment of act-evaluation.[1] It can say that an institution or law in a given society is just if and only if it expresses or is motivated by sufficient benevolence on the part of those who created it or sustain it in force. (By contrast, a direct utilitarian theory of justice would treat good consequences as the basis for describing a law as just.) All this, as I say, is but the barest sketch. But it gives us, I think, the basis for understanding how a theory of justice might allow moral responsibility in the door without any reference to or dependence upon considerations of free will.[2]

The plausibility of this sort of agent-based approach to social justice, however, can be challenged on the grounds that motivational assumptions seem hardly capable of putting sufficient constraints upon voters and or legislators to *prevent great miscarriages of justice*. In particular, how can our agent-based "justice as benevolence" guarantee that legislators and voters will not, say, pass laws that mandate twenty years in prison for ordinary car theft? If it cannot, then such views, by implying the possibility of *just* laws and punishments to that effect, are ultimately very implausible, and the idea that justice should be based in independent notions about desert would, in comparison, become more attractive. Agent-based justice as benevolence, however, may be able to answer such objections.

Think of what would have to be the case for justice as benevolence to allow a law providing for a mandatory twenty-year sentence for car theft to count as just or fair. How could a benevolent legislator think she was doing more good by instituting such a severe punishment rather than something milder? You may respond, What about recent "three strikes you're out" legislation concerning felonies? In such a case, the punishment introduced by legislation is very severe. But what about the motives of those who passed such legislation? Are they trying to do the most good they can for society or are they pandering politically to the

prejudices, fears, and resentments of constituents (or subject to such irrational attitudes themselves)? Are those clamoring for or instituting
such laws motivated by desires and attitudes that exhibit a concern for
their fellow citizens? I doubt that a view such as justice as benevolence
will force us into admitting the justice of "three strikes you're out" legislation or laws that make twenty-year prison terms mandatory for car
theft.

Of course, if legislators are sufficiently misinformed about the facts,
they may institute stronger (or weaker) penalties than they otherwise
would, but there are two cases here. If the legislators make strong benevolent efforts to learn the facts and are nonetheless thwarted, it may be
said that the penalties they institute are just. If they ignore the facts, then
their claim to genuine benevolence is commensurately undermined, in
which case the stronger laws they institute will not be just on an agent-
based account that appeals to the motive of benevolence.

It is difficult to imagine that legislators who bothered to inform themselves could think that legislating a mandatory twenty-year sentence for
car theft would benefit society more than other laws they might pass. To
that extent, an agent-based view such as justice as benevolence can argue
against those penalties and punishments that we intuitively regard as
undeserved or unmerited *without making any sort of independent appeal to
such notions.*

In that case, we are now in a position to see how an agent-based approach such as justice as benevolence supports ideas of moral accountability and responsibility. For it treats certain laws and punishments as
just (and fair) if they have the proper motivational provenance, and that
means that if, given the operation of just laws and procedures, someone
is justly sent to prison for a given term, then the person is in a morally
proper and valid fashion being held responsible or accountable for his
crimes. Does that not entail that, morally speaking, the person in question *is* responsible and accountable for his crimes? Any view that allows
a person to be justly punished for a certain crime and gives an account
of how this is possible has also vindicated the idea of moral responsibility or accountability in the bargain.

Notice too that we haven't just been speaking of punishments or
penalties that are mandated by some set of laws, but of punishments
and penalties instituted in accordance with *just* laws and (given the
agent-basing) motivated by the same *ideal or admirable* motives that
serve to justify moral evaluations generally. If benevolence is what
makes acts morally good, then laws and punishments also supported

on grounds of benevolence have been vindicated in terms of the same coinage, and there is every reason therefore to consider that vindication to be a moral one. An agent-based account that entitles us to hold people responsible for certain crimes has a good claim, then, to be a vindication of moral (and not just legal) responsibility in respect to such crimes, which accomplishes what we earlier set out to do.[3]

Moreover, a parallel justification extends to violations of norms or rules other than laws. If the rules of a tribe or even a family are based in benevolence, prescribed punishments or penalties for violations may be just, and the accountability for such violations has moral foundations. Modern, formal institutions are not, therefore, necessary to the kind of moral accountability or responsibility I am defending here; but it should also be clear that the theory I am offering does not allow us to speak of moral responsibility independently of norms or institutions and is thus somewhat narrower than our commonsense ideas about moral responsibility. Does the present view's inability to defend moral responsibility and just punishment in a "state of nature," however, count as a serious defect from the standpoint of moral theory?

This account also implies that convicting and punishing someone for a crime she didn't commit can be just, if just procedures were followed, but the evidence before a court was (inadvertently) misleading. This claim is not, in itself, particularly implausible; and neither does it automatically entail that such punishment is *deserved*, a conclusion that would be problematic. Scanlon, however, seems to be committed to just such a conclusion, claiming that "the only notions of desert which [my view] recognizes are internal to institutions and dependent upon a prior notion of justice: if institutions are just then people deserve the rewards and punishments which those institutions assign them" (Scanlon 1988, 188).

But this would appear to go too far; and I want to suggest, by contrast, that an agent-based virtue ethics should as much as possible avoid committing itself on questions of desert. If we can conceive social justice in agent-based terms and without bringing in the notion of desert, then the latter notion may be less crucial or basic to a theoretical understanding of morality than it was once thought to be; and because moral responsibility has now been defended without leaning on the concept of desert, perhaps that is all we need to say, for the moment, on this subject.

Having now seen that issues both of free will and of moral desert need not enter into a conception of just moral accountability, we need to consider, more specifically, how the approach taken here affects legal and moral issues concerning genetic predispositions to criminal behavior. Given what has been argued so far, questions about the just treatment of those with such predispositions do not turn fundamentally on whether or to what degree such people are free (able to control themselves and/or change). Rather, they turn on whether legislators and those they represent would be expressing or be motivated by (sufficiently) benevolent motives in recommending certain sorts of criminal penalties or certain sorts of educative measures (these alternatives are not exclusive) for such people.

Assume that we have identified people with genetic predispositions toward criminal behavior and that we (the electorate or their representatives – leaving aside, for simplicity's sake, the judiciary) are benevolent in the sense of seeking the public good, the general good of the country or other unit whose laws are at issue. Because we are worried about the effects of criminal activity on the security and quality of life in our country or state, there is reason – of a sort that tracks utilitarian reasoning but does so, so to speak, from within – for wanting to deter criminal behavior and incapacitate those who engage in it. This fact might lead one to suppose that a theory like justice as benevolence will be content to advocate fairly strong criminal penalties even for those who are predisposed toward and less able to resist illegal activities, thus risking accusations of inhumanity and callousness, which theories focusing on issues of self-control and free will in their discussions of legal justice attempt, and may manage, to avoid.

But justice as benevolence has a distinctive principled way of avoiding such harsh conclusions, and explaining why provides a deeper understanding of the nature of this and other agent-based theories of justice and moral responsibility. To focus on issues of retribution (criminal desert) and on the deterrent and incapacitating effects of criminal laws and punishments without due consideration of possible bad effects on (potential) criminals is to lack a generalized benevolence toward fellow citizens; the "throw away the key" attitude of many who advocate "three strikes" legislation seems incompatible with any kind of *generalized* or *impartial* benevolence. Utilitarianism, whatever its faults, at least takes the good of the potential criminal fundamentally into account in making political or legal judgments, and a virtue-ethical justice as benevolence (not surprisingly, given the similarities to utilitarianism) also provides ample room for such concern.

But the concern for (potential) criminals can and needs to take in more than the mere unpleasantness of incarceration for criminals. In particular, those with genetic predispositions toward criminal behavior may lack the full capacity for social self-control that others possess, and to that extent they and their lives are at risk in ways that are sometimes ignored in utilitarian and other discussions of criminal justice. The ability to conform oneself to society's norms is one element of mature adulthood: at a more advanced level and within a larger arena, it is analogous to the ability to tie one's own shoes or control one's bowels. But, as psychologists seem to agree, any child who cannot contain or control his excretory functions at the period of life when others are able to do this will inevitably feel shame and a lack of self-respect or self-esteem that can color many of his other endeavors and life experiences. Someone who lacks self-esteem or a sense of self-respect in important matters may find other things empty and not worth doing and/or lack the will and confidence to strive for what he does think worth doing or having (Rawls 1971, 440). Such a psychological state effectively undermines most of what normally is or can be good about a human life.

In any fairly decent society, law-abidingness constitutes, therefore, a basis for self-esteem that is clearly at risk among those with lesser ability to conform to social norms. (Of course, where the society is unjust or invidious in other ways, rebellion becomes a more likely and reasonable basis for self-esteem, and conformity can actually undermine self-esteem.) Thus, the goodness, the quality, of the lives of those with a genetic predisposition to criminal (non-self-controlled) behavior is at risk in a distinctive and crucial way. If they (are allowed to) fall into criminal behavior and recidivism, and their self-respect is undermined, then, quite apart from the unpleasantness and purely economic opportunity costs of being in prison and having a criminal record, they may lose the chance for worthwhile lives. Lacking hope and/or a sense of the real value or attainability of the things that make up most of what is good in people's lives, they will be among the worst-off members of their societies, although the coinage of this awfulness is more psychological than economic (shades of Plato's *Republic*).

Consider too that most of us know (or can readily be brought to see) all this. Thus, if legislators or others are serious about overall public good, about the good of fellow citizens generally, they have reason to worry about the effects of criminal legislation on those who may be less able (or less free) to conform themselves to criminal laws. Of course, there are also good reasons to make certain activities illegal because without a broad range of such laws the level of anxiety and destruction among the citizenry of a society would be intolerable. It is not clear that we could have, in the modern sense, a society, without broad criminal legislation. But the point is that concern for the good of all must also make us worry about those more likely to be caught in the net of any criminal legislation a society passes. The lives of most criminals just seem to be worse than most other lives one can lead, and so a high and general concern for the well-being of one's society should lead one to counterbalance criminal legislation with efforts to make criminality itself less likely.

Of course, this point is often made with regard to slums and ghettos. If criminality is more likely to occur in such environments, then a decent concern for those who live in them should make us want, and take steps, to improve those environments. And a similar point applies to genetic predispositions to criminality. If there are ways to educate or train those with such dispositions that make it easier for them to conform to social norms and, more generally, control themselves, then we ought to implement such education and training if we can do so without stigmatizing those who are to receive such education/training in such a way as to undermine their self-esteem from another direction. Obviously, there are delicate psychological, educational, and legal issues involved here, but my point is that any society that learns about genetic predispositions to criminal behavior will have moral reason to try to do something about them.

Given what we know (or think or strongly suspect) about the terrible lives led by most criminals, anyone, and in particular any legislator, who makes no effort to find out what can be (benignly or respectfully) done to lessen such predispositions or who, having found out, makes no effort to implement those findings, demonstrates a lack of generalized benevolence toward all the people of his society. Any penal legislation passed without such efforts will at best reflect a narrow kind of benevolence and thus count as unjust according to our agent-based virtue-ethical justice as benevolence.

And we can say, in addition, that those who pass laws in this fashion will have largely undercut their own moral or political right to punish those who violate them. For, as Allen Buchanan has pointed out, the language of social justice and of political or moral rights are by and large intertranslatable: one has a right to do something if and only if one can justly do it (Buchanan 1990, 227–252). So if it is unjust to institute criminal penalties without regard to their effect on those least able to conform themselves to such legislation, then there is at least a prima facie case for saying that those who institute the penalties have forfeited the right to apply them.[4]

In a way, too, what we have just been saying represents a kind of complement to educative or reformist theories of punishment and moral responsibility. For to justify punishment in terms of retribution, deterrence, or incapacitation is largely to ignore the well-being of those being punished, and the idea that criminal punishment should help to reform or educate the criminal at the very least suggests that the criminal is worth reforming, his or her life worth saving. Of course, education in prison also presumably redounds to the advantage of the law-abiding members of a society; education, deterrence, and incapacitation are all in their different ways *forms of crime prevention.* But the idea that punishment or prison should involve reform also has much, perhaps more, in common with the idea that we should work to mitigate genetic predispositions toward criminal behavior. To be sure, the treatment of such predispositions might involve certain forms of conditioning or other methods of treatment that hardly count as educational or, in the strictest sense, as ways of *reforming* criminal tendencies. But the idea of educative or reformist punishment does treat certain changes as important to the lives of criminals, and typically those who have advocated educative or reformist punishment have thought of the criminal not only as more morally virtuous but, as such, also as better off for such punishment.

But we need not suppose that virtue automatically (i.e., constitutively) makes one's life better in order to see that educative or reformist punishment can help to improve someone's life: we need only remember what was said earlier about self-respect and the ability to conform to social norms. If we were right earlier about the life prospects of those who fail to conform to social norms, that conclusion bears both on the attempt to mitigate criminal tendencies before criminality has occurred *and* on the attempt to reform those who have already broken society's rules. A generalized benevolence has reason (other things equal) to implement reformist forms of punishment, but also to try to prevent crime *before* it occurs – whether by eliminating the conditions of slums and ghettos or by discovering ways (that do not undermine self-respect) to treat or train those with criminal predispositions.

It is worth noticing how very differently considerations having to do with human free will enter into our agent-based virtue ethics of criminal responsibility from the way they are taken into account, for example, by most compatibilists. Thus, with regard to the particular issue of genetic predispositions, compatibilists typically worry about the *extent* to which such dispositions limit people's freedom. If it is more difficult for some people to conform to social norms according to compatibilists, there are limitations on how much they can justly be punished for breaking the law, and, on this issue as well as related others, questions of justice and moral accountability turn directly on questions of how free people are when they engage in criminal behavior.

But justice as benevolence cuts the Gordian knot of these metaphysical worries and considers only what it is like to concern oneself benevolently with the well-being of (all) fellow members of one's society in grounding the limits it seeks to impose on the severity of punishments for the freedom-impaired. The free-will related issues of self-control and diminished capacities enter (if they enter at all) only through the prism of such concern and not in the direct way in which they enter into the moral arguments of compatibilists.

In addition, there is a striking contrast between what justice as benevolence has to say about the moral responsibility of people with genetic predispositions (or other limits on their freedom of choice in violating the law) and what the so-called reactive-attitudes approach has to say on this subject. The latter, as developed by P. F. Strawson and others, treats issues of moral responsibility and just punishment as driven by facts about our natural or inevitable tendencies to react emotionally to certain sorts of antisocial behavior and thus too by facts about the reactive practices that such tendencies give rise to as a necessary precondition of any recognizably human social life (Strawson 1962, 1–25). All this, as with the present virtue-ethical view of things, tends to play down metaphysical issues of freedom, but the implications of the two approaches are, nonetheless, very different. Even when we know or suspect that a serial killer or child molester is driven by forces that are difficult or impossible to control, we tend to feel angry and punitive, and it is, therefore, difficult to see how a reactive-attitudes theory can morally insist upon humane treatment for such offenders. By contrast, a view that stresses benevolent concern for the well-being of all who live in a given society offers no moral foothold for punitiveness or other hostile reactive attitudes and is therefore in a very good position to make the argument for humane treatment.[5] Once again, however, that argument does not depend on free will or the lack of it, but rather takes off from considerations having to do with what we all take to be important elements of fulfilled, good human lives; and though considerations of the good life clearly also connect with what the reactive-attitudes approach has to say about the conditions of human social life, they do so much less directly and constitutively than in what I have been calling justice as benevolence.

Thus the approach taken here offers a distinctive way to treat issues of diminished freedom, whether they involve genetic predispositions of the sort we have focused on or related phenomena like ignorance of the law, irresistible impulse, provocation, and strict liability. It tells us that questions about free will do not, in their own right, affect issues of social and individual morality, but can do so, in effect, only by engaging with our fundamental and humane concern for the good of others. If such concern is the basis and substance of all morality, then it would be less than moral to allow a different concern about the metaphysics of freedom to dictate social policy.

NOTES

1. Actually, I am here suppressing certain complications in order to simplify the discussion. But I have elsewhere argued that the kind of benevolent motivation most appropriate to private life differs from that relevant in the public or political sphere; for private benevolence can understandably favor near and dear over people one does not know, but a public official does (morally) best to think and act more impartially, and so a benevolence that (in matters of national import and forgetting what we owe to other nations) solely seeks the good of one's country seems a more appropriate basis for assessing public laws and institutions. I have discussed such a more specific conception of social justice and individual morality in my "The Justice of Caring," but for purposes of the present sketch, talk about benevolence gives a sufficient idea of the sort of theory I have in mind.
2. Scanlon (1988) makes a rather similar point.
3. David Wasserman has pointed out to me that the present account of "holding responsible" is somewhat reductionistic. Most of us tend to think that holding a person morally responsible for something requires us to think of him (or her) as free and perhaps also to have certain "reactive attitudes" toward him, but on the view proposed here it is sufficient to believe that we may justly impose certain penalties on the person in question.
4. The qualification "prima facie" seems necessary, if one wants to allow a state some moral room to protect its population from heinous criminality that arises (partly) from the state's own (in)actions.
5. The present virtue-ethical approach does allow us to characterize certain criminals (e.g., child molesters) as moral monsters, and this is a kind of condemnation. But as I indicated earlier, such criticism presupposes a lack of metaphysical control on the part of the person thus characterized and so does not sit well with the kind of hostility or punitiveness that are so typical of human reactive attitudes.

REFERENCES

Buchanan, A. (1990). Justice as reciprocity versus subject-centered justice. *Philosophy and Public Affairs* 29: 227–252.

Hill, T. (1992). Kant's anti-moralistic strain. In *Dignity and practical reason in Kant's moral theory,* 176–195. Ithaca, N.Y.: Cornell University Press.

Rawls, J. (1971). *A theory of justice.* Cambridge, Mass.: Harvard University Press.

Scanlon, T. (1988). The significance of choice. In *The Tanner lectures on human values,* ed. S. McMurrin, 8: 149–216. Salt Lake City: University of Utah Press.

Slote, M. (1998). The justice of caring. *Social Philosophy and Policy* 15 (1): 171–195.

Strawson. (1962). Freedom and resentment. *Proceedings of the British Academy* 48: 1–25.

Chapter 12

Strong Genetic Influence
and the New "Optimism"

J. L. A. GARCIA

In "Freedom and Resentment," now regarded as a classic of twentieth-century philosophical analysis, Peter Strawson distinguished two schools of thought on the implications of universal determinism. Pessimists thought its truth would deprive the concepts of moral obligation and responsibility of "application" and imply that "the practices of punishing and blaming, of expressing moral condemnation and approval, are really unjustified." Optimists, in contrast, held that "these concepts and practices in no way lose their *raison d'être* if the thesis of determinism is true" (Strawson 1962). Strawson's choice of names is revealing. It seems to make little sense unless we assume not only the value of our familiar practices and discourse of moral responsibility but also the eventual intellectual triumph of universal determinism. Given all that, the only interesting difference is between those who gloomily warn that determinism's inevitable triumph dooms the moral practices to senselessness (pessimists), and those who cheerily claim there is no tension between the truth of determinism and our practices of holding people morally responsible (optimists).

This chapter derives from reflection on original and revised papers prepared for and discussed at the April 1995 meeting of the working group on genetic explanation and responsibility ascription. Throughout, I draw on my notes of the working group's discussion of the various papers, and therefore incorporate ideas originated by others. I do not wish to arrogate credit for other's ideas. However, neither my notes nor my memory is sufficiently complete or reliable to make accurate ascriptions of specific positions to particular discussants, and it seems better here to work simply with my own versions of the objections and ideas culled from the sessions. I freely acknowledge this chapter's substantial debt to those discussions for many of the points raised herein.

My own suspicion is that the sort of optimism Strawson had in mind appears to make sense only against the background of midcentury philosophical belief in the autonomy of ethics, the logical gap between facts and values, a sharp distinction between theory and practice, and the concomitant contemporary tendency to relegate morality to the realm of emotions, directives, choices, preferences, and deeds in a way that was thought to insulate it from being undermined by progress in scientific knowledge. We could carry on our social practices, pursue our personal ideals, conduct our emotional lives, and, in short, live our moral lives largely independently of what the world is like. To make the world more to our liking is the business of our practical reasoning and thus of our "ought" talk, but determining how things are was thought to be the radically different business of theoretical reasoning and "is" talk. This larger picture has come to strike many of us as rather too neat, indeed, simplistic, even if Strawson's optimists – now usually called "compatibilists" – are still going strong.[1]

Recently, some philosophers have adopted a stance similar to what Strawson called optimism in regard to new discoveries in genetics. They seem to think that, with the possible exception of a scientific demonstration that genetics renders certain forms of behavior utterly unavoidable, the assignment of ever greater regions of conduct to genetic control poses no real danger to our ordinary moral practices of holding people morally responsible, nor to the deep self-conceptions that inform them. My aim here is to examine some defenses of this new "optimism," focusing on whether we should be concerned about the impact of strong genetic influence on what we might call "negative responsibility" (i.e., on the justification of punishment, of blame, etc.), even if it stops a bit short of genetic *determination* of our behavior, and to raise some questions about it.

VAN INWAGEN

Peter van Inwagen holds that the increasing ascription of conduct to genetic influences would undermine human freedom, and with it moral responsibility, only if the influences are strictly deterministic.[2] What should we say about the situation where, as Marcia Baron puts it, "genetic factors . . . incline but do not necessitate?" In these cases, "the person with the genetic marker is not incapable of responsible moral agency. It is simply harder for him to refrain from the sorts of acts to which he is genetically disposed" (Baron, Chapter 8, in this volume).

Our focus here is on the question of strong genetic influence: could responsibility – specifically, blameworthiness – survive the scientific discovery that our behavior is almost entirely a matter of genetics? When what we know about an agent's genes makes it highly *unlikely* that she will restrain herself from acting in a morally objectionable manner, and even when what we know about her genes makes it quite *difficult* for her successfully to exercise such restraint, the influence of these genetic factors, van Inwagen maintains, does nothing to mitigate the agent's moral responsibility. This position is, in our terms, "optimistic," but is the optimism warranted? There are grounds for doubt.

Van Inwagen offers little in the way of a general view of mitigation, and the examples he offers suggest that his understanding of legitimately mitigating factors may be too narrow. Significantly, in several of the cases he offers what tempts the agents toward immoral conduct is some *moral* ground for action. (At least, such grounds are available to tempt them, although van Inwagen leaves it unclear which factors actually work in what ways in their deliberations.) "If [a] a mother steals because she and her children are starving, . . . if [b] I betray my country or the revolution (or whatever) under torture or [if (c) I betray it] because my family is being held hostage, . . . if [d] someone commits rape as the alternative of the murder of his family, that would be a mitigating circumstance" (van Inwagen, Chapter 9, in this volume).

Genetic influences, of course, do not operate like these exigencies. They do not provide the agent with moral grounds to act immorally in the way the temptations in (a), (c), and (d) appear to do (nor with prudential grounds, as in [b]), and so it may come to seem that they are not genuinely mitigating factors. However, many genuinely mitigating factors do not come by their powers of mitigation from counterposing opposed *moral* grounds for the agent to stray from the right course. Think only of commonplace cases where one causes harm or otherwise violates some moral duty because fatigued, tipsy, addled, overwrought, distracted, or enraged (where one is not at fault simply for having gotten into this state or remaining in it). In none of these cases of lessened responsibility, in contrast to all of van Inwagen's, is the mitigating factor something that also provides a strongly attractive *reason* for acting immorally.

Indeed, one of van Inwagen's supposed examples of a mitigating factor seems not to mitigate but completely to exonerate the agent. In his cases there are moral factors that can tempt the agent to immorality. In his case of a mother "stealing" food to feed her starving child, however, this seems too strong. Her taking food for the child sounds to me like "stealing" only in a conventional or legal sense, not morally, for it is unlikely she violates any morally legitimate property rights in such cases. Your right to dispose of your property as you see fit does not extend to keeping others from using it to meet their urgent need (unless, perhaps, it is already reserved for your own or another's comparable need). Traditional morality would not count the mother's action as stealing at all (not as the sin of stealing, at least). This example might be written off simply as a poor choice by van Inwagen, a mere slip. However, it lends support to our worry that he may be operating with an excessively narrow and overly moralized understanding of mitigation. To be sure, genetic influences on an agent's behavior will not resemble such morally counterbalancing factors. However, only an excessively restrictive and moralized account of mitigation would permit us to derive from this dissimilarity the conclusion that genetic factors cannot serve to mitigate moral responsibility.[3]

In any event, I think there are good reasons to doubt what we might call van Inwagen's all-or-nothing view of the impact that genetic influences might have on human freedom and moral responsibility. What I call the "all-or-nothing view" is the thesis that either genetic influences are deterministic and therefore, for incompatibilists like himself, completely destructive of moral responsibility, or those influences, no matter now strong, leave the latter generally undisturbed, if they fall short of causal determination.[4]

Van Inwagen claims that,

If a man contemplating rape is indeed able to refrain from acting on his present desire [whose presence, strength, or difficulty to resist is owing to genetic factors], if he is indeed able to refrain from committing the rape he is considering, then I do not see why the fact that he had had that desire should, afterward, provide him with any sort of excuse for what he has done. . . . If, moreover, the man contemplating rape is able to refrain from acting on his present desire, then I do not see how facts about the *source* of that desire can provide him any excuse if he decides to act on it. (van Inwagen, Chapter 9; emphasis retained)

This last question is somewhat tendentious, of course. The important issues concern *how difficult* it was for the agent to resist the desire. How strong was it? What desires, principles, or resolutions, were available to him to stand against it in his deliberations, and how strong were they? To what extent could he see and think clearly about his options? Insofar as the answers to these questions were affected by genetic (or, for that matter, environmental) factors in such a way as to render it that much more difficult for the agent to have acted other than he did, I think that they may mitigate his blameworthiness. Two reasons for this contention come to mind, beyond its intuitive appeal.

First, notice that this is the way we tend to think of positive desert, for example, praiseworthiness.[5] If someone (call her Tawana) has done me some good service, then I am inclined to be grateful to her. However, as it becomes clearer to me that her desires, feelings, and choices were all under the mechanical control of some California brain technologists, to adapt an example from J. M. Fischer, then my gratitude will likely, and quite reasonably, wane. Surely, I will think her less deserving of it (Fischer 1986). Perhaps Tawana wanted to help me, but both her help-ing me and her wanting to help were really their doing. More theirs than hers. She gets less credit for her behavior to the extent it was not really hers. If it would have been extremely hard for Tawana not to want to help, indeed, not to want quite strongly to help, then she is less deserv-ing of praise (or, at least, deserves less praise).

As I have elsewhere urged, for a person to deserve something good is for her to have been (or done something) so good in some way that she prima facie acquires rights to get some suitable good from others (Garcia 1986). When the California scientists make it extremely difficult for Tawana to avoid having that desire, then we are inclined to say she gets less credit for having it. Van Inwagen needs to offer some good rea-son for us to treat a genetic disposition much differently from Tawana's mechanically induced one, or some good reason to treat negative desert in a very different way from positive. I am unconvinced by the argu-ments offered to support a supposed asymmetry that makes positive re-sponsibility less dependent than negative responsibility on the freedom of the agent held responsible (Wolf 1980).[6] However, even if those ar-guments are sound, they would not support van Inwagen's contention against the complaint just lodged. To do that, he needs to show that neg-ative responsibility is *less* responsive than positive to an agent's dimin-ished control over her behavior, not that it is, as some asymmetrists con-tend, *more* responsive.

Second, the person who has a strong genetic inclination to do some-thing wrong and does it may feel just as strongly committed to the good and may exercise just as much responsible superintendence of her con-duct as does one who, lacking the genetic disposition, avoids the wrongdoing. This would be the case, for instance, in some of the mod-els of deliberative psychology under genetic dispositions that Patricia

Greenspan discusses (Greenspan, Chapter 10, in this volume). According to one such model, a common one, the genetic disposition to A strengthens the subject's inclination to A beyond the normal level.[7] However, it need not therein diminish her interest in doing other things (e.g., acting morally) that she knows to be incompatible with doing A, or even her interest in avoiding A as such. To that extent, then, such an agent (call her Carla), though she does wrong, may be just as interested in acting rightly, and committed to doing so, as one (call her Mercedes) who in fact does so. It is just that Carla also has a greater-than-usual (and, therefore, harder to resist) attraction to a particular type of wrongdoing, to which attraction she succumbs in the event. Even though she does wrong, then, Carla may expend no less effort to control her (abnormally strong) attraction to doing A than does Mercedes. She is just as conscientiously committed as Mercedes, and conscientiousness is a moral virtue.

More important, she may expend considerably *more* effort toward acting rightly than does a more typical wrongdoer (call her Carlita). If she does, then it is difficult to justify van Inwagen's view that Carla, despite the fact that it was much *more* difficult for her to avoid wrongdoing than it was for Carlita, nevertheless is no *less* to be blamed than is Carlita. According to van Inwagen, the greater difficulty Carla faces in trying to act well, based in her genetic inheritance, does nothing to mitigate her offense, nothing to render it less blameworthy. "What we can do, and what I believe we should do, is feel sorry for" people like Carla; but that's it (van Inwagen, Chapter 9).[8]

Consider this reason for rejecting van Inwagen's position. To be blameworthy is to deserve blame. I suggested that, when someone deserves something negative (such as blame), then she is so bad in some relevant way that some of her rights are vitiated in such fashion that purposely subjecting her to the bad thing need not violate them. If Carla and Carlita are equally blameworthy, then they must have been equally bad in the relevant way. Carla, however, is more attached to doing right than is Carlita, and virtuously strives more seriously to avoid wrongdoing. To that extent, she is, in the most pertinent way, less bad, and therefore less deserving of blame. In short, less blameworthy.

Van Inwagen offers some reasons against the thesis that blameworthiness, roughly and ceteris paribus, waxes and wanes with the ease of avoiding wrongdoing, just as – in a way I just described as intuitively appealing – praiseworthiness seems to wax and wane together with the ease of acting well.[9] He says, for instance, that even our intuitions are more selective than this, allowing that someone who grew up in a place where a certain crime is especially hard to detect, or easy to commit, or suited to the great value locally placed on daring, ingenuity, and the like might find crime harder to avoid but be no less blameworthy for that (van Inwagen, Chapter 9). In his cases, however, one crucial difference is that the defendant making this claim on her own behalf seems not to have tried to resist the appeal of wrongdoing. If she could show that she tried to resist the appeal of crime just as hard as a person of normal virtue would have, and nevertheless succumbed to temptation, it is not clear to me why she would not be on solid ground in asking to be held less blameworthy for her offense than is normal. Where this is not the case, I should conjecture, it may be that we are tacitly assuming that a person of normal virtue would have been sensitive to the bad effects the agent's environment was having on her desires, and would earlier have taken measures to counteract them or render herself less likely to give in to them.

This suggestion reminds us that the agent may be fully to blame for her action not simply when she failed to do enough to restrain herself at the time of acting – perhaps she tried as hard as any person of normal virtue would have at the time – but also when she did not take the reasonable advance precautions against wrongdoing that a morally virtuous person would have taken. This, then, brings us to those aspects of responsibility and freedom that go beyond the episode of acting itself to the agent's responsibility for her situation. (Should she have avoided banks, for example, recognizing her propensity to rob them?) More important for our purposes here, it raises the question of an agent's responsibility for the formation of her character.

I next consider the views of Patricia Greenspan and Michael Slote, whose "optimism" about the impact of strong genetic influence on our views of responsibility is grounded precisely on views of freedom and responsibility that focus on character and virtue.

GREENSPAN

I find much that is useful and insightful in Greenspan's writing. However, without fully endorsing them, she sympathetically explores two strategies for defending the new optimism's claim that strong genetic influence (and, perhaps, even genetic determinism, if there be such a possibility) could be reconciled with something pretty much the same as our ordinary moral practices of holding people responsible. She offers two suggestions, if I understand her: first, what moral responsibility requires of an action is not that it have been indeterministically free but that it reflect the agent's rationality along a familiar deliberative path; and second, that, in any case, we can understand why it would make sense to hold people responsible for their tendencies and dispositions, whether or not resulting from strong genetic influence, by looking to neglected elements of what we might call (though she does not) the "language game" of moral responsibility. Specifically, she directs our attention to feelings of anger and contempt, which she considers "retributive emotions." I will try to show why I think these strategies do too little to illuminate the necessary elements of our moral discourse and practice.

The claim that their rationality and their foundation in the agent's deeply held convictions, values, projects, and commitments suffice to render her and her actions fit for responsibility is common these days, but it is hard to see its persuasiveness.[10] One would think that someone's responsibility will crucially depend on how she came by those convictions. If that was all deterministic (or near deterministic), then it is hard to see why this kind of rootedness is enough. At least, this appeal does nothing either to satisfy or to undermine the moral concerns that motivate incompatibilists. Indeed, insofar as our convictions and other belief-forming psychological mechanisms operate deterministically, some have reasoned, that undermines the rationality of the beliefs.[11]

Greenspan's other claim about the ability of our moral discourse and practice to survive is more original and, perhaps, more interesting. I think, however, that her views of punishment, blame, and other moral responses are seriously flawed on several counts that make them untrustworthy in elucidating our ordinary moral thought.

First, Greenspan holds that "society . . . might be said to bear responsibility" for an agent's wrongdoing if society could have helped her avoid it by better responding to her genetic predispositions (Greenspan, Chapter 10). However, this raises questions. Is she entitled to this help? Is society obligated to provide it? If so, to whom is it so obligated? On what basis? In any case, why does not she (as agent) bear full responsibility, even if society also has some? After all, we should not think of responsibility as a zero-sum game, such that if society has some portion, then the agent must have less (than she would if society were not in the picture). Such a zero-sum understanding of moral responsibility leads to obvious absurdities. For instance, it would serve partially to exonerate a mob-connected professional executioner, who operates as part of a network or conspiracy, as compared with an isolated killer, who has no one with whom to share the blame and therein lessen the amount that accrues just to her. Moreover, mention of society only forestalls questions of freedom and responsibility. How can society be held responsible – in the sense of being blamed – if its response was itself merely a result of the pattern of genetic dispositions of the genetically shaped individuals who compose it?

Second, Greenspan concentrates on anger, scorn, and contempt as "versions of blame," which are "punitive." She even calls them "punitive reactions." However, this cannot be literal, for punishment is necessarily an action done intentionally and for a purpose. Emotional reactions are not like that, even if they sometimes have welcome deterrent effects. Moreover, real punishment – the act of punishing – is never justified simply by its expressing such emotional reactions. They must themselves be justified, and that depends, among other things, on whether their associated moral judgments are correct. There are difficulties in Greenspan's effort to show that those moral judgments can be seen as correct in the face of strong genetic influence. She thinks we can blame an agent for her action because her action reflects her character and priorities. This is true, though even actions that are *out* of character can be praised or blamed for the priorities, however temporary, they too express. But why should we follow Greenspan in saying that character is "something basic about who [someone] is," that it "constitutes a self-defining pattern," in cases where it is formed or sustained almost entirely by forces beyond the agent's own control? If "she can't help himself" is literally true and, moreover, in a context in which she could have never have intervened to become a different person, why should it be her character that defines her for moral purposes?

More important, why should we focus on character for criticism rather than the factors that shape character, if there is no serious opportunity for the agent to shape her character differently? What keeps that from being arbitrary? Why think moral criticism directed here anything more than absurd? What makes it any less misplaced than moral criticism directed at the other factors, such as the genes themselves or the chemicals that form them? Of course, people feel anger about all sorts of things, and often feel it with no moral criticism implicit.[12] We get angry about our mistakes or others', angry with software, angry over illnesses or bad weather, and so on. Feeling angry about something X may be common and even justified over a broad range of values for X. What

Greenspan needs to be justified, however, is not just anger about some X, but anger *with someone S;* and she needs that justified even when S cannot help doing what she does or being as she is. It is hard to see how Greenspan can get that. She seems to think such anger justified insofar as S's actions reflect her priorities and the rational reflection behind them, irrespective of whether they are beyond the agent's control at any time.

Unfortunately for her position, it is unclear what grounds Greenspan can offer for insisting that the agent is "a personal cause of wrong which is not erased by the acknowledgment that [s]he was unfree to do otherwise." In what way can she "deserve anger to the extent that she "slights" others" when this "slighting" is not free, not really under her control, not something she can avoid? To the extent that her behavior is like that of her genes and their constituent chemicals in being strictly determined, blaming her for it makes no more sense than does blaming them for it. It only appears to make more sense because she thinks, deliberates, and chooses, and those activities give the impression of freedom and control. Once that impression is ruled out as in strict determinism – or all but ruled out, as in strong genetic influence – the crucial difference disappears, taking with it the basis for holding the agent responsible.

Greenspan speaks of an ambivalence in our reactions in this connection: we want to blame but we also feel uneasy about doing that. However, I think she misunderstands the phenomenon. That ambivalence does not exist *within* our moral reactions, but *between* our moral urge to exonerate the agent and our largely prudential desire still somehow to deter and restrain the conduct and to express how unwelcome we judge its results to be. At that point, I think, Greenspan, despite the sincerity and ingenuity of her efforts, can no longer avoid resorting to a merely forward-looking justification for punishment and blame, based on their general beneficial effects.[13] With any such approach, however, we can ask *why* its good effects in other cases should justify blame in the significantly different case. It needs to be shown that what is missing in the special case is not just what was crucial for justifying blame ascriptions in the usual cases. And that is just what is at issue when we are talking about blaming people who have little or no control over what they do. To attend merely to good effects on society of blaming and punishing such people, as Greenspan sometimes seems to realize, is simply to evade the issue of whether those emotional, behavioral, and cognitive responses are morally justified and therefore just. To show that we act justly in blaming and punishing, it must be shown not merely that society benefits but that we treat even *the person punished* with due concern: with fairness, with respect for her dignity and status as human, and without any violation of her rights.

SLOTE

Michael Slote's form of "optimism," if that is what it is, is another approach that, like Greenspan's, centers on character. Unlike hers, Slote's discussion takes special care to attend to the welfare of the one punished. Let us assess its adequacy in responding to the demands of justice.

Slote suggests that "the idea of a good society can substitute for the idea of free will in providing a basis for the punishments, penalties, incentives, and rewards that are the hallmarks of moral and social accountability" (Slote, Chapter 11, in this volume). He hopes to use that idea to show we can be justified when we "think better or worse of people's motives or character quite apart from whether either ever was or is under their control." Indeed, "A serial killer may be judged morally vicious even if we believed he or she was *made* vicious" (Slote's emphasis). Because such "moral criticism and evaluation can occur and be valid in the absence of freedom and responsibility," he thinks we are then positioned to develop a new conception of responsibility out of the demands of social justice and, more to the point, in the face of strong genetic influence, perhaps even extending to some form of genetic determinism. Slote's view is complicated and only sketchily presented, but I try to articulate and raise questions about some of its principal contentions.

Here is Slote's reasoning, as I reconstruct it, making explicit its implications for the questions on which we focus here.

S1. The fundamental concepts of moral assessment are virtues, especially, benevolence.
S2. The true ascription of moral virtues and vices to someone does not depend on whether her will is free from strong genetic influence (or even determination).
S3. Someone can properly be held (and is) morally responsible, if she may justly be punished, blamed, and the like.[14]
S4. Someone is justly punished (or blamed) when "a just society via just procedures institutes and applies" laws that assign that punishment to her (Slote, Chapter 11, in this volume).

S5. "A society is just to the extent its members are benevolent . . . (toward one another)," and "an institution . . . in a given society is just if and only if it expresses or is motivated by [a] sufficient [level of the moral virtue of] benevolence on the part of those of who created it . . . or sustain it."

S6. Even if all members of a society are so strongly influenced by their genetic inheritance that they lack free will, some of them may still be so vicious morally and others of them may be so virtuous morally that the latter may establish and execute institutional practices whereby the former are justly punished.

S7. When that happens, the vicious are justly punished or blamed.

S8. Therefore, people can justly be blamed or punished, and moral responsibility thus preserved, even in the context of strong genetic influence.

S9. Still, those with strong genetic predispositions toward antisocial behavior should not be treated as criminals, where avoidable, because criminals have "terrible lives" and good will toward them dictates that such harm be minimized.[15]

This reasoning makes a strong and appealing cases for some measure of moral responsibility despite strong genetic influence. Because of S9, it is not immediately clear whether Slote endorses what I have called "optimism" about such influence. He may think that it does require significant changes in the way we respond to harmful behavior, calling for us to turn from criminalization and punishment toward prevention and rehabilitation. Nevertheless, I treat Slote's position as a form of "optimism" because his reasons for preferring prevention to punishment (which preference, by the way, even many retributivists would share) is not the backward-looking consideration that genetic predisposition mitigates, partially excuses, or exonerates the agent. Rather, it is apparently motivated only by the forward-looking desire to make the agent's life go a little better. That Slote himself ties his view to the rehabilitative, educative tradition on punishment indicates that he does not view genetic predisposition as calling for a different approach than what would otherwise be called for.[16] This factor does not seem to mark a crucial difference about the morality of punishment for Slote, and in that crucial respect his position is that of the "optimists." I regard Slote's position, then, as in part a defense of "optimism" and judge that it fails for reasons I will try to make clear.

While most moral theorists today would likely dispute S1, I think it correct. I share many of Slote's views about the centrality of the virtues within moral theory. Nor shall I raise questions about either S3 and S4, both of which seem to me plausible. What of S2? Slote appeals to Spinoza for its defense, but that will not impress those with incompatibilist leanings. Slote says that we may rightly brand the serial killer a "moral monster" and that when we do so, this "is no merely aesthetic judgment," for "in characterizing someone a moral monster we make a moral criticism without (necessarily) imputing blameworthiness or responsibility for having or developing the traits that ground that criticism."[17]

Perhaps some people do sometimes make moral judgments when they call others "moral monsters" and maybe they do sometimes call them such names without blaming them. What is counterintuitive in Slote's position is that he thinks they can do these things at the same time. Plainly, talk of people as moral monsters is parasitic on less metaphorical claims about the nature and depth of their wickedness. It is hard to see how these more literal claims could get purchase if no one was ever thought to be wicked in the central ways – the ones that, people ordinarily suppose, are her own fault, and which therefore justify blame, censure, punishment, and the like. For the same reason, it is hard to see how any elements of character count as moral virtues or vices in the total absence of freedom. Philosophers nowadays usually explicate the virtues and vices in terms of dispositions and tendencies. These terms are quite broadly applicable, however, in the sense that they can be applied to things of many different types. In moral contexts, they must be used in a much more restricted way, lest brute animals and even natural phenomena turn out literally to have moral virtues and vices in virtue of their tendencies and dispositions to bring people good or ill. People do say that hurricanes and wild beasts have dispositions and sometimes metaphorically call them "cruel," for example. However, we should not let this fact lead us into thinking such things have the sort of dispositions that can constitute character.[18]

Of course, this is too simple. The moral vice of malevolence, for instance, is not merely a tendency to cause people harm but a desire to cause it and a disposition to choose to cause it. Nevertheless, not just anything that could be called, in some sense, a desire or a choice will qualify. I may call a certain lioness (say one who has killed my loved ones) a malevolent beast. However, I know that she does not literally bear the moral vice, even if she can be said in some sense to have wanted and even chosen to kill her victims.[19] It seems reasonable to retain the commonsense view that only desires and choices properly tied to free will and not swallowed up into the causal chain can serve as the stuff of moral virtue and vice.

S5 is tied to Slote's interesting idea that we should "treat laws and punishments as just (and fair) if they have the proper motivational provenance," especially, benevolent motives – what he calls "justice as benevolence." I agree that benevolence is one of the few core moral virtues, perhaps the entire core, and that justice therefore needs to be grounded in it and understood in its terms. However, if that is to be plausible and to be saved from some of the errors that characterize competitive accounts, then both benevolence itself and its connection to justice need to be understood in more nuanced ways than those with which Slote here works. The topic is a large one, and my remarks will of necessity be inadequate to it.

First, benevolence. Slote's remarks suggest an understanding of benevolence not much advanced beyond that of the utilitarians. He pronounces laws just if they were motivated by "generalized benevolence toward the people of the society," as if that were enough and, in discussing how to test laws for justice, suggests we need to ask whether the legislators were "really trying to do the most good they can for society." One obvious problem for this is that it raises serious dangers of severe injustice, rather a drawback for any account of justice. Slote acknowledges the problem, but does little here to solve it.[20] I have elsewhere urged that benevolence is a moral virtue because of the importance of our wishing well each of the various people to whom we are connected in certain morally significant role-relationships. In that sort of roles-centered, virtues-based moral theory, the nature of the moral virtue of benevolence is, thus, to will each person what is good for her (Garcia 1993). That is one thing Slote's formula of "benevolence . . . toward all" might be taken to mean. However, the way he talks of "the public good" suggests he interprets universal benevolence quite differently, as an ultimate devotion to the group itself or, still more dangerously, as pursuit of benefits for its larger number.

That sort of majoritarianism is one possible "second-best" strategy when not all can be helped, but it is not the only one, nor, I think, the best. A more attractive alternative is a sort of moral "minimax" strategy according to which one minimizes the extent to which one departs from benevolence with respect to any affected party. That would provide a benevolence-based justification for at least one sort of familiar "deontological" restriction on the extent to which it is permissible to turn against some to benefit others.[21] What is important for our purposes is that such a view will indict any action, law, or punishment for offending against the virtue of benevolence if, in performing it, the agent treats anyone with inadequate goodwill. Note that the goodwill of one person, A, toward another, B, may be inadequate in more than one way. The most obvious way is when A just does not care (or does not care enough) about B's good. However, another way for A's benevolence toward B to be morally inadequate occurs when A is too selective in her goodwill, willing B some types of good but not others that are also condign and suitable to B's status and to her relationship with A. It is here, I think, where Slote's account fails in its ambition to secure the justice of punishment in the face of strong genetic influence. If we turn to the next consideration I raise about Slote's account, that of justice, we can see why it fails.

Slote does show sensitivity to the difficulty inherent in any simplistic identification of justice with the good of the greater number. Justice involves due care for *each* person, not just the majority, and Slote properly directs attention to the welfare and "self-esteem" of those who break the law. Still, even the act utilitarian allows that each person's welfare, and thus that of the lawbreaker, counts for something. Slote says that "benevolence is what makes acts morally good" but, to make that claim convincing, he must show that he can derive justice (and the other moral virtues) from benevolence. There is more than entrée to the discourse of justice: desert, rights, respect, dignity, to be sure, and perhaps – as we mentioned already – fairness and equality as well. Unfortunately, as I will show, Slote dismisses desert, mentions rights without following through on them, and essentially ignores the others. This dooms his effort to ground a meaningful and recognizable account of justice in his virtue of generalized goodwill.

On desert, Slote's account reduces it entirely to the assignments made by just institutions. But then even this spare conception of desert is laid aside and does no work in the justification of punishment. This account is inadequate both for its understanding and its marginalizing of desert. The understanding of desert is inadequate because, as Slote reluctantly acknowledges, it leaves no room for valid claims of pre- or extrainstitutional desert.[22] Yet an acceptable account of institutional desert must accommodate our views of what people deserve prior to and independently of social institutions. Similarly, the marginalization of desert by excluding it from the justification of punishment cannot work because a prime example of the importance of preinstitutional desert lies precisely in our intuition that undeserved punishment is inherently illegitimate. It is highly implausible to maintain that punishment could be justified independently of desert because, as the terms' suffixes indicate, desert is internal to such concepts as culpability, blameworthiness, liability, accountability, and responsibility.[23]

As for rights, Slote approvingly quotes Buchanan's claim that talk of social justice and talk of political and moral rights are interchangeable. Yet, having said this, he does nothing to identify the pertinent rights of the lawbreaker and show each is duly respected. For that matter, respect itself, like the human dignity that calls for it, warrants no discussion whatever in Slote's discussion.

This will not do. Slote cannot vindicate the justice of punishment and moral responsibility even in the face of strong genetic influence without troubling to show that, despite such influence, the lawbreakers nonetheless deserve punishment, that punishing them violates none of their rights. Nor can he do so without attending to the question whether the very concept of justice itself remains meaningful in a world of such influence, whether humanity retains the special dignity, respect for which grounds and generates human rights. I think a better approach to understanding justice in terms of the moral virtue of universal, but not unnuanced or undifferentiated, benevolence can find a proper role for the centrality of rights, while also explaining the importance for justice, of respect, dignity, and even desert. This is not the place to develop such an account, but I outline a few of its salient features.

The truly benevolent person not only wants to help others, she is concerned not to fail them. Thus, in addition to wishing others well, she takes special care not to fall so far short of benevolence as to be vicious. Because we can talk of my vicious behavior toward you in the language of duty – as a violation of my duties to you – and my duties to you can themselves be discussed in the discourse of rights (your rights against, or in respect of, me), this second-order concern lest anyone be treated viciously is, I think, the beginning of justice.[24] It is a concern that persons not be treated viciously by others, that is, a concern that no one's rights be disregarded. Benevolence is willing what is good, but there are many things good for people to have. Basic justice – the justice of the fundamental human rights – especially involves willing persons the goods that are distinctive of their status: deferring to them as beings with special dignity where that deference characteristically takes the form of specially valuing their welfare and such defining features of the person as the capacity, unless blocked, of developing a faculty of freely initiating and controlling her actions. This status calls for certain modes of response and delegitimizes others. It thus establishes a kind of moral relationship among persons, even persons personally unacquainted.

A just person is, as such, benevolent, then, as Slote affirms. But, *qua just*, a virtuous person is especially concerned that others have certain personal goods: first, the good of relationships with others that are fulfilled in the sense that persons take care to acquit themselves virtuously (at least without vice) in their dealings with one another. Second, the just person is especially concerned that persons have the goods that are definitive, characteristic, and constitutive of personhood, such as free choice and the respect to which the capacity for it entitles us.

Much more would need to be said to clarify, let alone begin to defend this approach. Even this much, however, allows us to see what Slote would need to do adequately to employ his motivation-based approach to ethical theory (toward which I lean) in defense of what I here call "optimism." He would need to show that we can continue meaningfully to regard human beings as endowed with special dignity and status, as having special rights founded thereon, and then he would need to show that we could properly continue to regard people as morally responsible – blameworthy, culpable, and the like – in the face of widespread and strong genetic influence over their antisocial (and otherwise noteworthy) conduct. This he does not even attempt, and there is no reason to think any such attempt would prove successful. It is hard to see what special value and dignity would reside in humanity once all human thought, response, deliberation, choice, and conduct were virtually swallowed up in the nexus of deterministic causes. Virtues are central to Slote's conception of morality, as I think they should be. Yet virtues and vices, it is usually thought, must be acquired or maintained at least in part through the agent's voluntary conduct. My taking a pill or being struck by a ray that makes me "love" you does not plausibly count as my acquiring a virtue.[25] But, again, it is hard to see what would remain of *voluntas* and its importance if chemicals really did call the shots. Nor is it apparent what adequate substitute is to be provided as a foundation for human rights.

In any case, the optimist has to show more than that morality would substantially survive such influence. Beyond that, she must more specifically show that and how blame and punishment would remain just. It is difficult to see how she can achieve either of these goals along the lines Slote lays down. For you to punish or blame me with full justification, it must be that I have behaved so viciously that you do not act contrary to the virtue of justice when you blame or punish me.[26] That is, your treatment of me is duly respectful, observes my rights, and therein responds adequately and appropriately to my dignity and the relationship between us that my dignity (and yours) establishes. Barring innocent mistakes, I then deserve what I get. No discussion that, like Slote's, dismisses the relevance and importance of desert, slights rights, and ignores human dignity and the respect it calls for, can be reasonably said to vindicate the optimist's claims that moral assessment and responsibility remain reasonable and morally justified once humans are seen as little more than puppets of their genetic inheritance.

WHY DO RESPONSIBILITY AND FREEDOM MATTER?

My view of the importance of responsibility for morality in general should already be clear. It may well be that neither we nor our actions can possess any significant moral features unless some actions (I mean concrete actions, act tokens) and some people can be blameworthy and otherwise morally responsible. The idea that we could jettison moral responsibility but retain morality's duties, rights, virtues, and so on – lose what some call "second-order morality" while keeping "first-order morality" – seems to me a delusion.[27] The moral importance of freedom itself is a more difficult matter to see clearly. At least since Mill's *On Liberty*, the view has become firmly entrenched in philosophy that anything roughly corresponding to what Slote calls "metaphysical freedom," whatever its exact nature, is quite irrelevant for politics. As ethics itself has increasingly been reduced to political philosophy, the impression has grown that freedom from determination or near determination by internal factors cannot really matter for morality either.

Most recent philosophical attention to human freedom has focused on political freedom, not "metaphysical." In the past few decades, philosophers have moved away from championing mere "negative freedom," as Isaiah Berlin called it – freedom from legal and other man-made constraints – and explored more "positive" and inclusive conceptions of freedom, often employing Kant's language of "autonomy." They hope thereby to give freedom and its importance a firmer foundation, by displaying it as more than merely an arbitrary grabbag of liberties to do this and that. Rather, they present freedom as a more unified business, whose importance derives from its role in generating, unifying, and expressing the individual's identity, even her "self."[28] However, their view that free choices are to be respected because they express the individual's self-creation runs into difficulty in explaining how the self gets this dignity that it (or its creation) bestows on choices, especially insofar as the self is a mere artifact or is almost entirely the result of genetic and environmental influences.

Some have built on this new understanding of political freedom as autonomy to reconceive freedom of will in a way that is decidedly less "metaphysical."[29] Harry Frankfurt, most notably, considers a person's will free when her self is integrated in such a way that her "first-order" desires (to do, or be, or have certain things) harmonize with her "second-order" desires about what sorts of desires to have. To counter the objections that there may be an infinite series or orders of desires, and that, to take one of Frankfurt's examples, the drug addict's self may be partly reintegrated if some higher-order desires join the first-order in being "pro-drug," he allows that the integrated self "identifies" itself with some higher-order desire "whole-heartedly." (See his essays in Fischer 1986 and in Fischer and Ravizza 1993.) His position is compatibilist or, at least, helpful to compatibilists in that it provides them an account of freedom that seems not to contradict universal determinism.

Whatever its other merits, surely, this conception of freedom of will does not capture the use of that term in framing and rebutting moral excuses. The reluctant drug addict, who acts from a desire she wishes she did not have, does not have an integrated self, and may even act against her own free will in some sense. However, no one could accept that whatever way in which her will is unfree would serve to excuse or exonerate her behavior. Certainly, it does not generate an excuse in the way that the claim might that she did not act of her own free will because she was under hypnosis.

My suspicion is that Frankfurt is onto something, but that "identification" does not capture it. I can make little sense of moral responsibility in the context of universal determinism, and even the considerably weaker thesis of strong genetic influence appears to me greatly to diminish blameworthiness and desert of either punishment or credit.[30] Bennett reminds us that the dream of morality's autonomy is just a dream, because any moral theory presupposes certain physical, scientific, and metaphysical claims (Bennett 1993). Commonsense morality (whether or not it is, as Derek Parfit thinks, itself a theory) does too, and the availability to many agents of choice that is neither determined, nor nearly determined, nor strictly a matter of chance may be one of them. If our choices are autonomous in Kant's sense of standing outside the causal nexus and therein manifesting our natures as beings not wholly parts of the natural, nomologically explicable order (if our choices reflect our status as "noumenal" beings, in his jargon), then perhaps we can begin to see why we and our choices merit the deference political liberals claim for them. If there is anything to morality, then, I think there must be less to genetic and other influences than many believe.

Cheerleaders for various supposed social and philosophical benefits of scientific investigation are notorious for inflating claims about what the sciences have proved and about what we must presuppose if we are to be scientific. So perhaps we should not be too worried, maintaining a justified skepticism before genetic (or environmental) boosterism. The adversary here is not real science but merely the metaphysical and epistemological extrapolations of the dubious scientism of some incautious enthusiasts.[31] Retaining my optimism about the prospects for what Strawson called "pessimism" (incompatibilism), I defer philosophical arguments to those more knowledgeable about both the relevant science and the relevant philosophy. (That deference does not require abandoning pertinent prephilosophical or other nonphilosophical beliefs, of course. For recent defenses of indeterminist freedom, see O'Connor 1995.)

In addition to this minimal freedom required for agency, though, there seems to be another type of freedom, which some have considered somehow higher or superior. Paul writes to the Galatians, "It was for freedom that Christ freed us!" (5:1) and continues, "You have been called to live in freedom, but not a freedom that gives rein to the flesh. . . . You should live in accord to the spirit and you will not yield to the cravings of the flesh. The flesh lusts against the spirit and the spirit against the flesh." The picture here, and elsewhere in Paul, is clearly of a divided self, and the text does seem to call for something like "identification," and "wholehearted" identification at that. Even if we are free from determinants, we may still be slaves, slaves to our own sinful inclinations. For him, to avoid this "lower" freedom, which gives rein to the flesh, we must side with the spirit and against the flesh, and must do so without reservations. The concept of identification, however, appears not fully to capture what he has in mind. Identifying is an action that agents perform and, as Gary Watson, Elenore Stump, and others have urged against Frankfurt, it may be arbitrary or ill-founded in ways that leave us quite unclear about why desires with which an agent identifies herself should *eo ipso* be taken more seriously than others. (See the editors' introduction and essays by Watson and Stump in Fischer and Ravizza 1993.) How does merely an agent's identification with it, even wholehearted identification, elevate a desire over foolish fancies, velleities, or idle wishes? I see no good answer to this.

Just as well, perhaps. As some philosophers talk of "identification" these days, the term expresses a chimera, the fantasy that a person can literally, by some act(s) of will, make herself an identity. Identity, however – what we basically are, and what persists through change, leaving us somehow the same (*idem*) despite the difference – is not something we come by so easily. The task of determining who we are is much more cognitive than constructive; our identity is much more something we discover than something we make. The deepest facts of human nature, of who we are, of how we came to be, and for what we exist, may very much matter for human freedom. Strawson thinks he knows the answers and, bowing before what he sees as the inexorable growth of scientific knowledge swallowing humanity up into the deterministic causal order, he considers "pessimists" those who think moral categories cannot survive such growth. Those "optimists" – as I have called them, adapting Strawson's language – who today confidently minimize the effects of strong genetic influence on our moral classification schemes and thinking follow in his footsteps.

There are other possibilities. Wojtyla, some decades ago, wrote that "freedom is on the one hand for the sake of truth and on the other hand it cannot be perfected except by means of truth" (quoted in Dulles 1995, 36). He thinks that we are made for freedom, but for the free choice of our genuine good. Freedom dignifies us, in this view, because it is a principal way in which we resemble God, who is free, and also (as Kant saw) a way in which we rise above the rest of earthly creation, which is not. However, this freedom is ill-used unless it operates to select in accord with what we really are and need. That is, genuinely free choice is realistic, facing up to our status as creatures and our need for God. Without God, we can get things we want but never be satisfied with them, for, as Wojtyla wrote later as pope, we then "detach . . . human freedom from its essential and constitutive relationship to truth" (*Veritatis Splendor*, sec. 4). This view of the self as morally responsible *qua* originating, undetermined cause, of course, is contrary to what some call "the Modern Scientific World View" (see Korsgaard 1996). It would need to be shown, however, just what scientific experiments – as distinct from philosophical claims and scientists' presuppositions – entail that we are not such causes.

Of course, any such understanding of humanity may prove incompatible with various assumptions commonly made in modern science, such as the causal closure of the physical world. However, again, it needs to be shown that scientific inquiry is impossible with anything short of unqualified endorsement of such a universal principle. More important, it needs to be shown what scientifically proven *results*, not just what common methods or assumptions of scientists' inquiry, entail such determinism (or, if there is such a thing, the near determinism that constitutes strong genetic influence). I am skeptical about that, but cannot claim the expertise needed to make the argument one way or the other. Note, however, that the issue is really one of metaphysics – the implications of science – rather than of science itself, as talk of a "World View" indicates. What I wish to emphasize is that ideas have consequences, and that we should not expect our moral beliefs, attitudes, and practices to be justified no matter what the world is like. The Strawsonian optimist's sunny assurances to the contrary seem to me a pipedream. If that is right, then the experienced moral world is itself a datum, an "appearance" that a scientifically adequate metaphysics will need to save. Given the strength, universality, longevity, and certainty of that moral world, I doubt there is good reason to think modern scientific results will or can decisively prove it illusory.[32] There is much that would need development and defense here, but philosophers' breezy assertions to the contrary appear to me so much bluff and humbug.

Autonomy is often counterposed to paternalism in both political philosophy and medical ethics. Paternalism, many have noted, is unbecoming just on the basis of the disturbing imagery embedded in its etymology. Neither my physician nor my society, after all, stands to me as father to child. However, autonomy is also scary in its root meanings. For the idea of everyone running about being and making a law for herself is unsettling in the extreme. In his later writings, Wojtyla instructively tames autonomy, insisting that we have nothing to fear from it, when (but only when) the autonomous person understands her autonomy as a "participated theonomy," that is, not as her making up a law for herself but as her making the natural law (i.e., the law of our nature and good as creatures) her own by internalizing it and adopting it as her personal rule of behavior.

Strawson endorses a picture of humanity that makes us entirely the creatures of our genes and environment. Then he labels "pessimists" those who think that, if this be the truth about us, then morality has no place, and "optimists" those who sunnily assure us this truth makes no difference. It has suited my purposes to follow and adapt his linguistic practice here. However, I have usually placed these terms between quotes to register my skepticism that the position of Strawson and his followers is real optimism, rather than learned self-delusion. I think the real optimists are those who think the metaphysical truth really does make a difference, but who find it elsewhere than where Strawson and his followers think it lies. Some are cocksure this will turn out to be wrong. I attempt no rebuttal here. My point has been to help clear the ground of the new Strawsonian "optimism," so as to enable us better to see that it is only some such truth as that Wojtyla affirms that could let us remain morally responsible. And make us free.

NOTES

1. The incompatibilist claim should be that the truth of universal determinism is sufficient to sink moral responsibility (or freedom of the will, depending on the type of compatibilism being rejected), but that its falsity is insufficient to rescue responsibility (or freedom). It is as hard to find a place for responsible and meaningfully free human action in a world where the only undetermined events are purely random ones at the subatomic level as it is to find them in a wholly deterministic universe. Although Stephan Korner, Alan Donagan, and other proponents of what might be called "quantum freedom" are correct in thinking that quantum mechanical indeterminacy makes room for moral agency to the extent that it contradicts the thesis of universal determinism, they need to show why it does not also immediately fill that space with a rule of chance no less destructive of moral agency. In an important unpublished paper, Jonathan Schaffer has helpfully suggested that the real issue is not whether the physical world is deterministic, but whether it is complete (in the sense that everything is physical), closed (in the sense that only events cause events), reducible (in the sense that every object and event has all and only the causal power of its most scientifically basic description), maximal (in the sense that all its features and entities derive from that description), and "blind" (in the sense that no event is irreducibly intentional). If all these features do hold of the physical world, then, Schaffer persuasively contends, the incompatibilists' root fears are realized, even in an indeterministic universe. (For a closely reasoned effort to cut libertarians off from the comfort of quantum indeterminism by questioning indeterminist accounts of quantum theory, see Loewer 1998) This does not undermine the general incompatibilist position, but it does suggest that the incompatibilist who sees hope in what quantum mechanics denies should also sense danger in what it affirms. (For two reviews of some of the recent literature on incompatibilism, see O'Connor 1993 and Clarke 1995.)

2. He seems to me to be correct that we can in principle make such ascriptions. I do not understand the position of those who would maintain that we could never tell the extent to which an item of someone's behavior was owing to genetic influences. "Lewontin ... has described the separation of behavioral variation into genetic and environmental contributions and the variation

between the two as 'illusory'" (Horton 1995, 38). Holding the other factors steady while we vary only the genetic influences should largely do the trick. Of course, we need to be wary of the simplistic ideas that there will be one gene controlling each interesting type of behavior (a gene for speech, say, though speech involves many dispersed anatomical structures) or that genes will control institutionally defined behaviors as such (a gene for stealing, or promising, or adultery).

3. Van Inwagen notes that these cases make immoral action more attractive (van Inwagen, Chapter 9). However, some common mitigating factors make such action harder to resist (but not more attractive) by lessening our capacity to summon the will to choose otherwise.

4. With van Inwagen, Baron, Greenspan, and others, I agree that this issue is not restricted to genetic influences. "All-or-nothing" views might also be counterposed to variable ones (whether gradualist or punctuated, as we might say) with respect to environmental influences on action.

5. I take positive desert to consist in deserving some form or expression of favor (such as praise), and negative desert, with which most of my discussion is occupied, to be deserving some form or expression of disfavor (such as blame or punishment) (Garcia 1986).

6. However, I do think philosophers often go too far when they treat one type of negative responsibility – blame – as simply the opposite of praise. Call this the strong symmetry thesis. (I intend a contrast with what we can call the weak symmetry thesis at issue in the text, that is, the thesis that positive and negative responsibility are comparable in the way they can be undermined by an agent's diminished capacity for self-control.) Without going into the matter here, suffice it to say that praise is, at bottom, an event in speech (even if uttered only subvocally and to oneself), whereas blame is essentially a state of emotions and judgments. (I simply assume here, pace Nussbaum 1994, and others, that emotions are not themselves reducible to judgments.) If I have praised you for many years, that means there were many occasions on which I made laudatory remarks. In contrast, if I have blamed you for many years, that means I have throughout a long duration held certain thoughts and feelings about you, dispositionally, at least.

7. Greenspan herself appears to be more interested in what some call an "inhibitory" model of genetic dispositions, according to which the disposition to A consists not in a subject's strengthened attachment to A-ing but in factors sharply diminishing beneath the norm her resources for withstanding the appeal of A-ing. In extreme cases, the subject's "will or character is deficient in the sense of not being up to the task imposed on all of us, the task of normal behavioral self-control" (Greenspan, Chapter 10).

8. Of course, van Inwagen owes us some explanation of why the facts that warrant our feeling sorry for someone do nothing to warrant our treating her less harshly. There are special circumstances where these remain unconnected, of course, as when one is acting in an official capacity subject to certain constraints, such as a judge, a referee, or a teacher. Van Inwagen's case, however, is nothing like them. It is one thing to say a judge should not rely on her personal feelings in applying society's standards by meting out pun-

ishments. It is quite another to say that the community should not appeal to their common feeling in setting those standards in the first place.

9. We should note that van Inwagen explicitly advances his own view as a rejection of what he acknowledges to be our intuitions.

10. Greenspan writes of this view, without explicitly endorsing it: "free will depends on how a gene . . . gives rise to its behavioral effect. If it works through normal channels, as it were – allows for normal exercise of choice or will in the agent – then genetic causation apparently . . . leaves free will intact" (Greenspan, Chapter 10).

11. So it appeared to C. S. Lewis. (See Lewis [1947] 1978. When Anscombe found fault in his reasoning there, he revised his argument, but not the general conclusion, in a second edition.) Recently, Alvin Plantinga has made a related argument, focusing on the claim that our belief-formation mechanisms allow of evolutionary causal explanation. His point is that this sort of causation can undermine justification, insofar as it may be that the mechanism with the greatest survival value may not tend to yield beliefs that are true (Plantinga 1993).

12. Greenspan also talks of "contempt . . . looking down on someone" as a form of blame that may survive seeing much or all of human behavior as subject to strong genetic influence. However, I should think any contempt for people (as distinct from contempt for their behavior) is always and intrinsically unacceptable morally, an offense against the virtues of charity and justice. So, to my mind, no question arises of justifying contempt for people.

13. "Retributive emotions such as anger are justified as appropriate or reasonable . . . in the sorts of cases where they play an important role in general terms as a way of modifying behavior. That they normally have this effect on failures of self-control is thus sufficient to justify them as emotional reactions in otherwise similar cases where they don't" (Greenspan, Chapter 10).

14. "Any view that allows a person to be justly punished for a certain crime and gives an account of how this is possible has also vindicated the idea of moral responsibility . . . in the bargain" (Slote, Chapter 11).

15. Slote seems to mean that making someone a criminal (treating her as a criminal) makes life go badly for her. There is some truth to this, of course, but it is odd for Slote to assert just this, without making the more important points that sometimes it is morally good (or, at least, not morally bad) to want and ensure that some people have lives bad in this way; and that a common reason for criminalizing some forms of behavior is that they tend to make their agent's life morally bad. Such facts as these motivate the pure and mixed retributivist approaches to the morality of punishment. Also, of course, we often punish someone in hopes of making her life go *better* through rehabilitation.

16. For his discussion of "educative/reformist theories of punishment," see Slote, Chapter 11.

We should note a complication here that will become more important in light of what I say later on. I fault Slote's account largely for failing in the "optimist's" project of showing that punishment is just despite strong genetic influence. Part of my criticism is that while Slote is concerned to emphasize the needs and welfare of the person punished, which justice requires, he attends only to the punishment's effects and not to the considerations of personal desert, human dignity, and rights that are central to justice (see Slote, Chapter 11).

One place where Slote's argument may engage justice is when he suggests that it may be unfair to punish people for offenses to which they may be genetically disposed. "Concern for the good of all must also make us worry about those more likely to be caught in the net of . . . criminal legislation." It is not clear just what sort of appeal Slote is making here, but one interpretation is that society would be unfair not to show special solicitude for those more likely to violate its rules. Fairness is often thought a consideration central to justice. I am perhaps more skeptical about fairness than is common in our Rawls-tinged age, but I do not try here to make the case that fairness is less essential to justice than usually thought. All we need note here is that this appeal to fairness does not lend support to the "optimistic" position that strong genetic influence leaves the morality of punishment and responsibility untouched. For insofar as fairness and justice demand that those with substantial genetic dispositions to antisocial conduct not be punished, strong genetic influence threatens to change our moral world, just as we "pessimists" affirm.

17. Berlin offers the view that, without meaningful freedom of choice, morality would have to be replaced with something like aesthetics: "If determinism were accepted, our vocabulary would have to be very, very radically changed. At best, aesthetics would have to replace morality. . . . Moral praise would have to take the form: if I praise you for saving my life at your own risk, I mean that it is wonderful that you are so made that you could not avoid doing this, and I am glad" (Berlin 1998, 60).

18. I once heard Laurence Thomas remind an audience that, though I can be sure that someone with a certain speech impediment will never call me "nigger," I do not trust him not to do so. I am relying on the wrong sort of disposition. Neither, we should add, does this disposition make the person trustworthy, for it is the wrong sort to constitute a moral virtue.

19. It is not unusual to talk of animals wanting such things. Talk of them choosing is less common, but it is not hard to think of cases where we might well say the lioness chose her victims in that she chose to kill these people (say, the weak, small, or hobbled ones) rather than those. Quite possibly, this talk is not literal. However, that fact need not undermine my response to Slote, and may even support it. Even if it implied that only what literally counts as a desire and/or a disposition to choose can count as a moral virtue (thus explaining why the lioness is not really morally vicious), it may well be that the reason that the relevant events in the lioness's mind do not count as choices is that they are decoupled from what Slote calls "metaphysical freedom." An unfree "choice" is not really a choice and, so, no disposition to make unfree "choices" can be a moral vice or virtue. (If, as many philosophers have thought, to desire is or includes having a disposition to make real choices, that would also imply that the relevant states of the lioness's mind do not count as desires. Indeed, if to have a mind is just to have, or to

have the capacity for, (real) desires, choices, beliefs, and the like, this line of thought leads toward the conclusion that the lioness has no mind at all. However, I am disinclined to go that far. Some of her states and events can count as mental, but they will be of a different and lower sort than their human counterparts.)

20. He allows of his own "agent-based virtue ethics" that it has "some difficulties accommodating ordinary deontology, but that is a problem shared with utilitarianism and consequentialism, and it is a problem all these theories may somehow be able to surmount" (Slote, Chapter 11).

21. For reasons of space and focus, I simply have to assume here the crucial thesis that turning against one person by, for example, intentionally harming her, constitutes a greater departure from benevolence (toward her) than does merely failing to help others (without intending them to go unhelped) constitute a departure from benevolence toward (any one of) them. For a defense of this thesis, see Garcia 1993.

22. "The theory I am offering does not allow us to speak of moral responsibility independently of norms and institutions and is thus somewhat narrower than our commonsense ideas about moral responsibility" (Slote, Chapter 11).

23. Perhaps Slote is tempted to exclude considerations of desert from his discussion of punishment because he recognizes that what he says about deserving would commit him to the absurd claim that a person can deserve blame and punishment for something she did not even do. He allows that his view commits him to the claim that "it can be just to convict and punish someone for a crime she didn't commit," if the just procedures were applied with innocent error, and adds, with a somewhat desperate-sounding qualification, that this position "does . . . [not] automatically entail that such punishment is deserved," conceding that this conclusion "really would be problematic" (Slote, Chapter 11). It would, indeed, and Slote gives no hint of a way of avoiding this absurd conclusion once his sketch of desert is applied to his conception of just punishment. He merely ducks the application. However, all an agent-based account of justice shows is that the state will not act unjustly in erroneously applying its procedures for punishment, not the stronger conclusion that it is just to punish even the innocent. Thus, it need not flirt with the unacceptable view that someone can deserve punishment for what she did not do. A better virtues-based view of deserved punishment holds that someone S deserves to receive punishment P from some authority A for doing some act V when S acted so viciously in doing V that A need not be acting unjustly to S in giving her P. That preserves the intuition that we deserve punishment only for what we do, while also maintaining the agent-based position that the fundamental categories of act evaluation are virtue concepts applying chiefly to the structure of the agent's motivation. For a criticism of other views that assign excessive powers to judges or umpires to change the facts of the world (making people guilty of crimes they never committed, making baseball runners safe even though they were touched by the tag, making pitches into strikes even though they never entered the strike zone, etc.), see Garcia 1987b.

24. In this way, justice resembles another second-order concern that is a virtue,

namely, conscientiousness. The best model for the virtue of justice is, I think, the virtue of conscientiousness. Attention to their essential similarity, as well as to some pertinent differences, is instructive. Conscientiousness is attention to one's duties, a concern to avoid doing wrong. As such, it is an essentially derivative, parasitic virtue. The acts and attitudes against which it sets one's heart must already be independently vicious in some way. (Otherwise, the conscientious need not bother about them.) Still, important differences distinguish justice from conscientiousness. (i) Conscientiousness focuses on the moral subject as agent, whereas justice focuses on her as moral patient, the one on the receiving end of action. (ii) Conscientiousness is essentially first person, concerning the agent with her own duties (viciousness), whereas justice is universalist, directing the just person's attention to anyone's rights against anyone else, that is, anyone's duties to anyone else. (iii) Conscientiousness focuses on the future, whereas justice pays little attention to temporal position. (iv) Conscientiousness stresses the way people are bound whereas justice stresses the need to keep people free in certain ways, especially, from others.

25. For discussion, see Aristotle's *Nicomachean Ethics* and Zagzebski 1996, sec. 2.5, pp. 116–125.

26. I presuppose here that your punishing me with full justification requires more than that, owing to some innocent mistake, you do not act unjustly in punishing me. It essentially involves my deserving to be punished by you.

27. So, I reject the claim (in Smilansky 1994) that the "substantive part of morality," concerned with right and wrong, virtue and vice, and the like, competes with the "accountancy part," dealing with accountability (responsibility) in such a way that eliminating the latter might actually benefit the former. True, an egotistical concern with maximizing the moral credit she gets might distract an agent from simply acting well, and even corrupt her, but it hardly follows that eliminating all credit and discredit is an "ethical advantage."

28. For an important and influential recent example, see Korsgaard 1996. In her May 1998 address to the American Philosophical Association's Central Division Meeting in Chicago, Korsgaard states more thematically and forcefully her view that a person's practical and moral identity is both invented and the ground of obligation.

29. It is interesting that Berlin, who did so much to redirect philosophers' attention to political freedom, rejects this sharp separation of political liberation and action from metaphysical freedom: "It seems to me paradoxical that some political movements demand sacrifices and yet are determinist in belief. . . . Can so many people be truly persuaded to face these dangers [e.g., of Marxist revolutionary struggle], just to shorten a process which will end in happiness whatever they may do or fail to do? This has always puzzled me, and puzzled others" (Berlin 1998, 60). Of course, a careful compatibilist will not allow that a certain outcome is guaranteed "whatever [people] may do." Rather, she will insist that it is guaranteed that they will do the things that cause that outcome. Still, it is difficult to see why people will bother to fight for what is assured. Of course, some hold a moral doctrine according to which what matters is not the result of one's actions, but how and why

one acts (although this is less characteristic of Marxists than of their Christian adversaries). But even for many of the latter – I here exclude more traditional Calvinists – how and why one acts must be up to one's free choice, rather than determined by past states of the world in conjunction with natural causal laws.

30. To be honest, we incompatibilists should admit that we cannot do very much to explain moral responsibility or the working of free agency even in an indeterministic universe (O'Connor 1993). Our view must be that there are limits to forms of explanations that exclude reliance on irreducible intentional phenomena, because reality is different from that imagined by theorists who limit events to causal determination and randomness. It is a mistake for incompatibilists to concede that these limits to irreducibly nonintentional explanation indicate that there is something real and important missing from their picture of reality, and an even greater mistake for them to try to fill in what is missing. That effort leads to the "incoherence" with which Nagel and others have taxed positive incompatibilist accounts of moral responsibility.

31. See Korsgaard's efforts to carve out a sphere for the practical and moral despite her devastating (and unsupported) concessions to the enormous metaphysical implications she sees in what she unashamedly calls "the Modern Scientific World View" (Korsgaard 1996, esp. 94ff.).

32. Taking it as a given, of course, does not relieve us of responsibility ultimately to give some reply to Nietzsche's contention, as, moral nihilist, that "there are no moral phenomena at all, but only a moral interpretation of phenomena" (Nietzsche 1966, 85).

REFERENCES

Bennett, J. (1993). The necessity of moral judgments. *Ethics* 103: 458–472.

Berlin, I. (1998). My intellectual path. *New York Review of Books*, May 14, 53–60.

Clarke, R. (1995). Recent work: Freedom and determinism. *Philosophical Books* 36: 9–18.

Dulles, A. (1995). John Paul II and the truth about freedom. *First Things* (August–September): 36–41.

Fischer, J. M. (Ed.) (1986). *Moral responsibility*. Ithaca, N.Y.: Cornell University Press.

Fischer, J. M. and Ravizza, M. (1993). *Perspectives on moral responsibility*. Ithaca, N.Y.: Cornell University Press.

Garcia, J. L. A. (1986). Two concepts of desert. *Law and Philosophy* 5: 219–235.

    (1987a). Goods and evils. *Philosophy and Phenomenological Research* 4: 385–412.

    (1987b). Constitutive rules. *Philosophia* 17: 251–270.

    (1993). The new critique of anti-consequentialist moral theory. *Philosophical Studies* 71: 1–32.

Horton, R. (1995). Is homosexuality inherited? *New York Review of Books*, July 13, 36–42.

John Paul II. (1993). *Veritatis Splendor* (Encyclical Letter). Washington, D.C.: United States Catholic Conference.

Korsgaard, C. (1996). *Sources of normativity.* Cambridge: Cambridge University Press.

Lewis, C. (1978). *Miracles.* New York: Macmillan.

Loewer, B. (1998). Freedom from physics: Quantum mechanics and free will. *Philosophical Topics* 24: 91–112.

Miller, D. (1976). *Social justice.* Oxford: Oxford University Press.

Nietzsche, F. (1966). *Beyond good and evil.* Trans. W. Kaufmann. New York: Vintage.

Nozick, R. (1981). *Philosophical explanations.* Cambridge, Mass.: Harvard University Press.

Nussbaum, M. (1994). *Therapy of desire.* Princeton, N.J.: Princeton University Press.

O'Connor, T. (1993). Indeterminism and free agency. *Philosophy and Phenomenological Research* 63: 499–527.

(Ed). (1995). *Agents, causes, and events.* New York: Oxford University Press.

Plantinga, A. (1993). An evolutionary argument against naturalism. In *Faith in theory and practice*, ed. E. S. Radcliffe and C. J. White. Chicago: Open Court.

Scanlon, T. M. (1988). The significance of choice. *Tanner lectures on human values,* vol. 7, ed. S. Murrin. Salt Lake City: University of Utah Press.

Smilansky, S. (1994). The ethical advantages of hard determinism. *Philosophy and Phenomenological Research* 54: 355–363.

Strawson, P. F. (1962). Freedom and resentment. *Proceedings of the British Academy* 48: 1–15. Reprinted in Fischer and Ravizza 1993: 45–66.

Wolf, S. (1980). Asymmetrical freedom. *Journal of Philosophy* 77: 151–60. Reprinted in Fischer 1986: 225–240.

Zagzebski, L. (1996). *Virtues of the mind.* Cambridge: Cambridge University Press.

Chapter 13

Genetic Predispositions to Violent and Antisocial Behavior: Responsibility, Character, and Identity

DAVID WASSERMAN

This chapter explores the meaning and moral significance of claims that particular genetic features predispose a person to violent or antisocial behavior. It argues that credible evidence of genetic influence is unlikely to have straightforward implications for moral responsibility. While such evidence may encourage observers to hold the agent less responsible for his aggressive or impulsive actions, it may also tempt them to see the agent as essentially impulsive or aggressive. It is, if anything, likely to deepen the ambivalence we already feel toward individuals who appear disposed to crime and violence.

There is a perennial tension in the criminal law between the desire to punish dangerous, recalcitrant offenders more severely, because they are more difficult to deter and less susceptible to reform, and the uneasy recognition that the conditions that make them more dangerous may also make them less responsible.[1] This tension is often seen in terms of a conflict between utilitarian and retributivist approaches to punishment, with the former appearing to favor stronger deterrents or longer incapacitation in the face of greater recalcitrance, and the latter appearing to favor weaker punishment in the face of diminished responsibility. The tension is more complex, however, because there is also a conflict among retributivist intuitions.

On the one hand, evidence that a person has a violent or antisocial predisposition may seem powerfully mitigating. It may invite the suspension of blame and its replacement with a more custodial or thera-

I am grateful for the comments I received on this paper from members of the April 1995 working group on genetic explanation and the ascription of responsibility, for extensive written comments from Alan Strudler, Marcia Baron, and Pat Greenspan, and for editorial guidance from Arthur Evenchik.

peutic attitude. Even if it does not exempt the offender from legal responsibility, it may suggest that he lacked opportunities for self-control that most of us have, or that self-control was far harder for him than for most of us. On the other hand, we punish recidivists more severely not only because they appear to be more dangerous, but because their recalcitrance appears to provide stronger evidence of an antisocial character than any single offense, however serious. We judge the offender not only for his acts, but for the character they reflect.[2] Evidence of a bad character or disposition may increase the agent's liability to punishment, and increase his punishment if he is liable.[3]

The claim that the most prolific or dangerous offenders differ genetically from the rest of us may reinforce *both* retributive strains, suggesting a volitional disability that makes blame inappropriate and a permanence that invites blame. Whereas claims of genetic influence may help to fill in a deterministic account of the agent's conduct or challenge assumptions about his freedom to do otherwise, they may also suggest that violent and antisocial impulses are deeply embedded and defining features of the agent's character.

As additions to the roster of social and biological factors that constrain choice and affect behavior, genetic factors will tend to be mitigating, all the more so because the agent has no responsibility for them. But as apparently constitutional features of the agent's character, genetic factors may work against mitigation, because their presence may be taken to reveal that the agent's behavior arose from an antisocial character or had its source in something essential to him. If the agent acted in character or was necessarily predisposed to violent or antisocial behavior, it may be difficult to treat him as a victim of bad moral luck and soften our condemnation on that basis.

Claims of genetic predisposition thus appear to have contrasting relevance for what Gary Watson (1996) calls "the two faces of responsibility": accountability and attributability. The claim that someone is genetically predisposed to violent or antisocial behavior may suggest that he has had a much harder time than most of us in conforming his behavior to norms of decency, thereby making him less accountable. At the same time, such a genetic predisposition appears to endow him with a violent or antisocial disposition and thereby to make the behavior fully attributable to him, as an expression of his underlying character. Attribution alone, Watson contends, is not enough to underwrite punishment or other sanctions; we are only warranted in imposing sanctions on the agent if he is accountable for his acts – if he had a reasonable opportunity to avoid engaging in them. Attribution may not be necessary for sanctioning either – an agent may be held accountable, and sanctioned, for conduct that is not fully attributable to him, in the sense that it is impulsive or out of character, as long as he can be said to have had a fair opportunity for avoidance. But our formal and informal sanctioning processes do take account of the extent to which the agent's conduct is attributable to him, in at least two ways. The more difficult it is to attribute the conduct to the agent, the less characteristic the conduct appears, and the more searching the inquiry into fair opportunity may be. And if the (variable) threshold for accountability is reached, the agent may be more severely condemned or sanctioned for more characteristic conduct – conduct that appears to be more fully attributable to him.

In the second section of this chapter I consider claims that genetic factors reduce accountability in the same ways as other biological and social factors; in the third section I consider claims that they strengthen the attribution of violent or antisocial behavior to the agent in ways that other biological factors do not. Assessing both sorts of claims require some understanding of what it means to claim with scientific credibility that a person is genetically predisposed to such behavior, so in the first section I examine how genetic predisposition claims are understood by the researchers who make them. The variety of ways in which genes may affect behavior, and the indirect and contingent ways in which they are most plausibly claimed to do so, suggest the difficulties of generalizing about the relevance of genetic-predisposition claims to the ascription of moral and legal responsibility.

THE CHARACTER OF GENETIC
PREDISPOSITION CLAIMS

One popular conception of a genetic predisposition to crime and violence (although perhaps not so popular as some critics suppose) imagines a person "born to kill," devoid of sympathy or remorse, requiring only slight insult or incentive to inflict lethal harm. This is the conception reflected in the character of Rhoda Penmark in the 1950s play and movie *Bad Seed*, a conception assailed by critics of human behavioral genetics.

Although Rhoda Penmark is a caricature, and a dated one, her portrayal reflects three assumptions about genetic sources of a predisposition to violent and antisocial behavior that are still widely held: that predispositions with genetic sources exert more direct and complete control over behavior than predispositions acquired solely through a harsh upbringing or social environment; that they endow those who have them with a fundamentally antisocial character; and that they are less mutable than predispositions with exclusively environmental sources, resistant to standard forms of therapy and social control. None of these assumptions is supported by credible research on genetics and human behavior.

The type of genetic influences that researchers expect to find are far less powerful and invariant, and potentially more mutable, than those to which Rhoda Penmark was subject. What behavioral geneticists generally mean when they claim that a person is genetically predisposed to a given kind of behavior is, in part, that he has genetic features that make him more likely to engage in that behavior in certain environments. Gottesman and Goldsmith (1994, 72) offer a further qualification, speaking of predispositions for behavioral tendencies:

Notions such as "genes for crime" are nonsense, but the following kind of notion is reasonable: There may be *partially* genetically influenced *predispositions* for certain behavioral tendencies, such as impulsivity, that in certain experiential contexts, make the probability of committing certain kinds of crimes higher than for individuals who possess lesser degrees of such behavioral tendencies.

Two features of plausible claims of genetic predisposition to violent and antisocial behavior complicate their relevance for moral responsibility and criminal liability. First, genes may predispose to criminal or violent acts in a variety of ways. A person's genetic endowment may make him generally susceptible or resistant to all forms of social influence, or it may contribute to specific traits conducive to violent or antisocial behavior in some circumstances – for example, a high energy level, a tendency to novelty-seeking, or a low anxiety level. With respect to the features alleged to predispose people to violent and antisocial behavior more specifically, the prevailing suspicion among contemporary researchers is that genetically based neurochemical deficits cause people to act impulsively by preventing them from "turning off" their arousal (Raine 1993, ch. 4; Stoff and Cairns 1996). Other researchers speculate that some people are genetically disposed to habituate quickly to punishment, making them harder to discipline, so less able to exercise self-restraint.[4] Still others propose low intelligence as a mediating variable, because it limits gainful employment and interferes with the rational assessment of consequences. It may also be possible that some people are "constitutionally" volatile: easily and intensely aroused

(Raine 1993). And perhaps some people lack the biological underpinnings for sympathy or empathy, so that they are unable to feel another's pain in a way that would constrain them from inflicting it. There are sharp debates about whether the proposed mediating variables, from neurochemical deficits to low intelligence, are actually subject to significant genetic influence, but that is not an issue on which I focus here.

A second, related feature of any genetic predisposition is its contingency. Because any genetic effect on behavior is mediated by an array of amniotic, somatic, and environmental variables, as well as interactions with other genes, the actual commission of criminal or violent acts by a person with a given genetic constitution is highly contingent. A person predisposed to crime or violence may never commit a criminal or violent act, and the same predisposition may favor courageous or heroic actions in some settings. As Gottesman and Goldsmith (1994, 72) speculate, "the energetic mesomorph who is unafraid to enter dark alleys might be a successful mugger or a decorated patrolman, depending on neighborhood heroes/heroines in childhood, attitudinal, cultural, and other chance factors that are easily imagined." Genetic predispositions with different causal pathways or developmental roots are obviously subject to different contingencies, and the degree of contingency may be far greater for some pathways than others. But even if people who are constitutionally less constrained by allegiance to public norms or the threat of punishment are more likely to engage in a wide array of conduct deemed criminal, Gottesman and Goldstein's example suggests that it will always be a highly contingent matter whether such underlying traits produce criminals or heroes.

These two features of plausible claims of genetic influence – their varying causal pathways and their contingent effects on behavior and character – complicate the mitigating or aggravating role they are thought to play in revealing volitional disabilities or in grounding negative character judgments. Obviously, there can be no plausible claim of a strictly genetic determinism; genes alone do not determine behavior. Perhaps less obviously, the fact that genes affect behavior indirectly and contingently leaves considerable doubt about what it would even mean to have a genetic predisposition to violent and antisocial behavior.

It is clearly not enough that genes play some causal role in such behavior; genes indisputably contribute to the anatomical and physiological prerequisites for almost all violent and antisocial conduct, but that hardly makes all agents who engage in such conduct genetically predisposed to it. Nor is it enough that genes make some agents more likely than others to engage in violent or antisocial conduct in a given environment. And it would also be misleading to treat a general genetic susceptibility to any environmental influence as predisposing a child to violent or antisocial behavior, even if he grew up in a neighborhood where bad influences were so pervasive that that susceptibility put him at "high risk" of such behavior. There are similar doubts about whether genetic factors that play a more specific role in the genesis of criminal and antisocial acts can be said to predispose the agent toward them. Can a genetic constitution that makes a person less intelligent be said to predispose him to violent or antisocial behavior if he is more likely to engage in such behavior only because of the limited opportunities for gainful employment or social acceptance available to him? If so, how can we distinguish this from a genetic constitution that affects a person's skin color, subjecting him to lifelong discrimination that forecloses many or most alternatives to criminal conduct?

Rather than classify genetic predispositions by their causal roles, we might look at the range of environments in which a given genotype is associated with violent or antisocial behavior; at what behavioral geneticists call its "reaction surface." If a given genotype were associated with a greater likelihood of such behavior in every environment, however harsh or benign, we might regard it as predisposing. But without any basis for classifying environments, let alone any knowledge of how people react in all but a small range of environments, this approach is not really feasible. And even if we could survey the relevant environments, we would have to decide how to classify genotypes with irregular reaction surfaces – for example, genotypes associated with more violent and antisocial conduct in some environments, but less or none in others.

Moreover, even if we had a satisfactory account of what it means to have a genetic predisposition to violent or antisocial behavior, such predispositions would be likely to have limited relevance to "disposition" in the ordinary sense of character or personality: a person may be genetically predisposed to violent or antisocial acts without having a violent or antisocial disposition in this sense. Thus, for example, a generally sweet, friendly, mild-mannered person with a genetic incapacity to control his occasionally hostile impulses might be regarded as predisposed to violent and antisocial behavior, but not as having a violent or antisocial character.

In the next two sections, I explore these complications in assessing the mitigating and aggravating potential of genetic predisposition claims. I argue that the apparent relevance of those claims for responsibility and blame often rests on a mistaken or oversimplified model of genetic influence.[5] I begin by considering how genetic predisposition evidence might bear on what many philosophers and lay people regard as the critical question in the ascription of moral responsibility: whether the agent acted freely, in the sense that he could have done otherwise, or in some other sense.

GENETIC PREDISPOSITIONS AND THE CAPACITY TO DO OTHERWISE

Claims of genetic influence may raise doubts about the agent's freedom to do otherwise by challenging the assumption that the agent is a distinct source of action, less subject than events in the physical world to causal laws. Patricia Greenspan (Chapter 10, in this volume) describes the challenge in the following terms:

> Genetic explanation . . . would seem to plug the agent . . . right into a deterministic chain of events, by way of a causal account of personality formation. The appeal to genetic causes of character may look worse for free will, then, because it co-opts the conceptual apparatus of a standard kind of incompatibilist defense of free will.

But the idea of the will as exempt from causal laws has been under attack for centuries. If the attacks now seem more credible, it may only be because the biological character of genetic explanations seems to improve the prospects for a comprehensive determinism at a molecular level: it now seems possible to analyze the agent's decision making in the same molecular terms as events in the external world. This appearance may be deceptive, however: the discovery of a neurochemical "substrate" for psychological activity may only sharpen the debate about the extent to which the former can explain or subsume the latter.[6]

Evidence of genetic predisposition, however, might temper blame even if it did not make determinism more plausible or advance deterministic explanations of specific behaviors. Such evidence might challenge the agent's responsibility in two ways: first, by revealing that his misconduct did not arise from normal deliberative processes but through impulses that overwhelmed or bypassed those processes; second, by revealing or suggesting that self-control was much more difficult for the agent than for most of us.

Although the first would be more clearly extenuating than the second, only a small fraction of violent and antisocial behavior, genetically influenced or not, appears to result from a complete bypass or disruption of the agent's deliberative processes. Even if genetic sources were found for psychotic delusions, epileptic seizures, irresistible impulses, and other conditions that induced violence but negated responsibility, those conditions account for a negligible proportion of the violent and antisocial behavior that occurs in any society. And the discovery that such conditions had a genetic source would hardly be necessary for extenuation. Critics of behavioral genetics expect it to do little more than gild the lily in this manner. They predict that any genetic contributions found to have major effects on behavior will be the work of mutations that cause major dysfunctions in a very small number of people – "the biochemical equivalent of hitting a subject on the head with a club" (Balaban 1996, 87).

Evidence of genetic predisposition might be more likely to play a mitigating role in revealing or suggesting that the agent faced a significantly greater cumulative burden of self-restraint than most of us do. The potential for mitigation would be clearest if the agent were genetically predisposed to violent or antisocial behavior by heightened arousal, so that he was routinely beset by the kind of powerful impulses that most of us experience only in very unusual circumstances. At an extreme, genetic predisposition evidence might make it appear that *some* aggressive or impulsive conduct by the agent was inevitable. As Greenspan (1993, 38) explains:

It might be argued that even if determinism is false, on the assumption that it is hard but *possible* for an agent incapable of acquiring a good temper as a general trait to control aggression on a given occasion, the temperament of such an individual does rule out an overall record of controlled response. The psychological pressures on him are intense enough to ensure that he will commit a violent act at one time or other, though when and how are subject to variation of a sort that strict determinism rules out.

Even if lifetime self-restraint were not literally impossible for the agent, he would bear a much heavier lifetime burden of self-restraint than most of us do, a burden that would serve to mitigate the occasional, almost inevitable lapse. Evidence of a genetic predisposition would give his claim of more frequent and intense psychological pressure an objective basis and would tend to absolve him of responsibility for that burden.

But this may not be the way that genes typically predispose. Recent neurogenetic research has focused on abnormalities in the inhibitory,

not the excitory system. As Greenspan suggests (Chapter 10) this research appears to implicate weak inhibition, not heightened arousal:

The impulsive criminal has the same impulses as the rest of us, at least initially,
on this account; he simply lacks our capacity to suppress them, or to let them
pass unsatisfied. He's no more disturbed at suppressing them, but just can't
manage it . . . because he lacks the requisite supply of serotonin. . . . The current
research model essentially takes the "mania" out of kleptomania – the suggestion of stirred up counterforces to self-control, something on the order of a demon fighting oneself – and substitutes personal inadequacy, a shortage of the
means of self-control.

On the assumption that we could characterize an impulsive disposition in the way Greenspan does,[7] the question remains of how it would
affect our appraisal of the agent's behavior and character. If he literally
could not suppress his impulses, if he could exercise no deliberative
control, we should not hold him responsible. But this does not seem to
be what Greenspan is suggesting. It is not that he is incapable of suppressing his impulses, just less capable than most of us.

Is someone who faces ordinary impulses with a reduced capacity to
resist them less deserving of mitigated punishment than someone who
faces unusually strong impulses with a standard capacity for self-control? On the one hand, we are more inclined to extenuate the misconduct of someone who struggles heroically against strong impulses than
of someone who blandly yields to routine ones, even if the latter lacks
the "internal resources" to put up a good fight. This may be in part because it is easier to see the former as a victim of bad moral luck, an issue I take up in the next section.

Greenspan's account, however, might also challenge the assumption
that the agent yielded to his impulses because he identified with or endorsed them and was, in that way, responsible for his impulsive behavior. His posture might appear more acquiescent than approving. I am
not suggesting that neurogenetic research will yield this more sympathetic picture, but rather that such agent's conduct is likely to be subject
to a variety of more or less charitable or extenuating interpretations.

In conclusion, although genetic factors may well lie behind some of
the more extreme conditions associated with violence, such as certain
types of epilepsy and psychosis, evidence of a genetic contribution
would, for exculpatory purposes, be largely redundant. Genetic factors
might well have a more significant mitigating effect in suggesting that
certain people face a greater cumulative burden of self-restraint than

most of us or, perhaps more plausibly, face the standard gamut of goads and temptations with a weaker capacity for resistance. Such evidence of genetic akrasia, however, would hardly negate liability; the vast majority of violent crimes are committed by people with weak impulse control. But such evidence would challenge the assumption that the agent failed to resist his impulses because he endorsed them, or had a callous disregard for the harm he would cause by yielding to them. Moreover, genetic evidence might have some mitigating effect in challenging the assumption that *repeat* offenders displayed a callous insensitivity to the interests of other people. It would suggest a less reprehensible lack of self-restraint.

GENETIC INFLUENCE, IDENTITY, AND IDENTIFICATION

If evidence of genetic predisposition can mitigate blame by suggesting an usually heavy burden of, or a limited capacity for, self-control, it can also aggravate blame, by suggesting that the offending behavior was characteristic of the offender. In Watson's (1996) terms, it may strengthen the attribution of bad conduct to the agent even as it reduces his accountability for it. Evidence that the agent's impulses had a *genetic* source might suggest that he was not acting "out of character" in committing the offense, denying him the mitigation, or at least sympathy, he would enjoy if those impulses were seen as external to him.

Not all forms of genetic predisposition would have this aggravating tendency. In particular, a genetically based incapacity to control one's impulses would reveal, if anything, that it was in the agent's character to act out of character. Greenspan's impulsive offender may have a sweet, amiable disposition, with fewer, or weaker, hostile and aggressive impulses than most of us. Even if he can be said to be predisposed to violent and antisocial behavior, in that his lack of self-control makes it likely that he will engage in such behavior in a wide range of environments, that weakness does not endow him with a violent or antisocial character.

Some mechanisms of genetic influence, however, such as a rapid habituation to aversive stimuli, might contribute to the formation of a character that is more recognizably antisocial, if such habituation blunted the development of the capacity for remorse that somehow evolves from the fear of punishment in the course of a normal upbringing.[8] Although the effect of such genetic features would appear to place responsibility for the agent's character beyond his control, their role in its formation also make it hard to see him as a victim, as he might be seen if his character could be attributed to environmental sources. Thus, Watson (1996, 240) observes

> our ambivalence towards the vicious criminal who was himself a victim of an abusive childhood. . . . His conduct is attributable to him as an exercise of his "moral capacities." It expresses and constitutes his practical identity, what he stands for, what he has made of his life as he found it. . . . What gives rise to our "pity" are concerns about fairness. Facts about his formative years give rise to the thought that the individual has already suffered too much and that we too would probably have been morally ruined by such a childhood.

To the extent that a genetic predisposition can emerge without such an abusive childhood, it will not offer the same countervailing mitigation.

There are three aspects of predisposing genetic factors that appear to work against mitigation. The first is their allegedly constitutive role in the agent's identity. Although Watson's vicious criminal could have acquired a better character, if, say, he had been placed in foster care shortly after the abuse began, it might seem that an agent whose violent or antisocial character is attributable to a genetic predisposition could *not* have acquired a better character – that his bad character is essential to him in a way that the bad character of Watson's criminal is not. The second aspect is onset. During the formative years in which his character is shaped, Watson's criminal is already a person with some distinguishing qualities that elicit sympathetic identification, if not affection; we can see his abusive childhood as deforming his character. In contrast, someone with a predisposing genetic factor appears to be a "naked subject" during the time when that factor shapes his character; it is difficult to see his genes as doing something to *him*. The third aspect is the manner in which genetic factors appear to operate. Watson's child is not merely made vicious; he is made vicious by abuse. He is a real victim, and the tragedy of his life is that he goes on to victimize others in a similar way. In contrast, the genetically predisposed individual does not acquire a vicious character by vicious human agency. His genes may have an adverse effect on his character, but they do not operate in a way that can be said, except in the most figurative sense, to victimize him.

These contrasts, however, are exaggerated, based on misconceptions about the role genes play in the development of the organism and the identity of the person. The first overstates the dependence of the agent's identity on predisposing genetic factors and the traits to which they contribute; the second conflates "genetic" with "congenital"; the third treats predispositions with genetic sources as distinct kinds of predisposition, acting independently of, rather than through, environmental conditions. In fact, there is a great deal of uncertainty about the relationship of genetic constitution to personal identity: current research suggests that genetic factors have greater influence on behavior later in life, and it is generally recognized that their influence largely consists in the way they mediate environmental influences, including abusive treatment by other people. In light of the modest and complex role that genes play in shaping character and constituting identity, there is no reason why agents predisposed to violence by their genes should not, in general, elicit the same kind of ambivalent response as Watson's environmentally disposed criminal.

Genetically Influenced Predispositions and Personal Identity

One feature of plausible claims of genetic predisposition discussed in the first section – environmental contingency – virtually guarantees that the actual display of violent or antisocial conduct by a predisposed agent involves an element of moral luck. Had the agent not had the bad luck to encounter an "environmental trigger," he might not have engaged in this violent or antisocial act. However common such triggers may be, the agent might have avoided them.

Yet to the extent that the agent's character as well as his actions is the object of moral appraisal, the role of moral luck seems to recede. Having a violent or antisocial genetic predisposition does not seem to be a matter of bad moral luck in the same way as encountering an environmental trigger. Although some philosophers speak of "constitutive moral luck," bad luck in one's constitution may not elicit the same pity or warrant the same mitigation as bad luck in one's experience or personal history.

It might be thought that (1) to the extent the agent's disposition is genetically based, she could not have existed without it, and (2) if the agent could not have existed without the predisposition, we cannot regard her, and she cannot regard herself, as a victim of bad moral luck in having that predisposition. While I am not sure if (2) is true, I believe that (1) is false: the same individual might have existed without almost any genetically based predisposition. Altering the genetic features and developmental conditions that yield the predisposition would not necessarily alter the identity of the zygote or of the person who developed from it. And although the predisposition may sometimes be so integral to the person that *she* could not have existed without it, that will depend on how early and how pervasively the predisposition takes hold, not on its genetic origins.

Assume that the genetic source of the agent's predisposition is a specific point mutation (though most behavioral predispositions are likely to involve several genetic loci, we can safely ignore the complexities of gene interactions for present purposes). If the only way the agent could have escaped that mutation would be to have been conceived from a different pair of gametes, then, under what Bernard Williams (1990) calls the zygotic principle (ZP), the person without the point mutation would not have been he. But conception from a different pair of gametes is not the only counterfactual possibility; we can, for example, imagine that his predisposing mutation was repaired by prenatal germline therapy or altered prenatally by exposure to radioactivity.

These alternative possibilities limit ZP to a claim about one end point on a spectrum of genetic alterations, where a change in the zygote's genotype can only be achieved by replacing one of the gametes that formed it. Even without gamete substitution, however, the identity of the zygote, and of the person who develops from it, may be altered by comprehensive genetic changes.[9]

On the other end of the spectrum, however, a single genetic alteration, such as the "correction" of a point mutation, would not appear to alter the identity of the zygote or of the person who developed from it, however significant it might be phenotypically.[10] But how, between these end points, do we decide if genetic changes alter identity?[11] Does identity through genetic change depend on the number of genes altered, the magnitude of the molecular changes, the number of traits, the importance of the traits, or the magnitude of their phenotypic changes? And how would we measure magnitude or importance?

Fortunately, we do not need to resolve these complexities to reject the claim that this person could not have existed without an antisocial predisposition. Even if the agent could not have existed without the point mutation, he might have existed without the somatic, amniotic, or other circumstances in which the mutation would yield the predisposition. This contingency makes it reasonable to regard the agent's bad luck in developing an aggressive or sadistic disposition as a misfortune in his history, not, or not only, in his constitution.[12]

Genetic versus Congenital Factors

A second reason why genetic influences might be thought to place the person's character or disposition outside the scope of moral luck is that those influences are assumed to be at work from the time of conception or early gestation, so that the individual will already have been shaped by them when he is born or becomes conscious. But this assumption is mistaken, because genetic influences vary widely in their onset. And the onset of the influence matters more than its source in the genome or the environment.

This point can be made with a comparison of four "case histories." These cases are deliberately oversimplified, in order to exaggerate the differences between genetic and environmental influences on antisocial dispositions. If anything, the role played by time of onset would appear even greater if those differences were more realistically rendered.

1. As a result of harsh and abusive parenting, D is from the start an ag-gressive child, his juvenile delinquency and adult criminality merely an extension of his earliest sandlot behavior.
2. As a result of a point mutation in a gene regulating monamine oxi-dase, D is from the start an aggressive child, his juvenile delinquency and adult criminality merely an extension of his earliest sandlot be-havior.
3. D is a playful, amiable, and vulnerable child his first five years. Then, as the result of harsh and abusive parenting, he becomes increasingly aggressive, developing into a juvenile delinquent and adult criminal who manifests no trace of the playful, amiable, and vulnerable child he once was.
4. D is a playful, amiable, and vulnerable child his first five years. Then, as the result of a point mutation in a gene regulating monamine oxi-dase, he becomes increasingly aggressive, developing into a juvenile delinquent and adult criminal who manifests no trace of the playful, amiable, and vulnerable child he once was.[13]

We are, I think, less inclined in cases 3 and 4 than in cases 1 and 2 to see the child as essentially violent and antisocial, since it is easier in 3 and 4 than in 1 and 2 to see these dispositions as imposed or inflicted on him. In 3 and 4, we can see a playful, amiable, and vulnerable child over-taken by aggressive impulses; in 1 and 2, the child makes his debut as aggressive and sadistic.[14] The *onset* of the behavior-shaping process ap-pears to matter more than its genetic or environmental origins. This point can be amplified by comparing versions of cases 2 and 4 where the point mutation is inherited with versions where it arises from the child's accidental exposure to radioactivity. Despite the fact that the predisposition is not inherited in the second version, and is arguably as environmental as the parental abuse in cases 1 and 3, the extent to which the predisposition seems integral to the person is no different than in the original versions of 2 and 4.

An analogy to genetic disease may be instructive. We recognize a discrete loss or disruption where the onset of a genetic condition is postnatal, as in Huntington's disease. In contrast, it is more difficult to find a discrete loss or disruption where the condition is congenital, as in Down syndrome. Clearly, this is not because Down syndrome is any more genetic than Huntington's disease (in fact, Down syndrome is a *chromosomal* condition that generally results from the process of gamete formation, so it is not inherited) or any less mutable; it is because it is harder to see the child with Down syndrome as "afflicted" by a condition she was born with, and which may be integral to character and personality – we can hardly imagine her not being retarded, or see her retardation as something alien to her. Even if it makes biological sense to regard Down syndrome itself as disrupting the normal course of mental development, it is difficult to see it as something happening to the child: she is the product, rather than the victim, of the disruption.

Similarly, even if genetic research decomposed the process by which the antisocial disposition emerged in case 2, so that we saw the mutation as disrupting the mechanisms by which people normally develop or achieve socialization and impulse control, it would still be difficult to regard the congenitally aggressive child as afflicted by his disposition. He was always aggressive; there is no kinder, gentler segment of his history or aspect of his character that we can see as afflicted or overwhelmed by these harsher impulses.[15] But this might also be true for the genetically "normal" child subject to early, pervasive abuse.

Victimized by One's Genes?

The preceding section presented a false, or at least exaggerated contrast: between people made violent by their genes and people made violent by their upbringing. But genetic mutations and abusive parenting are not two distinct ways in which people can acquire violent dispositions. Rather, an abusive upbringing acts on a genetic, amniotic, and early developmental background. Children respond in very different ways to abusive parenting; we can add siblings in cases 1 and 3 who become extremely passive or timid in response to their parents' abuse. There might well be some genetic difference that helps explain why one sibling responded with violence, the other with passivity. The genetic feature that contributed to the one child's violent response would not make him less a victim of abusive parenting; indeed, any relevant genetic difference between the two siblings would weaken or block the attribution of their divergent responses to a conscious decision or voluntary choice.

The children in cases 2 and 4 are caricatures, much like Rhoda Penmark, becoming violent without environmental provocation. There are, of course, many environmental triggers for violent acts besides parental abuse. (Indeed, some developmental scholars now claim that peer-group pressure is a more important influence than parental behavior.) Some violent acts will be reenactments of the agent's own experience of abuse; many, perhaps most, will have a less poignant genesis. Even if we are not inclined to regard a genetic predisposition to violence as a burden, like abusive parents, indifferent teachers, or a bad neighborhood, it is unlikely that people predisposed to violence by various genetic factors have faced less hardship or endured less suffering in their childhoods than those who commit violent crimes without those particular predisposing factors. There would, perhaps, be a negative correlation between genetic predisposition and mitigating hardship if violence had an etiology similar to the etiologies of cancer for which there are varying degrees of genetic predisposition – with people who are less predisposed requiring greater environmental insult. But this seems like a simplistic and highly implausible model for violence.

Genetic Influence, Identification, and Empathy

Watson (1996, 240), however, suggests another source of mitigation that may be blocked by evidence of genetic predisposition: the conviction that "we too would probably have been ruined by . . . a childhood" like that of the abused criminal. Evidence that the criminal was genetically predisposed to violence in a way we are not would undermine that conviction, by making it appear unlikely or impossible for "us" to have been ruined in quite the same way.

It may be that even if the criminal could have existed without his genetically based antisocial disposition (e.g., because that disposition would not have emerged with gentler parenting), there may be no circumstances under which *I* could have acquired that antisocial disposition, and therefore no way I can coherently claim that "there but for the grace of God go I." Watson formulates (1988, 278) the metaphysical constraint on our identification in these terms: "To make sense of a counterfactual of the form, "If i had been in C, then i would have become a person of type t," C must be supposed to be compatible with i's existence as an individual (i must exist in the possible world in which C obtains)."[16]

The question is whether this formal constraint limits the empathy or identification I can have with "a person of type t." Clearly, as Watson observes, I cannot see the difference between myself and a brutal killer as a matter of moral luck if it is not possible for me to have become like him. But as Ferdinand Schoeman argues (1988, 300–313), there are at least two ways in which the conviction that "there but for the grace of God go I" can mitigate. First, the more easily I can imagine myself in the agent's situation, the more strongly I may be inclined to leniency, because I would want to be treated leniently in those circumstances. But, as Schoeman (1988) insists, this provides no moral basis for varying degrees of leniency. Even if I cannot extend leniency at all without at least a scintilla of empathy, the agent hardly deserves less leniency simply because I find it harder to imagine myself in his situation.[17]

The conviction that "there but for the grace of God go I" can also mitigate by making morally relevant features of the agent's situation more salient. But although the conviction that I could have gone the same way may make me more sensitive to the constraints, pressures, and temptations the agent faced, it is not clear why I cannot appreciate those psychological factors without that conviction.

Philosophers disagree about whether we can imagine the logically impossible; however, I do not need to imagine that I *am* someone it is logically impossible for me to be in order to appreciate the constraints, pressures, and temptations he faces. Of course, my appreciation may be severely limited or distorted by differences in temperament, history, and world view.[18] It may be much harder to project myself into someone else's circumstances than it appears, and I may inevitably import many of my own characteristics, so that my identification will be impoverished or distorted to the extent I differ from the person I identify with. But this will be so whether or not it is logically impossible for me to have acquired a similar temperament, history, or world view. I may be unable to appreciate the grim fatalism or paranoia of a close friend who returns from the front lines, even though it was only the luck of the draft lottery that spared me from acquiring a similar outlook. Conversely, when I read *The Brothers Karamazov* I may be able to identify or empathize quite strongly with people whose temperament, history, or world view it would have been impossible for me to acquire.

GENETIC PREDISPOSITIONS
AND SELF-IDENTIFICATION

A final way in which genetic predisposition evidence may be relevant to the agent's responsibility is through its effect on his own attitude or moral posture toward his violent or antisocial impulses. Several writers on moral responsibility, notably Harry Frankfurt (1988) and Gerald Dworkin (1970), argue that the agent's identification with, or endorsement, of his desires affects his moral responsibility for the actions motivated by them. Even if the agent is not free to refrain from acting upon his desires – that is, if he could not have done otherwise even if he rejected or disavowed those desires, he "takes responsibility" by endorsing or identifying with them.[19] We may blame the agent for yielding to desires he does not endorse, as in the case of impulsive violence, but we regard the actions as less fully his own, and in that respect regard him as less responsible, than when he acts on desires that he accepts as his own.[20]

In considering how genetic predisposition may bear on the agent's identification with his desires, it is useful to make a threshold distinction between actual genetic influence on the agent's conduct and the agent's belief that his conduct is under genetic influence. While the former is unlikely to have any general effect, the latter may have significant but opposing effects, inviting as well as discouraging the agent's identification.

With regard to the former, it might be argued that desires or impulses that had a source in a genetic abnormality would be especially susceptible, or especially resistant, to identification or endorsement. On the one hand, the agent might be more inclined to endorse or identify with such desires because of their familiarity. It may be that, in general, the earlier in the agent's development a desire emerges, and the more often it is aroused, the more fully the agent will identify with or endorse it. But as I previously noted, genetic influences may operate very late in development, while environmental influences may operate at a very early stage.

On the other hand, the desires arising from a genetic abnormality might be less continuous or coherent with the rest of the agent's desires, or they might present themselves to the agent in a way that is particularly resistant to identification. Joel Feinberg makes a similar suggestion about familiar compulsions like kleptomania. While denying that such impulses are literally compelling or irresistible, Feinberg suggests that they are often "senseless in the special way that permits us to speak of them as incoherent" (1970b, 287). Not only are they not self-interested in the usual sense, but they cannot even be regarded as expressing the agent's interests. They are as opaque to him as they are to observers.

However apt a characterization this may be of familiar manias, there is no reason to think that such manias are subject to strong genetic influences, or that desires and impulses that are subject to strong genetic influences have the disconnected and opaque character of manias. As Greenspan (Chapter 10) observes, research in human behavioral genetics suggests, if anything, that the way in which genes typically predispose to violent or antisocial behavior is by limiting the capacity to control desires, not by generating especially strong or opaque desires.

Whether or not an agent is more likely to identify with or endorse violent and antisocial impulses that in fact have a genetic source, the agent's belief that they have such a source may well affect his identification or endorsement. That belief might strengthen the agent's inclination to identify with or endorse those impulses, by making them seem a more integral and less mutable part of his character. The agent may well embrace the "genetic essentialism" that seems to pervade popular culture, and regard his genetic makeup as his basic or essential nature.

On the other hand, the agent's belief that his antisocial impulses have a genetic source might work against his identification with those impulses. He might regard a genetic diagnosis as externalizing the impulse, by locating its source in something objective and physical, as if his doctors had traced it to a tumor or brain lesion. He might regard a genetic abnormality as a burden inflicted on him by his parents, and disavow the impulses he associates with that abnormality as a way of asserting his autonomy. The treatment of his genetic abnormality as a medical condition might also encourage him to see his violent or antisocial impulses as symptoms of a disease, not as an expression of his "true character." And if there were any prospect of weakening his antisocial impulses or strengthening his self-control with drugs or therapy, he might come to see his predisposition as a temporary affliction.

The agent is not compelled, of course, to disavow impulses attributed in part to genetic causes, any more than he would be compelled to disavow his creativity or compassion if they were shown to have genetic sources. But if he has a kinder and gentler side, or higher-order regrets about his impulses, the discovery or claimed discovery of a genetic source for those impulses may encourage him to withhold his identification.

In general, genetic predisposition evidence is likely to have a highly equivocal effect on the agents' identification with, or endorsement of, his violent or antisocial impulses. If such identification strengthens an attribution of antisocial conduct to the agent, genetic predisposition evidence will once again complicate rather than resolve issues about the agent's responsibility.

CONCLUSION

Researchers are unlikely to discover genetic predispositions that will compel us to significantly revise our general practices of blaming or our judgments about the blameworthiness of specific agents. Any genetic contribution to violent or antisocial behavior, or to a violent or antisocial character, is likely to be indirect and contingent, mediated by an array of developmental and social variables. For that reason, credible claims of genetic influence should not radically alter the picture we have of the forces that shape and constrain human agency and character.

If anything, such claims reinforce both of our conflicting tendencies toward agents who are predisposed to violent and antisocial behavior. The discovery of genetic sources for such predispositions may make them look more oppressive, and involuntary, but it may also make them appear more integral to the agent's character. Both of these impressions are exaggerated, but it is unlikely that claims of genetic predisposition will be received with less ambivalence when they are better understood.

NOTES

1. In suggesting a long-standing tension between these two tendencies, I am well aware that the first is much stronger in the current political climate. Both the public and the criminal justice community are preoccupied with the specter of the repeat violent offender, thought to be responsible for a high proportion of crime, and to be resistant to any sanction besides prolonged incarceration or execution. Many recent initiatives in prosecution and sentencing have been designed to identify and incapacitate such offenders. Claims that some individuals are genetically predisposed to crime and violence are likely to lend scientific credibility, and the promise of di-

agnostic refinement, to these approaches. I address these prospects in Wasserman 1995.

2. The role of character in criminal liability is, however, a matter of considerable debate. George Fletcher, for example, is among the contemporary legal scholars who contend that character plays a central role in our system of excuses: "the distinguishing feature of excusing conditions is that they preclude an inference from the act to the actor's character" (1978, 799). Fletcher, however, seems confused about what it means to make an "inference . . . to the actor's character." He claims that the question is "whether a particular wrongful act is attributable either to the actor's character or to the circumstances that overwhelmed his capacity for choice" (1978, 801). But these alternatives are far from jointly exhaustive: in the usual sense of "character," a wide range of angry, grumpy, spiteful, impulsive, and whimsical actions may be "out of character" without being attributable to circumstances that overwhelm the capacity for choice, and those actions are typically unexcused. What seems required for most excuses is that the acts result from a departure from, or breakdown in, normal deliberative processes, so that they cannot be said to be a product of the actor's wrongful choice. But many out-of-character actions do not involve such a breakdown or departure; the fact that they are uncharacteristic does not make them unchosen.

3. For example, although we do not exempt people from liability merely because they were not predisposed to act as they did, we do exempt them if they were not predisposed and their acts resulted in part from the state's instigation. R. A. Duff (1993, 350–359, 363–364) provides other examples of how character judgments enter into criminal liability and defenses.

4. The distinction between general susceptibility and specific predisposing features may break down here, because rapid habituation to aversive stimuli could play both roles, making a child generally resistant to environmental influences and specifically predisposing him to crime or violence.

5. If those notions are widely held, of course, such speculation may accurately predict changes in our reactive attitudes even if it does not justify them.

6. Another way that claims of genetic predisposition may make determinism look more credible is by making causal influences appear more straightforward and mechanistic. An analogy may be helpful here. Getting a "head" from the toss of a (head-) biased coin not only looks more likely, but more fully determined, than getting a head from the toss of a fair coin. Although the two events are equally random or nonrandom, the toss of the fair coin seems more random because the coin lacks a "predisposition," and the congeries of physical forces shaping its flight are largely beyond the observer's ken. If the outcome reflects the predisposition, it may look more determined. But again, this is just an appearance: the outcome is no more determined in the toss of a biased coin than it is in the toss of a fair coin. This appearance may be misleading in another respect: it treats a predisposition as "biasing" the agent toward violence, a metaphor less apt for some underlying mechanisms than others.

7. It is not clear how the results of genetic research could support or confirm

this characterization of impulsivity. What would count as evidence that the impulsive person did not experience stronger impulses? Perhaps her synaptic discharges will be no greater than average in those settings where the relevant impulses arose. But we could hardly assume that the strength of the experienced impulse varied in any simple way with the strength of the electrical discharge. We cannot simply read off the phenomenology of impulsive behavior from its neurophysiology. Researchers could attempt to correlate self-reports with synaptic discharges, but, to the best of my knowledge, this has not been done. In any case, such a correlation would merely be suggestive. It may be that scientific evidence cannot confirm or refute an "internal" account like Greenspan's, but merely render it more or less plausible or felicitous.

8. As Greenspan (personal communication) suggests, the character assessment may be more complex if such habituation prevents the development of the internal resources for self-control, but not of a (behaviorally inert) capacity for remorse.

9. This is certainly what Williams thinks (1990, 171): "Suppose there were a feasible and legal programme for polishing up people's DNA so that their offspring had enviable characteristics. A is annoyed that he is less clever, handsome, etc., than B, whose parents took the trouble and had the resources to enter the programme, while A's did not. A cannot coherently think that if his parents had entered the programme, he would have been more clever and handsome; he would have just been a non-existent person with a smarter sibling."

10. Although at least one philosopher, Noam Zohar (1991), had argued that it would.

11. Jim Stone (1994, 292) frames the issue this way: "Freeing a complete genetic code of interfering DNA is . . . compatible with [the fetus's] survival, even if we alter every cell in her body. Also, plugging a very small genetic gap that prevents the developmental path from proceeding is compatible with the fetus' survival, even if we alter every cell in her body. . . . On the other hand, a genetic deficit so vast that we must, by genetic engineering, determine the features and functions of the outcome, involves so many genes, implicated in the development of so many traits, that we are neither destroying or saving an embryo, but creating one" (292).

12. There may, however, be a subtler reason to see gene expression as falling outside the scope of moral luck – not because this agent's genes could not have expressed themselves differently, but because the conditions affecting the expression of his genes are typically not conditions to which he is subject. They are not part of his personal history, but part of the molecular etiology of his condition. I am not sure, however, why this should prevent the occurrence of those conditions from being regarded as a matter of moral luck. The agent may be unlucky, even if the locus of his bad luck is at the molecular level. What is left of the suggestion is merely the claim that the agent is not the victim of human agency, a claim I take up later.

13. If this fourth case seems too fanciful, consider a mother who described the

earliest expressions of aggression and destructiveness in her previously sweet and vulnerable male children as "testosterone poisoning."

14. Although, as Greenspan (personal communication) suggests, the contrast between the congenital and later onset cases may fade as the child becomes an adult, who will in either case have had a violent disposition for most of his life.

15. Of course, many people who act violently or aggressively will have kinder, gentler aspects of their character. I have eliminated those aspects in these examples to focus on the role of onset and highlight the significance of having a kinder, gentler past. As Greenspan suggests, the significance of onset may not be as great for agents with such kinder, gentler aspects – for example, impulsive agents who have sweet, amiable dispositions, marred by occasional fits of temper.

16. Watson (1988) does not distinguish the questions of whether (1) I could have become a person of type t and (2) a person of type t could have avoided becoming a person of that type. These are both questions of moral luck, but quite distinct ones, since (2) could be true when (1) is false: it may be the case that X could have avoided becoming a person of type t under some circumstances even if I never could have become a person of type t under any circumstances.

17. I have benefited from Greenspan's comments in clarifying this point.

18. It might be argued, somewhat analogously to the way Marya Schechtman (1990) argues against the possibility of identity-creating memories that are not identity-presupposing, that one could not feel the pressures and temptations a person faced without being significantly like that person. But one can fail to be significantly like another person even if one could have become quite similar to him, so this barrier to sympathetic identification may be present regardless of whether it has a constitutional source.

19. Although this approach has been criticized as deferring rather than resolving the question of the agent's moral responsibility, because his identification or higher-order approval may itself be causally determined, its emphasis on the agent's posture toward his own impulses seems to capture some of the phenomenology of blaming. An agent who regrets having, and acting on, violent or antisocial impulses, or who regards those impulses as alien to him, may seem less responsible for yielding to them than an agent who welcomes those impulses and regards them as integral to his character or identity. Of course, we might expect the difference in posture to make some difference in how frequently or fully the agent yielded to those impulses, and regard the agent's professions of regret or estrangement as disingenuous if they did not seem to constrain his conduct at all.

20. In Watson's (1996) terms, the agent's posture toward his effective desires may bear on whether we attribute the resulting actions to him, regardless of whether we hold him accountable for them. Watson regards such attribution as reflecting an "aretaic" conception of responsibility, which does not involve moral condemnation. We can attribute desires and actions to the agent without condemning him for them; conversely, we can hold him ac-

David Wasserman

countable, and condemn him, for acting on desires that we do not attribute to him; that he and we do not regard as fully his own. Other philosophers, such as Jorge Garcia (Chapter 12, in this volume), doubt that responsibility can be bifurcated in this manner, and that we can attribute base desires and reprehensible actions to a person without condemning him for them.

REFERENCES

Balaban, E. (1996). Reflections on Wye Woods: Crime, biology, and self-interest. *Politics and the Life Sciences* 15 (1): 86–88.
Duff, R. A. (1993). Choice, character, and criminal liability. *Law and Philosophy* 12: 345–383.
Dworkin, G. (1970). Acting freely. *Nous* 4 (4): 367–383.
Feinberg, J. (1970a). Crime, clutchability, and individuated treatment. In *Doing and deserving*, 252–271. Princeton, N.J.: Princeton University Press.
(1970b). What is so special about mental illness? In *Doing and deserving*, 272–292. Princeton, N.J.: Princeton University Press.
Fisher, J. A. (1994). Why potentiality does not matter: A reply to Stone. *Canadian Journal of Philosophy* 24 (3): 261–280.
Fletcher, G. (1978). *Rethinking criminal law.* Boston: Little Brown.
Frankfurt, H. (1988). Identification and wholeheartedness. In *Responsibility, Character, and the Emotions: New Essays in Moral Psychology*, ed. F. Schoeman, 27–45. Cambridge: Cambridge University Press.
Gottesman, I., and H. H. Goldsmith. (1994). Developmental psychopathology of antisocial behavior: Inserting genes into its ontogenesis and epigenesis. In *Threats to optimal development: Integrating biological, psychological, and social risk factors*, ed. C. A. Nelson, 69–104. Hillsdale, N.J.: Erlbaum.
Greenspan, P. S. (1988). Unfreedom and responsibility. In *Responsibility, character, and the emotions: New essays in moral psychology*, ed. F. Schoeman, 62–80. Cambridge: Cambridge University Press.
(1993). Free will and the genome project. *Philosophy and Public Affairs* 22: 31–43.
Raine, A. (1993). *The psychopathology of crime: Criminal behavior as a criminal disorder.* New York: Academic Press.
Schechtman, M. (1990). Personhood and personal identity. *Journal of Philosophy* 1990: 71–92.
Schoeman, F. (1988). Statistical norms and moral attributions. In *Responsibility, character, and the emotions: New essays in moral psychology*, ed. F. Schoeman, 287–315. Cambridge: Cambridge University Press.
Stoff, D., and R. Cairns (Eds.). (1996). *Aggression and violence: Genetic, neurological, and biosocial perspectives.* Mahwah, N.J.: Erlbaum.
Stone, J. (1994). Why potentiality still matters. *Canadian Journal of Philosophy* 24 (3): 281–292.
Wasserman, D. (1995). Science and social harm: Genetic research into crime and violence. *Report from the Institute for Philosophy and Public Policy* 15 (1): 14–19.
Watson, G. 1988. Responsibility and the limits of evil: Variations on Strawsonian theme. In *Responsibility, character, and the emotions: New essays in moral*

psychology, ed. F. Schoeman, 256–286. Cambridge: Cambridge University Press.

(1996). Two faces of responsibility. *Philosophical Topics* 24: 227–247.

Williams, B. (1990). Who might I have been? In *Human genetic information: Science, law and ethics,* ed. D. Chadwick, G. Bock, and J. Whelan, 167–179. London: John Wiley and Sons.

Zohar, N. (1991). Prospects for "genetic therapy" – Can a person benefit from being altered? *Bioethics* 5: 275–288.

Index

adaptation: in the ancestral environment, 179–81, 186, 188–94; selection pressures and, 16, 175, 181–8, 186–8, 190–2; versus adaptationism, 175; violent behavior and, 179–83, 185–7. *See also* evolutionary psychology

adoption studies, 6, 7, 73. *See also* heritability; twin studies

agent-causation, 246–7, 263–6. *See also* causation; free will

aggression, 11–12, 126, 179–80; MAO mutation and, 11–12. *See also* impulsivity

aging, memory loss in (metaphor), 250–1, 253–4

alcoholism, genetic studies of, 10

allelic association studies, 103–4

American Society of Human Genetics (ASHG), 39

analysis of variance. *See under* quantitative genetics

antisocial personality: behavioral heritability and, 7; correlations v. causes of, 39–41, 40f; DSM-III-R description of, 161–2; DSM IV description of, 151–2; DSM-III-R description of, 161–2; as polygenic trait, 102

Aristotle, *Nicomachean Ethics*, 247, 251–2, 254–6, 256 n.4, 262

autonomy. *See* free will

Avery, L. C. Bargmann, and H. R. Horvitz, *odr-7* study, 89

Bad Seed, The (film), 305–6, 318

Bailey, J. M., et al., 99

Bargmann, C.: neurobiological experiments with *C. elegans*, 87–91, 97, 105

Baron, Marcia, 18, 201–23 (Chapter 8), 274

behavior: adaptation and, 175, 179–83, 185–7; defined, 86; the genome as a "plan" for, 170–1, 174–7, 187–8; instinctual or species-specific, 84; "internal" vs. "external" causes of, 233–5, 238–40, 241 n.4, 246–8, 252, 254; social classification of, 12, 26–9, 42–4. *See also* behavioral genetics; violent behavior

behavioral genetics, 5–6, 12–16, 104–5, 106; classification issues in, 26–9, 38, 42, 43–4, 201–2; controversies in, 1–5, 8–10, 25, 43–5; quantitative trait loci (QTL) studies, 103; ranking of interactive causes in, 33–8, 43; reductionism in, 31–3, 43; replications of experiments in, 99–100. *See also* genetic predisposition; heritability studies; neurogenetic/ neurobiological research

behavioral mutants: *C. elegans* mutants, 86, 88–9; Drosophilia mutants, 93, 94–5t; genetic explanations and, 95–8

bell-curve (normal or Gaussian distribution), 149–50, 213

benevolence, justice as, 17, 262–3, 266–8, 270–1, 282–3, 286–9, 297 n.16, 303

Bennett, J., 291

Benzer, S., 93

Berlin, Isaiah, 290
biological v. genetic factors, 51, 173, 203
biologizing: of athletics, 158–60; of criminality, 142–3, 148, 157–8; of social behavior, 203
bipolar disorder, genetic studies of, 10
blameworthiness. *See* legal responsibility; moral responsibility
Boyd, R., and P. J. Richerson, 184
Brandon, R., 81
Breggin, Peter R., *Talking Back to Prozac,* 159
Brenner, Sydney, 85–6, 87
Brunner, H. G., 10, 101, 126, 164, 190, 316
Buchanan, Allen, 268–9, 288
Buss, D., 177

Carlton, L. R., 101
causation (notion of), 38–42; linear approximation, 35–6; reductionism in assigning causes, 31–3, 43. *See also* genetic causation
C. elegans: behavioral mutants, 86, 88–91, 105–7; male mating behavior, 91–2; species description, 86–7
Chalfie, M., and J. White, 86, 87, 91
character: genetic predisposition and, 253, 254, 304, 308, 323 n.2; identity and, 18, 30–1, 293, 312–17, 320–1; moral luck and, 314, 318–20, 324 n.12; virtues and flaws in (Aristotelian perspective), 247, 251–2, 254, 261, 262, 264–5, 265 n.4, 271 n.5; and volitional impairment and, 252, 253, 255. *See also* free will
Charcot, Jean-Martin, 146–7
Churchland, P., and T. Sejnowski, 97
City of Hope (research program), 146–7
classification issues: defining antisocial traits, 141–2, 151–2, 155, 161–3; defining behavior, 12, 26–9, 42, 43–4, 201–2; defining criminality, 5, 12–13, 19–20 n.1, 126–7, 156–7, 201–2; taxonomies of "human kinds," 154–5, 158, 160, 162, 163–4
cognitive science, 124
Colbert, H., 89–90, 105
compatibalism: critique of (Garcia), 279, 283–90; generally, 18, 251,

277–83, 308, 310, 311, 321; philosophical foundations of (Slote), 17, 259–63
common sense or folk psychology, 177–9
Copeland, Peter, 36–7, 98
Cosmides, L., and J. Tooby, 181
crime and criminality: assumptions about, 405; biologizing of, 142–3, 148, 157–8, 160–3; definitional issues re: 19–20 n.1, 126–7, 141–2, 201–2; genetic heritability studies of, 7, 218 nn.1 and 2, 225–8, 233–4; genetic v. environmental/social causation of, 205–10; pathologizing of, 125–31, 136–8; as a "problem," 144, 145, 164–5, 193–4; reactive heritability and, 128–9, 132–8, 157, 188; social construction of, 5, 12–13, 26–9, 42, 43–4, 156–7. *See also* genetic predisposition; violent behavior
criminal justice: accountability and attribution in, 303–5, 312, 325–6 n.20; "diminished capacity defense," 208, 210–11; educative or reformist theories of, 269; genetic markers as evidence, 19, 206–7, 228–31, 254–5; heat-of-passion defence, 211–12; insanity defense, 207–8, 209; legal precedents, 209, 212; punishment in cases of genetic markers (hypothetical cases), 17, 303–5, 309, 311–14; reform or prevention programs advocated, 218, 269; the social contract and, 261–2; treatment of individuals with genetic markers, 17, 303–5, 309, 311–12, 313–14. *See also* legal responsibility; retribution

Daly, Martin, and M. Wilson, *Homicide,* 177, 180–2
Darwin, Charles, 171
definitional issues. *See* classification issues
DeFries, J. C., 101
depression, genetic studies of, 248
determinism: environmental, 76; genetic, 26, 31–2, 83, 109, 173, 243, 321; moral agency/free will and, 16, 244–8, 252–3, 273, 290–5. *See also* free will
developmental genetics, 56, 61, 76–7, 138 n.5

DeVries, J., 105
Diagnostic and Statistical Manual of Mental Disorders: DSM-III-R, 161–2; DSM-IV, 151–2
diseases, genetic, 10, 38, 317; Huntington's disease, 101, 317; PKU, 58
Dostoevsky, Fyodor, *The Brothers Karamazov*, 320
Douglas, Mary, 145
Drosophilia, 92; behavioral mutants 93, 94–5t, 96–7; courtship mating behavior of, 93–7, 94–5t, 112 n.10; *fru* (fruitless mutation), 94t, 96–7
drugs: addiction to, 130–1, 159–60; to treat disease or condition, 118, 159–60
"Dutch family" study, 10–11, 101, 126, 164, 190, 316. *See also* MAO mutation
Dworkin, Gerald, 320

ELSI (Ethical, Legal, and Social Implications Program), 3
environmental factors: in altered or new environments, 58–9, 66–70, 67f, 69, 77 n.9, 122; criminal behavior and, 120, 234–8; defining and classifying, 72–5, 308; heritability and, 57–8, 66; in natural (vs. controlled) environments, 60–1, 71–2; phenotypic variance and, 90, 120; protective (in case of genetic predisposition), 217–18; versus genetic factors, 51–3, 72–5. *See also* genetic causation; interactionism; social contingencies
environmental variance, 47; genetic variance and, 48–9, 51–5, 64, 68, 70–1. *See also* phenotypic variance
epilepsy, genetic studies of, 146
Ethical, Legal, and Social Implications (ELSI) Program, National Center for Human Genome Research, 3
ethology, 84
eugenics, 2, 3, 149–52
evolutionary psychology, 16, 153, 169, 172–7. *See also* adaption

Feinberg, Joel, 321
Ferveur, Jean-François, 97
folk psychology, 177–9
Frankfurt, Harry, 291, 256 n.2, 320
free will: the assumption of agent-causation, 246–7, 263–5, 290–1; compati-
balist view of, 245–6, 300–1 n.29; determinism and, 16, 244–8, 252–3, 264–5, 273, 283, 285, 290–45; genetic predisposition and, 201–4, 209–10, 213–14, 230–6, 244, 269–71, 290–2; metaphysical conception of, 290–1; moral responsibility and, 276–7, 294–5; self-control and, 201–4, 209–10, 213–14, 230, 244, 269–71; Spinoza on, 259–60, 262–3, 289–90, 322. *See also* character; compatibalism; incompatibalism; moral responsibility

Garcia, Jorge, 17, 18, 273–302 (Chapter 12)
"gay gene," 100–1, 145
gene-environment interactions, 54, 50–5, 57–9, 73–5, 112 n.4, 192–4; developmental genetics and, 76–7; formula expressing, 120. *See also* interactionism
gene-gene interactions, 90t
"Genetic Basis of Complex Human Behaviors, The" (Plomin et al.), 40f, 169
genetic causation: complexity of, 8, 102, 109, 187, 192–4, 225; direct versus indirect, 41–2; polygeic, 108; versus analysis of variance, 39–41, 40f, 43. *See also* causation (notion of); genetic predisposition
genetic determinism, 26, 31–2, 83, 109, 173, 243, 321. *See also* determinism; genetic causation
"Genetic Dissection of Complex Traits, The" (Lander and Schork), 103
genetic engineering, 324 n.11
genetic explanations: behavioral mutants and, 95–8; causal-mechanical approach (genetic mechanisms), 81–3, 104, 107, 180; conceptual assumptions, 13, 35–6, 313–14; deductive nomological model, 81, 83, 104, 110; explanatory limits, 117, 118–19, 123, 124; mechanical explanations and, 79–80, 111–12 n.3; sociopolitical implications of, 117. *See also* behavioral genetics; neurogenetic/neurobiological research
genetic predisposition: "all-or-nothing view" of, 276–7, 294–5; in asympto-

genetic predisposition *(cont.)*
matic individuals with, 35–6, 201–2, 214–18; defined, 306–7; free will or self-control and, 201–4, 209–10, 213–14, 230, 269–71, 290–2; gene-environment interaction in, 117–18; an individual's possible identification with, 155–6, 157, 163, 16–17, 312–13, 320–2, 325 n.20; mechanisms and pathways to, 16, 119, 131–2, 188–92, 307; problems and benefits of testing for, 214–18, 222 n.26, 283, 315; "strong genetic influences" model of, 18, 273, 275, 276–8, 284, 310, 315, 322; therapeutic treatment of, 315; versus congenital factors, 316–17. *See also* free will; genetic causation; heritability
genetic variance: environmental variance and, 48–9, 51–5, 64, 68, 70–1. *See also* heritability; phenotypic variance
Gibbard, Allan, 16, 142, 153, 169–97 (Chapter 7)
Goodman, Nelson, 155
Goodwin, F. K., 160
Goring, Charles, 149–52
Gottesman, I., and H. H. Goldsmith, 306, 307
Greenspan, Patricia, 18, 243–58 (Chapter 10), 277–83, 308, 310, 311, 321
Greenspan, R. J., 93, 97–8

Hacking, Ian, 15, 141–67 (Chapter 6), 155
Hall, Jeffrey C.: studies of *Drosophilia* courtship behavior, 93–7, 94–5t, 96–7
Hamer, Dean, 36–7; studies of male sexual orientation, 98–101, 102, 110, 111
Hamilton, W., 175
heritability, 7–8, 39, 47, 187; defined, 39, 54, 118, 119, 138 n.3, 169–70; developmental processes and, 76–7; heritability studies, 1–2, 9–10, 38–40, 103, 121–2; limitations of heritability studies in determining, 7–8, 121–4; social or environmental intervention and, 57–8; studies of, 1–2, 9–10, 38–40, 103, 121–2; versus inheritance, 7–8, 14, 48, 54, 57, 169
Herrnstein, Richard, 150; and C. Murray, *The Bell Curve*, 213

Hinckley, John, 207
homicide: diminished-capacity defense, 208, 210–11; genetic studies of, 177, 180–2; heat-of-passion defense, 211–12; insanity defense, 207–8; variation in causes of, 193
Human Genome Project, 3, 85

identical twins. *See* twin studies
identity (personal), 18, 30–1, 293, 312–13, 314–17, 320–1
impulsivity, 101, 160; issue of moral agency and, 18, 211–13, 296 n.7, 306–7, 321, 323–4 n.7; neurogenetic research on, 11–12, 244, 248–51, 310–11
incompatibalism, 246–7, 289, 294–5, 295 n.1, 309–12, 319
inheritance: Mendelian theory of, 80–1, 107; versus heritability, 14, 48, 54, 57, 169
inhibition, lack of. *See* impulsivity
insanity defense, 207–8. *See also* criminal justice
interactionism, 33–8, 43. *See also* gene-environment interactions
intermittent explosive disorder (IED), 161–2
IQ (intelligence quotient): criminality and, 149–51, 221 n.15; debate over genetic contribution to, 58–9, 69, 74, 183; heritability of, 7; MAO mutation and, 11

Jencks, C., 30, 39
Jensen, C., 58

Kandel, E. R., J. H. Schwartz, and T. M. Jessel, *Essentials of Neuroscience and Behavior*, 105
Kanner, Leo, 156
Kant, Immanuel, 261–2, 290–2
King, M. C., 100
Kitcher, P., 81, 104
Kruglyak, L., 99–100
Kuhn, Thomas, 104, 145
Kupferman, I., 84

Lakatos, Imre, 144–6, 147
Lander, Eric S., and L. Kruglyak, on replicating experiments, 99–100; and

N. J. Schork, "The Genetic Dissection of Complex Traits," 103
learning: and plasticity of organisms, 90, 98; resistance to, 306–7
leech, neurobiological studies of, 91–2
legal responsibility, 13; insanity defense, 207–8, 209; moral responsibility and, 205–6, 210, 261–6, 303–5, 322; precedents establishing, 208–13; volitional component of, 18, 207–8, 256 n.2, 262–3, 289–90, 322. *See also* moral responsibility; retribution
LeVay, S., 98; and D. Hamer, 100
Lewontin, R., 57–8
Linnoila, M., et al., 160
Locke, John, 154–5
Lockery, S., 91
Lombroso, Cesare, 148
Lorenz, C., 84

McClearn, G., 105
McGuffin P., 40f, 169
Malcolm X, 189–90
MAO mutation, 9–10, 316; and aggression, 11–12. *See also* impulsivity; serotonin levels
markers, genetic. *See* genetic predisposition
"mean gene," 145
"medicalization" of social behavior, 203
Mendel, Gregory: theory of inheritance, 80–1, 107
Millard v. State, 209
Mill, John Stuart, *On Liberty,* 290
molecular genetics, 8, 15, 111; MAO mutation, 9–10, 316. *See also* neurogenetic/neurobiological research
moral luck, 314, 318–20, 324 n.12
moral responsibility: assignment of, 264–5, 278–9, 289, 296 n.6; cases of genetic predisposition considered, 16–19, 201–4, 209–14, 242 n.9, 252–7, 275–9, 309–12, 319, 322; criminals as "moral monsters," 260, 285–6; degree of control over actions and, 241 n.4, 246–8, 255, 278, 282; determinism compatible with, 273–4, 279–83, 284, 289–90, 294; free will in relation to, 276–7, 294–5; legal responsibility and, 205–6, 210, 261–6, 303–5, 322; onset of behavior as a factor in, 254, 316–17;

social contingencies and, 18, 254, 307, 316, 322. *See also* character; determinism; free will; legal responsibility
Morgan, T. H., 92

Nagel, E.: analysis of mechanical explanations, 79
National Center for Human Genome Research, 3. *See also* ELSI; Human Genome Project
National Institutes of Health (NIH), 3. *See also* ELSI; Human Genome Project
"nature versus nurture" controversy, 30–1, 33, 47, 84
neurogenetic/neurobiological research, 1, 8, 84–6, 91–92, 98; *C. elegans* (Bargmann), 87–8; command neurons, 84; of *Drosophilia* (Hall), 93–7, 94–5t; embryogenetic development, 90t; multifunctional neurons, 90t; mutation studies, 88–91, 105–7; neuroanatomy, 97; neurochemical substrate of behavior, 33, 306–7; rules relating genes and neurons, 85, 89, 90t, 102; single-gene behavior links (rare), 91. *See also* evolutionary psychology; MAO mutation; serotonin levels
neurosis, genetic studies of, 7
Nisbett, R., 183

Owen, M. J., 40f, 169

Paternalism, 294
Paul, Galatians 5:1, 292
Pearson, Karl, 149–52
People v. Tanner, 209
pharmacology. *See* drugs
phenotype: behavioral, 83; defined, 51–2; genetic contribution to, 10–11, 51–5, 59–60, 60f, 119, 123–4; phenotypic traits, 77 n.2
phenotypic variance: complexity of factors in, 54–5; environmental contribution to, 90, 120; formulas expressing, 71–2, 120; in humans, 57–8; linguistic (of native languages), 123–5; within vs. between populations, 76. *See also* genetic variance; gene-environment interaction
Plomin, Robert, et al., 39, 40f, 102, 105, 169

Popper, Karl, 144
Powledge, T., 142
Prozac, 159
psychology: of criminal behavior, 119,
125–7, 171–4; DSM-III-R description
of antisocial personality disorder,
161–2; DSM IV description of antisocial personality disorder, 151–2; psychological traits, 69–70. *See also* neu-
rogenetic/neurobiological research
public policy, 170
punishment. *See* criminal justice; retribution

quantitative genetics, 14, 47, 72;
analysis of variance (ANOVA),
48–9, 55–6, 57, 58–63, 77 n.1 and 8;
developmental processes not considered in, 76–7; distinguishing genetic
from environmental variance in,
48–9, 51–5, 64, 68, 70–1, 73–5; environmental alteration and, 58–9,
66–70, 67f, 69, 77 n.9; formulas expressing variance, 49, 50, 53, 54,
64–5, 68–9; variance v. causation,
39–41, 40f, 43. *See also* environmental
variance; genetic variance; phenotypic variance
quantitative trait loci (QTL) studies,
103
Quetelet, Adolphe, 153

racism, 4, 74
Railton, P.: on the ideal explanatory
text, 104, 111
Rawls, John, 262
reactive heritability, 128–9, 132–8, 157,
188
recidivism, 163–4, 304, 322–3 n.1
reductionism, 31–3, 43. *See also* genetic
determinism
Regina v. Newell, 212
retribution, 17, 16, 251–6, 256–7 n.5,
259–61, 266, 269, 270–1, 280–2, 297
n.13, 303; reactive attitudes, 16,
252–3, 255, 270. *See also* moral responsibility
Rice, G., et al., 99, 100
Risch, N., 99
Roberts, State v., 209
Rose, Steven, 43–5
Rutter, M., 142

Salmon, W., 81–2; on the causal-
mechanical approach, 81–3
Saunders, A. R., 99
Scanlon, Thomas, 262, 265
Schaffner, Kenneth, 15, 79–16 (Chapter
4), 142, 154. *See also* genetic explanatory model (GE)
schizophrenia, genetic studies of, 7
Schoeman, Ferdinand, 319
Schork, N. J., 103
Sejnowski, T., 97
Sengupta, P., et al., 89–90, 105
serotonin levels: aggressive behavior
and, 126, 159, 160, 185, 204, 248–51,
254; CSF 5-HIAA metabolite, 11. *See
also* MAO mutation
sexual orientation studies, 98–101, 110,
111
Slote, Michael, 17, 279, 283–90, 259–71
(Chapter 11), 297 n.16, 299 n.23
Sober, Elliott, 14, 38, 47–78 (Chapter 3),
111; on the additivity thesis, 66–70,
77 n.9
societal contingencies, abusive parenting, 18, 254, 316; criminal behavior
and, 120, 128–38, 157, 234–8;
in defense or mitigation of criminal
behavior, 18, 190, 254, 307, 316, 322;
triggering a genetic predisposition,
218. *See also* environmental factors
sociobiology. *See* evolutionary psychology
Spinoza, *Ethics*, 259–60
State v. Roberts, 209
Steele, C. M., 184, 188
Strawson, Peter F., 16, 252, 255, 270–1,
273–4, 292, 294–5
Stump, Elenore, 292
Symons, D., 177

Taylor, Kenneth, 14, 15, 38, 117–39
(Chapter 5), 157
Tooby, J., 181
Tourette, Gilles de La, 146
Tourette's syndrome, 146
twin studies: of behavioral conditions,
7; of criminality, 164–5, 202; critique
of, 75–7; monozygotic (MZ) twins
reared apart, 63–6, 70, 121; monozygotic (MZ) v. dizygotic (DZ) twins,
68–70, 77–8 n.10, 102, 121. *See also*
adoption studies; heritability

Uniform Crime Reports, 156

Van Fraassen, B., 104, 110–11
Van Inwagen, Peter, 17, 225–42 (Chapter 9), 274–9
variance. *See* environmental variance; genetic variance; phenotypic variance
"Violence Initiative," 304
violent behavior: adaptation and, 179–83, 185–7; changing classifications of, 26–9; constitutional volatility, 306–7; genetic heritability research on, 4, 10–13; inhibition of, 180–1; intermittent explosive disorder (IED), 161–2; questions regarding issue about causality, 29–33, 35, 39–43, 147, 172–5, 188, 192; selection pressures of ancestral environment and, 16, 153, 165, 179–83, 185–7, 188–94. *See also* aggression; antisocial personality; impulsivity
Virkunnen, M., and M. Linnoila et al., 160–2, 164†

virtue ethics (Aristotelian), 247, 251–2, 254, 261, 262, 264–5, 265 n.4, 271 n.5
volitional component of legal responsibility, 18, 207–8, 256 n.2, 262–3, 289–90, 322

Wachbroit, Robert, 1–21 (Chapter 1), 14, 25–46 (Chapter 2)
Wasserman, David, 1–21 (Chapter 1), 18, 271 n.3, 303–27 (Chapter 13)
Watson, Gary, 292, 304–5, 312, 318, 319, 325 n.20
White, Dan (trial of), 210–11
White, John, 86, 87, 91
Williams, Bernard, 315
Williams, G., 175
Wilson, Margot, 177, 180–2
Wimsatt, W., 81
Wojtyla, K., 293–5
Wood, W., 87

XYY karyotype, 8–10, 208–9, 218 n.1 and 2, 220–1 n.15